D0358702

ADVANCES IN CREEP DESIGN

ADVANCES in CREEP DESIGN

The A. E. Johnson Memorial Volume

Edited by

A. I. SMITH and A. M. NICOLSON

National Engineering Laboratory, East Kilbride
Glasgow, Scotland

HALSTED PRESS DIVISION
JOHN WILEY & SONS, INC.
NEW YORK

Distributed in the United States and Canada by
Halsted Press Division
John Wiley & Sons, Inc., New York

ISBN 0 470 80172 7

LIBRARY OF CONGRESS CATALOG CARD NUMBER 72-1357

WITH 235 ILLUSTRATIONS AND 23 TABLES

Printed in Great Britain by Galliard Limited, Great Yarmouth, England

LIST OF CONTRIBUTORS

M. M. Abo El Ata
University of California, Berkeley, California, USA

J. F. Barnes
Engine Research Department, National Gas Turbine Establishment, Pyestock, Farnborough, England

L. Berke
Department of Aeronautics and Astronautics, Stanford University, Stanford, California, USA

J. M. Clarke
Engine Research Department, National Gas Turbine Establishment, Pyestock, Farnborough, England

B. Crossland
Professor of Mechanical Engineering, Department of Mechanical Engineering, Queen's University, Belfast, N. Ireland

U. Edstam
Chalmers University of Technology, Gothenburg, Sweden

H. Fessler
Department of Mechanical Engineering, University of Nottingham, England

W. N. FINDLEY
Professor of Engineering, Brown University, Providence, Rhode Island, USA

I. FINNIE
Professor of Mechanical Engineering, University of California, Berkeley, California, USA

P. A. T. GILL
Department of Engineering Science, University of Durham, England

R. M. GOLDHOFF
Metallurgy Applied Research, Materials and Process Laboratory, Power Generation Division, General Electric Company, Schenectady, New York, USA

A. GRAHAM
Department of Mechanical Engineering, City and Guilds College, Imperial College of Science and Technology, London, England

G. R. HALFORD
National Aeronautics and Space Administration, Lewis Research Center, Cleveland, Ohio, USA

J. HENDERSON
Computer Stress Analysis and Creep Division, National Engineering Laboratory, East Kilbride, Glasgow, Scotland

N. J. HOFF
Professor of Aeronautics and Astronautics, Stanford University, Stanford, California, USA

T. C. HONIKMAN
Department af Aeronautics and Astronautics, Stanford University, Stanford, California, USA

J. HULT
Professor of Strength of Materials, Chalmers University of Technology, Gothenburg, Sweden

L. M. KACHANOV
Professor, Faculty of Mathematics, Leningrad State University, Leningrad, USSR

A. KRISCH
Max-Planck-Institut für Eisenforschung, Düsseldorf, Germany

E. C. LARKE
Consultant, Research and Development Department, Imperial Metal Industries (Kynoch) Limited, Witton, Birmingham, England

F. A. LECKIE
Professor of Engineering, University of Leicester, England

I. M. LEVI
Department of Aeronautics and Astronautics, Stanford University, Stanford, California, USA

S. S. MANSON
Chief, Materials and Structures Division, National Aeronautics and Space Administration, Lewis Research Center, Cleveland, Ohio, USA

F. K. G. ODQVIST
Emeritus Professor, Royal Institute of Technology, Stockholm, Sweden

R. OHTANI
Department of Mechanical Engineering, Kyoto University, Kyoto, Japan

R. J. PARKER
Research and Development Department, Imperial Metal Industries (Kynoch) Limited, Witton, Birmingham, England

R. G. PATTON
Department of Mechanical Engineering, Queen's University, Belfast, N. Ireland

YU. N. RABOTNOV
Member of the Academy of Sciences of the USSR, Professor of Mechanics, Moscow State University, USSR

W. SIEGFRIED
Institut Battelle, Geneva, Switzerland

W. J. SKELTON
Department of Mechanical Engineering, Queen's University, Belfast, N. Ireland

A. I. SMITH
Head of Computer Stress Analysis and Creep Division, National Engineering Laboratory, East Kilbride, Glasgow, Scotland

J. D. SNEDDEN
Computer Stress Analysis and Creep Division, National Engineering Laboratory, East Kilbride, Glasgow, Scotland

D. A. SPERA
National Aeronautics and Space Administration, Lewis Research Center, Cleveland, Ohio, USA

P. STANLEY
Department of Mechanical Engineering, University of Nottingham, England

S. TAIRA
Professor of Mechanical Engineering, Department of Mechanical Engineering, Kyoto University, Kyoto, Japan

C. H. A. TOWNLEY
Central Electricity Generating Board, Berkeley Nuclear Laboratories, Berkeley, Gloucestershire, England

CONTENTS

SECTION I—THEORETICAL ANALYSIS

SECTION III—CREEP AND DESIGN OF COMPONENTS

SECTION IV—REVIEW OF A. E. JOHNSON'S RESEARCH

EDITORS' FOREWORD

The importance of allowing for creep in the design of certain components which operate under sustained loading at elevated temperatures has long been recognised, and two leading British laboratories, *i.e.* Metro-vics and NPL Engineering Division, undertook analytical and experimental work in this field as early as the 1930s. Creep under multi-axial stress systems is a difficult and tedious subject for experimental investigation and is not readily tractable mathematically, but it is universally recognised that no investigator has made a greater contribution to this subject than the late Dr A. E. Johnson, and it is therefore especially fitting that leading research workers in this field should have contributed papers to this Memorial Volume, giving their latest experimental and theoretical results on creep under multi-axial stress conditions and applications in design. In view, however, of the longtime significance of Dr Johnson's work and the relevance, to design, of the experimental and theoretical results contained in the contributions to this Volume, it was considered appropriate to entitle the Volume 'Advances in Creep Design'; it is hoped that it will provide research scientists and designers with a convenient review of the current stage of development of the subject.

Recent research in this field falls into several recognisable sections and the chapters of this book are arranged accordingly. In several of the leading academic schools of engineering science, theories of creep and rupture under simple and multi-axial stress systems are being developed and may lead to more rational procedures of design for creep conditions; in the first five chapters leading Russian, Swedish and British experts, of international repute, summarise their latest ideas in this field. Rabotnov develops a general theory of rupture at stress concentrations. Kachanov considers the theory of creep rupture and Odqvist of rupture under three-dimensional stress systems; the latter is based on the concepts of Dr Johnson and Kachanov. Leckie reviews concepts and theorems relating to creep deformation, including the reference stress technique, in structural design. Hult proposes a general theory of creep design.

In the second section of the Volume there are nine papers concerned primarily with aspects of creep properties relevant to engineering design for elevated temperature service. Goldhoff describes the effect of periods of unloading on creep strains giving rise to recovery, *i.e.* dimensional recovery. Larke and Parker attempt to mathematically analyse data on the three stages of creep deformation and derive equations for each stage; their results may be relevant to extrapolation for design purposes. Crossland correlates tension and torsion creep properties and illustrates derivation of constant-load tensile creep properties from torsion tests, using a strain-hardening theory. Graham discusses the validity of the constitutive equations for creep under multi-axial stress—in particular the effect of the development of anisotropy, during creep, on tertiary creep and rupture. Henderson reports results of a study of complex stress relaxation, *i.e.* creep under conditions of constant total strain. Siegfried reviews cracking in notched test pieces and includes a discussion of precipitation and embrittlement due to structural changes during creep. Manson shows how creep may interact with low-cycle fatigue properties. Krisch demonstrates that recognisable metallurgical structural changes can affect long-time creep properties. Findley describes complex-stress creep properties of plastics.

In the third section of the Volume several authors discuss creep analysis and its application to the estimation of the creep deformation of specific components. In the papers by Taira and by Finnie the prediction of complex stress creep is considered, particularly in relation to creep deformation and rupture of tubular components. Fessler treats the case of creep in thick rings under internal pressure. Clarke analyses creep in thick wall tubes under mechanical and thermal loading—a condition simulating cooled, hollow gas-turbine blades—and shows that the stress redistribution causes the overall strain behaviour to approach that for the mean axial stress and the mean radial temperature. Hoff presents an example of creep buckling, the particular component analysed being a flat rectangular plate under compression on two edges. The increasing use of computers in stress analysis of components for creep is reflected in Townley's paper which describes the application of computer techniques to analysis of discs, tubes, welded tubes under static conditions and rotating discs under cyclic loading.

The final paper reviews briefly the contribution of Dr Johnson to research on creep and fracture, particularly under multi-axial stress conditions, and the application of his results to the analysis of the creep of components. A bibliography of Dr Johnson's published papers is appended.

The task of detailed literary editing was undertaken by the NEL Publications Section, and grateful acknowledgment is made of the work of the late Cdr Gibb.

BIOGRAPHICAL NOTE ON DR A. E. JOHNSON

Arthur Edward Johnson was born in Manchester in 1908 and educated at Stand Grammar School and the School of Engineering at Manchester University. He graduated B.Sc. with 1st Class Honours in 1928, and M.Sc. in 1929, and he gained the Fairbairn Engineering Prize in his year. Shortly after joining the staff of the Engineering Division, National Physical Laboratory, in 1929, he became one of the team engaged on creep research under the late H. J. Tapsell. About 10 years later he commenced the study of creep under multi-axial stress systems—a subject in which he gained an international reputation.

At the NPL, and from 1952 at the National Engineering Laboratory, Johnson developed the highly-refined experimental techniques required to obtain data on the stress–strain–time–temperature relationships of engineering materials subjected to creep under multi-axial stress conditions. Over a period of a quarter of a century, by experiment and analysis, he systematically determined the laws of creep deformation and fracture under complex stresses, and his published papers—some 70 in number—are the most important references on this subject. Several gained for him prizes from the Institution of Mechanical Engineers—an Institution Prize (1945), a Water Arbitration Prize (1952) and a James Clayton Prize (1965). During the Second World War he studied the elevated-temperature design of jet engine discs and a detailed paper based on this work led to the award in 1957 of the Howard Prize for Mechanical Motive Power of the Royal Society of Arts. Johnson was awarded the degrees of M.Sc. (Tech.) in 1947 and D.Sc. in 1950 by Manchester University. He was a Member of the Institution of Mechanical Engineers and a Fellow of the Institution of Metallurgists.

In 1961 Johnson visited the main centres (both academic and industrial) in the USA engaged on research on complex-stress creep and fracture of metals and structures for discussions with the leading American workers in this field. He was promoted to Senior Principal Scientific Officer on Special Merit in 1962, and shortly afterwards was selected by the then Department of Scientific and Industrial Research for the Wolfe Award for 'the most outstanding contribution to the research work of the Department' during the year. Johnson was a member of the Organising Committee and a Session Chairman of the Conference on Thermal Loading and Creep in Structures and Components, held by the Institution of Mechanical Engineers in 1964 on behalf of the Joint British Committee on Stress Analysis.

Johnson was shy and retiring by nature and a man of simple tastes. His way of living had, perhaps, something of the ascetic about it, for he felt no need of the creature comforts which other men find help to ease the burden of life. His barely furnished bachelor flat contained not a single non-technical book, no television and no radio or other 'diversions'. He lived alone and, tragically, he died alone. And yet, in his daily life, he was not lonely; his work was his constant companion. Every evening and every weekend he was hard at work on his technical papers either in the laboratory or at home. Yet he was a most amiable and interesting companion to the few who really got to know him, a tolerant and painstaking adviser to those who sought his advice, and he was by no means devoid of humour. In his early years at NPL he was a well-known rugby player; in later years his sole recreation was walking and his daily routine included a walk of several miles irrespective of the weather.

In paying a tribute to Johnson's work and achievements, one can perhaps do no better than to slightly paraphrase words used in *The Engineer*'s obituary article in 1874 on Sir William Fairbairn, the first British engineer to undertake experiments on the creep of metals—'No man, living or dead, has done so much to make mechanical engineering, in one important branch, so nearly perfect. . . . His name is inseparably connected with his researches into the strength of metals under complex-stress conditions'. Johnson died suddenly of a heart attack on the night of Wednesday, 28th October, 1965, and it is a remarkable coincidence that he was buried little more than a few paces from the grave of Fairbairn in the grounds of Prestwich Parish Church, Manchester; and so Johnson now rests, in good company, in the shadow of the church in which he sang, as a choir boy, some fifty years ago.

SECTION I
THEORETICAL ANALYSIS

Chapter 1

CREEP RUPTURE UNDER STRESS CONCENTRATION

YU. N. RABOTNOV

SUMMARY

Starting from experimental results an attempt is made to construct the simple equations of creep taking into account the formation of internal microcracks leading to rupture.

The stress distribution for the stage of creep preceding rupture is believed to be approximately the same as in an ideally plastic body obeying a certain generalised yield condition. Using this assumption for plane strain, the problem of prediction of the lifetime is reduced to integration of the hyperbolic system of differential equations.

Some simple examples of the application of the method are given: pure bending of a bar and tension of a strip with double circular notch.

NOTATION

a, b	Constants dependent on temperature.
h	Depth of beam.
K	Ratio of length of bars.
m	Effective stress concentration factor.
p, q	Parameters defined on p. 8.
q_i	Structural parameter.
q_s	Static value of structural parameter.
S	Arc-length of characteristic line.
T	Temperature.
t	Time to rupture.

x, y, z Coordinates.
x $\frac{1}{2}\varphi(\zeta)$.
β Constant in eqn (8).
ε Creep strain.
$\dot{\varepsilon}$ Creep strain rate.
ε_* Creep strain at rupture.
ζ Non-dimensional coordinate of neutral axis.
η Non-dimensional coordinate of crack front.
θ Angle between x axis and second principal stress direction.
λ Factor dependent on t.
ξ z/h (non-dimensional form of z).
σ Direct stress, hydrostatic stress component.
σ_e Effective stress.
σ_1 Maximum tensile stress.
σ_r Radial stress.
σ_s Arbitrary constant.
σ_θ Hoop stress.
τ Maximum shear stress.
φ Function.
ψ, ψ_i Function.
ω Relative number of microcracks.

Subscripts
0 Zero time.
1 Bar 1, or stage 1.
2 Bar 2, or stage 2.

Superscripts
. Instantaneous value or rate.
* Rupture value.
n Exponent of the power law of damage accumulation.

INTRODUCTION

The first attempts to formulate a mathematical theory of creep of metals valid up to the moment of rupture and permitting the prediction of the real lifetime of a structure were undertaken only recently. In this connection one should emphasise that the general problem of fracturing of solid bodies seems to be one of the central problems of modern mechanics and the above-mentioned attempts must be considered as constituent elements of the future general theory. In the field of strength under high temperature conditions, the situation seems to be somewhat clearer, than, for instance, in the theory of plastic rupture at normal temperature. In fact, the characteristic feature of creep rupture is the so-called 'embrittlement'. The creep process is accompanied by the formation of cavities in the grain boundaries, giving rise to the formation of intergranular microcracks. The growth of microcracks leads to the formation of a macrocrack spreading through the body. The rupture of modern heat resisting alloys normally occurs at very low strains and therefore the analysis of the state of stress based on the assumption of geometric linearity of the equations remains valid till the moment of formation of the macrocrack. Direct experiment leads to a relatively simple criterion of rupture expressed in terms of stresses.

A very general assumption permitting formulation of the constitutive equations of creep is that the creep rate at a given instant is determined by the stress, the temperature and the structural state of the material, the last being defined by a certain number of structural parameters q_i [1]†:

$$\dot{\varepsilon} = \varphi(\sigma, T, \dot{q}_i) \tag{1}$$

The rates of change of these structural parameters should be described by a system of the kinetic equations:

$$q_i = \psi_i(\sigma, T, q_s) \tag{2}$$

One of these parameters, which will later be denoted as ω, can be taken as a measure of damage; it is the relative number of cavities or microcracks. When ω reaches a certain critical value, say $\omega = 1$, the local rupture occurs. The simplest theory of creep, describing the behaviour of the body up to the moment of rupture will be based on the assumption that ω is a single structural parameter, affecting the creep rate. Then the creep equations become

$$\dot{\varepsilon} = \varphi(\sigma, T, \omega), \qquad \dot{\omega} = \psi(\sigma, T, \omega) \tag{3}$$

Some tentative expressions for the functions φ and ψ were considered in ref. 1. One of the simplest assumptions is

$$\dot{\varepsilon} = a\left(\frac{\sigma}{1-\omega}\right)^n, \qquad \dot{\omega} = b\left(\frac{\sigma}{1-\omega}\right)^K$$

† Numbers within square brackets correspond to the references listed at the end of each chapter.

the values of a and b being dependent on the temperature. If $n = K$, $\varepsilon = \varepsilon_* \omega$, $\varepsilon_* = a/b$ and the rupture occurs when the strain ε reaches the fixed value ε_*, depending only on the temperature, then the equations take the form:

$$\dot{\varepsilon} = \varepsilon_* \left(\frac{\sigma}{1-\omega}\right)^n, \qquad \dot{\omega} = \left(\frac{\sigma}{1-\omega}\right) \tag{4}$$

Dots denote the differentiation with respect to the modified time defined in such a manner that $b = 1$. At constant stress the time to rupture t_0 will be found by integrating eqn (4) and putting $\omega = 1$, namely:

$$t_0 = \frac{1}{(n+1)\sigma^n}$$

The characteristic feature of the eqns (4) is that they describe the third stage of creep, neglecting the first stage, that is the strain hardening. This is the case of so-called short time creep when the temperature is high enough and the strain hardening is negligible.

EFFECTIVE STRESS CONCENTRATION FACTOR

A very simple example will be considered first to illustrate the basic ideas and introduce the notion of the effective stress concentration factor. Three bars are connected in parallel as shown in Fig. 1, the cross-section of bar 1 being unity and of each of the bars 2 being $\frac{1}{2}$.

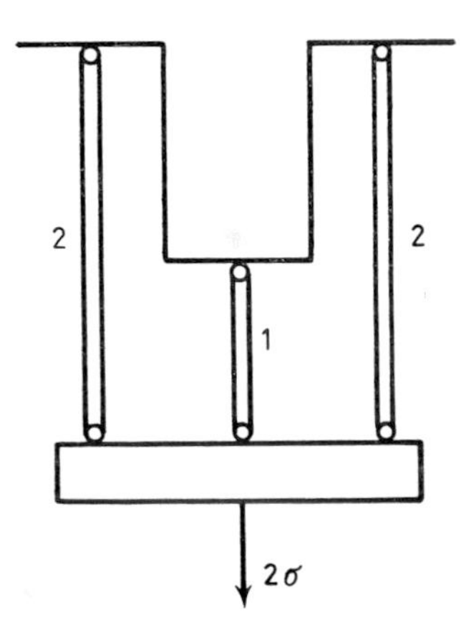

FIG. 1. Arrangement of bars.

The equation of equilibrium is

$$\sigma_1 + \sigma_2 = 2\sigma$$

The equation of compatibility is

$$\varepsilon_2 = K\varepsilon_1$$

where K is the ratio of lengths of the bars. For an ideally plastic structure, the stresses in the bars are equal: $\sigma_1 = \sigma_2 = \sigma$. The time to rupture under this assumption will be

$$t_0 = \frac{1}{(n + 1)\sigma^n}$$

The real time to rupture being t, the value of

$$m = \left(\frac{t_0}{t}\right)^{1/n}$$

can be denoted as an effective stress concentration factor. The definition of the time to rupture initially requires some convention. In this example the rupture of bar 1 occurs first at the time t_1, but the bars 2 continue to bear the load during some time t_2, the total time to rupture of the structure being $t_1 + t_2$. The lifetime can be estimated whether as $t = t_1$ or as $t = t_1 + t_2$. To solve the above stated problem two approaches will be used.

Steady state creep solution

The stress distribution under the conditions of steady state creep will be

$$\sigma_1 = \frac{2}{1 + K^{1/n}}\sigma, \qquad \sigma_2 = \frac{2K^{1/n}}{1 + K^{1/n}}\sigma$$

If this distribution remains valid up to the moment of rupture of the first bar, t_1 will be defined as follows

$$\frac{t_1}{t_0} = \left(\frac{1 + K^{1/n}}{2}\right)^n$$

The integration of the creep damage equation for the second bar gives the value of $\omega_2 = \omega_2{}^*$ at $t = t_1$:

$$1 - (1 - \omega_2{}^*)^{n+1} = \left(\frac{2K^{1/n}}{1 + K^{1/n}}\right)^n \frac{t_1}{t_2} = K$$

After the rupture of the first bar, the total load is carried by the bars 2 and therefore $\sigma_2 = 2\sigma$. The integration of the same equation permits the value of t_2 to be found:

$$\int_{\omega_2^*}^{1} (1 - \omega_2)^n \, \mathrm{d}\omega = 2^n \sigma^n t_2 \quad \text{or} \quad \frac{t_2}{t_1} = \frac{1 - K}{2^n}$$

The total time to rupture being $t_1 + t_2$, the effective stress concentration factor will be determined as follows:

$$m = \frac{2}{[(1 + K^{1/n})^n + 1 - K]^{1/n}} \tag{5}$$

Exact solution

The equations of damage accumulation for each bar will be

$$\dot{\omega}_1(1 - \omega_1)^n = \sigma_1{}^n, \qquad \dot{\omega}_2(1 - \omega_2)^n = \sigma_2{}^n \tag{6}$$

When $\dot{\varepsilon}_1 : \dot{\varepsilon}_2 = \dot{\omega}_1 : \omega_2$, then $\omega_2 = K\omega_1$. Now the value of $\sigma_2/(1 - \omega_2)$ can be expressed in terms of σ and ω_2, namely:

$$\frac{\sigma_2}{1 - \omega_2} = \frac{\sigma}{p - q\omega}, \qquad p = \frac{1 + K^{1/n}}{2K^{1/n}}, \qquad q = \frac{1 + K^{-[(n+1)/n]}}{2}$$

The second of the eqns (6) can be rewritten as follows

$$\dot{\omega}_2(p - q\omega_2)^n = \sigma_n$$

At the moment of rupture of the first bar, $\omega_1 = 1$ and $\omega_2 = K$. The integration of the last equation gives:

$$\frac{t_1}{t_0} = \frac{1}{q}[p^{n+1} - (p - qK)^{n+1}]$$

When $t > t_1$, $\sigma_2 = \sigma$ as in the first case and the lifetime of the second bar alone will be determined in the same way, but now at $t = t_1$, $\omega_2 = K$ instead of $\omega_2 = \omega_2{}^*$. It is easy to prove that

$$\frac{t_2}{t_0} = \frac{(1 - K)^{n+1}}{2^n}$$

Now the value of the effective stress concentration factor will be determined by the formula:

$$m = 2:\left\{\frac{(1 - K^{1/n})^{n+1}}{1 + K^{1+(1/n)}}\left[1 - \left(1 - \frac{1 + K^{1+(1/n)}}{1 + K^{(1/n)}}\right)^{n+1}\right] + (1 - K)^{n+1}\right\}^{1/n} \qquad (7)$$

The formulae of eqns (5) and (7) give the same result $m = 1$ at $K = 1$ and $m = 2^{1-(1/n)}$ at $K = 0$. For the intermediate values of K, eqn (7) gives lower values of m than eqn (5). So, for instance, when $n = 3$, the values of the effective stress concentration factor given by eqn (7) are:

$K =$ 0·2	0·4	0·6
$m =$ 1·115	1·041	1·010

This example shows that the acceleration of creep in the third stage, due to the damage accumulation, leads to considerable stress redistribution and therefore the simple analysis based upon the assumption of the equal stress distribution, as in the case of ideal plasticity, permits a reasonably good estimate of the real lifetime of the structure to be obtained.

RUPTURE CRITERION FOR COMBINED STRESS

When considering a general case of the combined state of stress, the first problem is to formulate an adequate rupture criterion. The investigations of Dr Johnson showed that to a certain degree of precision the time to rupture will be defined by the maximum tensile stress. The experimental work of Dr Johnson was followed by investigations in the USSR by a number of authors, and among them by Sdobyrev who undertook a series of experiments on some heat resisting steels and refractory alloys [2]. In most of these experiments, tubular specimens in combined tension and torsion were used.

The analysis of these experiments and of a number of experimental results published by other investigators, including the data of Dr Johnson, showed that the time to rupture depends not only on the maximum tensile stress, but also on the shear stress. Sdobyrev proposed the following empirical formula for the effective stress:

$$\sigma_e = \beta\sigma_1 + (1 - \beta)\sigma_0 \qquad (8)$$

Here, σ_1 is the maximum tensile stress, σ_0 the stress intensity and β is approximately 0·5. In uniaxial tension, σ_e is the tensile stress and therefore the usual stress–time to rupture diagram can be used to estimate the lifetime of any body under combined stress conditions. In practice, the creep is always accompanied by stress redistribution and the value of σ_e at any point varies with time. Therefore the prediction of the time when the first macrocrack appears must be based on some principle of the cumulation of damage. Usually the influence of the stress redistribution on the lifetime is not great; for turbine discs for instance the correction needed does not exceed 3–4 per cent of the stress, calculated on the assumption of steady state creep.

Under the conditions of plane stress, the ratio σ_e/σ_1 varies, at $\beta = 0{\cdot}5$, as follows:

$$1 \geq \frac{\sigma_e}{\sigma_1} \geq 0{\cdot}93 \quad \text{when} \quad \sigma_2 \geq 0$$

$$1{\cdot}36 \geq \frac{\sigma_e}{\sigma_1} \geq 1 \quad \text{when} \quad \sigma_2 \leq 0$$

In turbine discs both the principal stresses are as a rule positive and the difference between σ_e and σ_1 does not exceed 7 per cent. Therefore the use of the σ_1 criterion of rupture seems to be justified. In other stress conditions, for example in the plane state of strain, the difference may be more considerable.

It is well known that in some cases the notched specimen gives a greater time to rupture than the unnotched one having the same cross-sectional area. This supporting effect of the stress concentration cannot be explained on the basis of the maximum tensile stress criterion of rupture.

ALTERNATIVE COMPLEX STRESS FRACTURE CRITERION

Johnson *et al.* [3] pointed out that, for some materials, the creep rate under biaxial stress depends on the stress intensity while for other materials this dependence remains valid only during the first and second stage of creep, the creep rate in the third stage being dependent on the maximum tensile stress. One can believe that, for the materials of the first group, the influence of the damage on the creep rate is negligible and the time to rupture can be predicted by the theory of Kachanov [4]. The dependence of the creep rate on maximum tensile stress for the materials of the second group shows that the formation of macrocracks makes the material weaker in the corresponding direction.

Supposing that the preponderant orientation of microcracks is normal to the σ_1 direction ($\sigma_1 > \sigma_2 > \sigma_3$), the effective stresses defining the creep rate will be

$$\frac{\sigma_1}{1-\omega}, \quad \sigma_2, \quad \sigma_3$$

The magnitude of the creep rate will be defined by a certain combination of these stresses, corresponding to the stress intensity or maximum shear stress, the creep rate distribution obeying the associated flow rule. The kinetic equation of damage can now be written as follows:

$$\frac{\partial \omega}{\partial t} = \psi(\sigma_e)$$

Here the equivalent stress is given by eqn (8), where the value of σ_1 is replaced by $\sigma_1/(1-\omega)$. The exact solution of particular problems of rupture, based on the use of the proposed system of equations, seems to be rather difficult. One example of the solution of this kind for the rotating disc is given in ref. 5.

CREEP FRACTURE OF BAR IN BENDING

It is well known that the stress distribution in creep does not usually differ very much from that in an ideally plastic body. On the other hand, the example given above shows that the assumption of the equal stress distribution permits a reasonable estimate of the lifetime to be obtained. A more general assumption is that the stress components at any instant satisfy the following equation:

$$F\left(\frac{\sigma_1}{1-\omega}, \sigma_2, \sigma_3\right) = \lambda \sigma_s \tag{9}$$

Here σ_s is an arbitrary constant and the factor λ, depending on t because ω is a function of t, will be found from the conditions of equilibrium. At $t = 0$, $\omega = 0$ and eqn (9) turns into the ordinary condition of ideal plasticity. Assuming the power law of cumulation of damage, the corresponding equation will be

$$\frac{\partial \omega}{\partial t} = \beta {\sigma_e}^K \tag{10}$$

As an illustrating example, the problem of fracturing of a rectangular bar in pure bending will be considered. The cross-section of the bar is shown in Fig. 2.

The stress in the upper part of the section being tensile, the neutral axis moves downwards and its position at the time t is defined by the coordinate

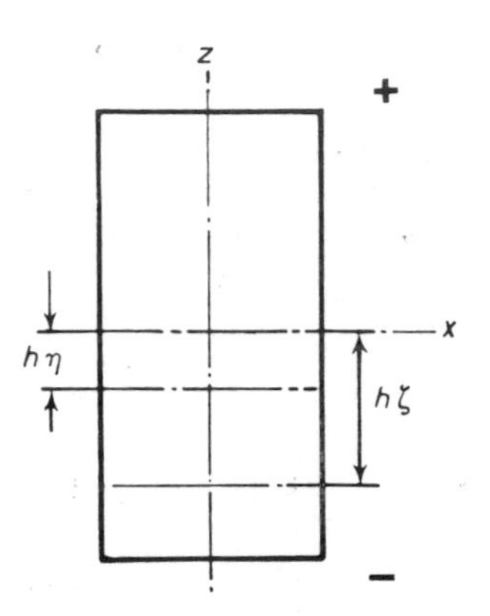

FIG. 2. Cross-section of beam.

$z = -h\zeta(t)$, $\zeta(0) = 0$. If σ^+ is the tensile stress at $z = -h\zeta$ and σ^- the compressive stress at $z < -h\zeta$, then, according to eqn (9)

$$\sigma^+ = \sigma(1 - \omega), \qquad \sigma^- = -\sigma, \qquad \sigma = \sigma(t)$$

In the first stage of the process $\omega < 1$ at any z. Equating the resultant stress to zero:

$$\int_{-\zeta}^{1} \sigma(1 - \omega)\,\mathrm{d}\xi - \int_{-1}^{-\zeta} \sigma\,\mathrm{d}\xi = 0 \tag{11}$$

where $\zeta = z/h$, and the moment equilibrium equation results in

$$\int_{-\zeta}^{1} (1 - \omega)\xi\,\mathrm{d}\xi - \int_{-1}^{-\zeta} \xi\,\mathrm{d}\xi = \frac{\sigma_0}{\sigma} \tag{12}$$

The equation of the damage cumulation will be

$$\dot{\omega} = \frac{\partial\omega}{\partial t} = \sigma^K \tag{13}$$

Differentiating eqn (11) with respect to t and using the condition $\omega = 0$ at $\xi = \zeta$ gives:

$$\dot{\zeta} = \frac{1}{2} \int_{-\zeta}^{1} \dot{\omega} \, \mathrm{d}\zeta$$

But $\dot{\omega}$ is a function of t only, therefore

$$\dot{\zeta} = \tfrac{1}{2}\dot{\omega}(1 + \zeta) \tag{14}$$

and, after integration

$$\omega = 2 \ln (1 + \zeta) + \psi(\xi)$$

The function $\psi(\xi)$ equals zero when $0 \leq \xi \leq 1$, at $\xi = -\zeta$, $\omega = 0$ therefore

$$\omega = 2 \ln (1 + \zeta) \qquad 0 \leq \xi \leq 1$$

$$\omega = 2 \ln \frac{1 + \zeta}{1 - \xi} \qquad -\zeta < \xi < 0 \tag{15}$$

When ζ reaches the value $\zeta_1 = e^{\frac{1}{2}} = 1{\cdot}649$, ω becomes unity for all positive ξ, which means the crack spreads suddenly in the upper half of the section. Differentiating eqn (12) and eliminating $\dot{\omega}$ with the aid of eqn (14) gives the following equation

$$\dot{\zeta}(1 + \zeta) = -\left(\frac{\sigma_0}{\sigma}\right)$$

and, after integration,

$$\frac{\sigma_0}{\sigma} = \tfrac{1}{2}[3 - (1 - \zeta)^2] \tag{16}$$

Now the integration of the system of eqns (13), (14) and (16) allows determination of the duration t_1 of the first stage

$$t_1 = \frac{1}{\sigma_0{}^n} \frac{1}{2^{n-1}} \int_1^{\zeta} \frac{[3 - (1 - \zeta)^2]^n}{1 + \zeta} \, \mathrm{d}\zeta \tag{17}$$

Now the second stage of the rupture process will be considered, when the crack moves in the domain of the negative ξ and the position of its front is determined by the coordinate $\xi = -\eta$. As formerly, differentiation of the equilibrium equations gives:

$$\dot{\zeta} = \tfrac{1}{2}\omega(\zeta - \eta)$$
$$\zeta(\zeta - \eta) = \left(\frac{\sigma_0}{\sigma}\right) \tag{18}$$

At $t = t_1$ and $0 \geq \xi \geq -\zeta_1$, $\omega = 1 - 2 \ln (1 - \xi)$ from eqn (15). The increment of ω being dependent on t only and ω being a monotonic function of t, then

$$\omega = \varphi(\zeta) + 1 - 2 \ln (1 - \xi)$$

At $\xi = -\eta$, $\omega = 1$ therefore

$$\varphi(\zeta) = 2 \ln (1 + \eta)$$

Now the value of $x = \frac{1}{2}\varphi$ will be taken as an independent variable and $\eta = 1 - e^x$ introduced into the first eqn (18). The resulting equation will be

$$\frac{d\zeta}{dx} = \zeta + 1 - e^x$$

The integral of this equation satisfying the initial condition $\zeta(0) = \zeta_1$ is

$$\zeta = (1 + \zeta_1)\, e^x - 1 - x\, e^x$$

Now ζ and η are expressed in terms of an auxiliary parameter x. This solution remains valid till the moment t_2, when $\eta = \zeta$; at this moment $x = \frac{1}{2}$ and $\zeta_1 = \zeta_2 = (1 + \zeta_1)^2 - 1 - \frac{1}{2}(1 + \zeta_1) = 0{\cdot}896$. The duration of the second stage can now be determined by integrating eqns (13) and (18). It can be shown that the value of $t_2 - t_1$ is small compared with t_1.

At $t > t_2$ the third stage of the process begins, ending at $t = t_3$, when $\zeta = \zeta_3$ and so on. The exact solution of the problem requires the consideration of an infinite number of steps each of them ending at $t = t_n$ and $\zeta = \zeta_n$. When n tends to infinity ζ_n tends to 1 and the time to rupture will be equal to $\lim (t_n)_{n=\infty}$.

A solution of this kind would have a rather theoretical interest. From the practical point of view the state of the bar when the crack is spread over half of the section must be regarded as inadmissible and therefore the value of t_1, defined by eqn (17), can be considered as a real estimate of the lifetime of a bar in bending.

GENERAL CASE OF PLANE STRAIN

The above described approximate method of solving creep rupture problems can be applied to the general case of plane strain. Equation (9) assumes in this case the following form:

$$\frac{\sigma_1}{1 - \omega} - \sigma_2 = \lambda \sigma_s \qquad (19)$$

Putting $\sigma_1 = \sigma + \tau$, $\sigma_2 = \sigma - \tau$, eqn (19) can be rewritten as follows

$$\tau = m\sigma + b, \qquad m = -\frac{\omega}{2 - \omega}, \qquad b = \lambda\sigma_s \frac{1 - \omega}{2 - \omega} \tag{20}$$

The limit condition, eqn (20), has the same form as that used in statics of soils; the difference is that the parameters m and b are functions of time. For a fixed value of t, the equilibrium equations and the condition (20) form a hyperbolic system which can be integrated using the method of characteristics. The angle between the axis of x and the second principal stress direction being θ, one can put:

$$\sigma_x = \sigma - \tau \cos 2\theta, \qquad \sigma_y = \sigma + \tau \cos 2\theta, \qquad \tau_{xy} = \tau \sin 2\theta$$

The standard procedure allows the equations of characteristics to be found:

$$\frac{dy}{dx} = -\frac{(2 - \omega)\sin 2\theta \pm 2\sqrt{(1 - \omega)}}{\omega + (2 - \omega)\cos 2\theta} \tag{21}$$

and the relations along the characteristics:

$$\pm \frac{\sqrt{(1 - \omega)}}{2 - \omega} \frac{\partial \sigma}{\partial S} + \tau \frac{\partial \theta}{\partial S} = \frac{\sigma + \frac{1}{2}\lambda\sigma_s}{(2 - \omega)^2} \frac{\partial \omega}{\partial n} \tag{22}$$

In the last equation n is a normal direction to the corresponding characteristic.

The tangent of the angle which the characteristic makes with the second principal direction is $\pm\sqrt{(1 - \omega)}$. When $\omega = 1$, the characteristics coincide, forming a crack. In practice near the stress concentration the large rates of creep are localised in a limited region owing to the intense crack formation, while in the remaining part of the body the creep rates are negligible. The scheme described idealises this situation; the creep accompanied by cracking is supposed to occur only in a finite plastic domain, surrounded by the rigid material. The boundary of this plastic region is formed by characteristics changing their position with changing ω. When, at a certain point, ω reaches the value $\omega = 1$, the angle between the characteristics becomes zero and the plastic region is continued by the crack as is shown schematically in Fig. 3.

When Sdobyrev's criterion of fracture is used, eqn (10) has the following form:

$$\frac{\partial \omega}{\partial t} = \left\{\left[1 - \beta\left(1 - \frac{\sqrt{3}}{2}\right)\right]\lambda + (1 - \beta)\frac{\sigma_2}{\sigma_0}\right\}^K \tag{23}$$

In this equation, t is the non-dimensional time defined in an appropriate manner.

The numerical procedure of solving the stress concentration problems can be described as follows. Equations (21), (22) and (23) are replaced by the finite difference equations; the distribution of ω at a given value of t being known, the first two equations can be used to construct the field of characteristics. Then the new distribution of ω at the moment $t + \Delta t$ will be found with the aid of eqn (23) and the new set of characteristics calculated in the same way. For any t, the value of λ will be found from the condition of equilibrium. When, at a certain point, the value of ω reaches

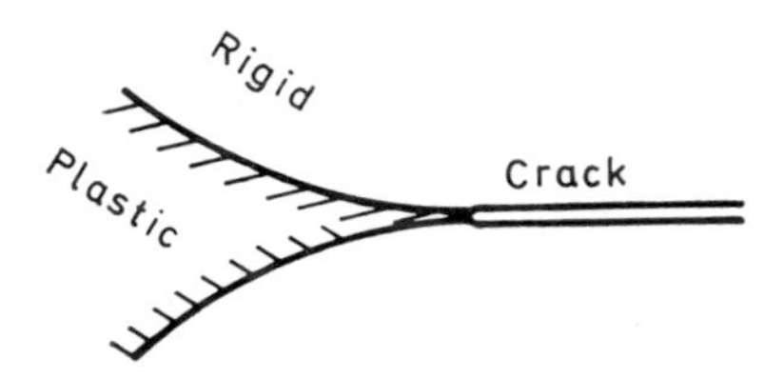

FIG. 3. Plastic region at crack tip.

the value $\omega = 1$, the system becomes parabolic and the plastic region will be continued as a crack. In the initial state, $\omega = 0$ everywhere and the field of characteristics corresponds to the solution of the ideally plastic problem.

As a simple example we shall consider the axisymmetric stress distribution as in the case of the thick-walled tube loaded by uniform pressure. The equations of equilibrium and compatibility are:

$$\frac{\partial \sigma_r}{\partial \xi} + \frac{\sigma_r - \sigma_\theta}{\xi} = 0$$

$$\frac{\sigma_\theta}{1 - \omega} - \sigma_r = \lambda \sigma_s$$

where ξ is the non-dimensional radius.

Putting $\sigma_r = \lambda \sigma_s (S - 1)$, then $\sigma_\theta = \lambda \sigma_s (1 - \omega) S$ and the equation of equilibrium gives:

$$\frac{\partial S}{\partial \xi} = \frac{1 - \omega S}{\xi} \tag{24}$$

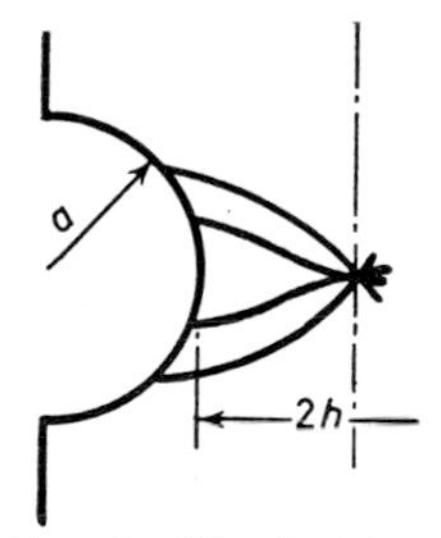

FIG. 4. Notched bar.

The equation of damage accumulation is:

$$\frac{\partial \omega}{\partial t} = \lambda^K \left[(1 - \beta)S + \frac{\sqrt{3}}{2}\beta \right]^K \tag{25}$$

The value of λ is determined from the equilibrium condition:

$$\lambda \int_1^\infty S(1 - \omega)\, d\xi = \text{const} \tag{26}$$

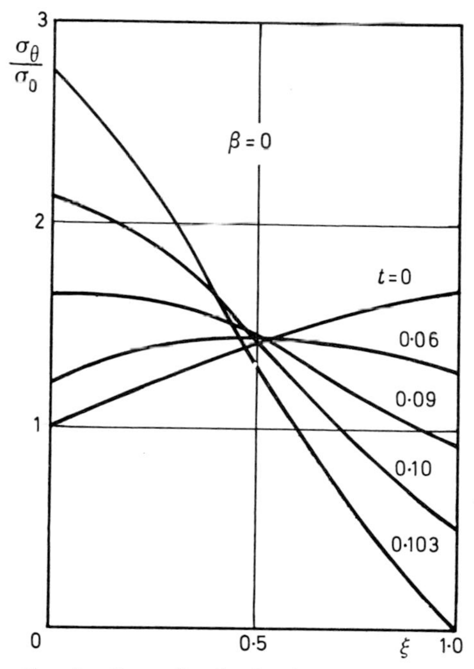

FIG. 5. Stress distribution in notched bar.

Tension of a rectangular bar with circular notches, as shown in Fig. 4, gives another example of application of the solution of this type. If the material is ideally plastic, the plastic zones are bounded by logarithmic spirals, having a point of intersection on the axis of symmetry. The corresponding state of stress is considered as the initial state. In the next stage of the process, when $\omega \neq 0$ and the characteristics are no longer orthogonal, the state of stress remains axisymmetric. In Fig. 4 the boundary of the plastic zone is shown at the initial moment when $\omega = 0$, and at the moment $t = t^*$ when $\omega = 1$ on the axis of symmetry and the angle between the characteristics becomes zero.

The time t^* will be considered as the time to rupture. The results of numerical integration of eqns (24), (25) and (26) for $K = 3$, $h/a = 1$ are given in ref. 6. It was found that $t^* = 0{\cdot}103$ when $\beta = 0$, and $t^* = 0{\cdot}187$ when $\beta = 0{\cdot}5$. The time to rupture of the unnotched specimen was equal to $t^* = 0{\cdot}0947$. Therefore, the supporting effect of the notch is equal to 1·024 at $\beta = 0$ and 1·25 at $\beta = 0{\cdot}5$. The distribution of the stress σ_θ for different values of t when $\beta = 0$ is plotted on Fig. 5. The negligibly small supporting effect in this case (2·4 per cent) is probably due to the error of numerical integration; when at $\beta = 0{\cdot}5$ this effect is considerable.

CONCLUSIONS

The problem of the influence of stress concentration on the lifetime of structures at high temperature is difficult. One can believe that at present it is not possible to propose a reasonably simple and adequate method valid for different materials and conditions. The approach used in this chapter can be useful in cases when the region of intensive creep and damage accumulation spreads across the structural member and the influence of the surrounding elastic field can be neglected. At lower temperatures, the zone of damage is more localised and the stress distribution strongly depends on the elastic deformation.

In principle, the calculation of the stress field can be achieved using the isochronous creep curves and some numerical methods.

REFERENCES

1. Rabotnov, Yu. N. (1963). 'On the equation of state for creep.' *Progress in Applied Mechanics* (The Prager Anniversary Volume), pp. 307–315. London and New York: Macmillan.

2. SDOBYREV, V. P. (1958). 'Long-term strength of alloy E1-437B under complex stresses' (in Russian). *Izv. Akad. Nauk SSSR Otd. Tekh. Nauk* (4), 92–97. English translation: *NEL TT* 652. East Kilbride, Glasgow: National Engineering Laboratory.
3. JOHNSON, A. E., HENDERSON, J. and MATHUR, V. D. (1962). *Complex Stress Creep, Relaxation, and Fracture of Metallic Alloys*. Edinburgh: HMSO.
4. KACHANOV, L. M. (1960). *Teoriya polzuchest'* (Theory of Creep). Moscow: Fizmatgiz. English translation: BISHOP, E. (1967), Boston Spa: National Lending Library for Science and Technology.
5. RABOTNOV, YU. N. (1966). *Polzuchest' elementov Konstruktsiy* (Creep of Structural Elements). Moscow: Nauka.
6. RABOTNOV, YU. N. (1967). 'Non-stationary creep with a power law of strain hardening' (in Russian). *Inzh. Zh. Mekh. Trerdoge Tela*, **3**, 66–71.

Chapter 2

SOME PROBLEMS OF CREEP FRACTURE THEORY

L. M. KACHANOV

SUMMARY

This chapter deals first with the influence of ultimate stress on the brittle rupture time under creep conditions; then the problem of the propagation of a creep crack is studied.

NOTATION

A, n	Constants in creep rupture law.
B	Coefficient in eqn (1).
b	Half-width of strip.
C	Coefficient in eqn (11).
E, ν	Elastic constants.
h	Half-height of unfractured rod.
k	Coefficient in eqn (17).
I_m	Moment of inertia.
M	Bending moment.
m	Creep index.
N	Intensity coefficient.
P	Force on strip.
R	Radius.
t	Time.
U	Deformation work.
u	Normal to failure front.
x, y	Coordinates.

α, β	Constants in crack growth equation.
γ	Surface energy.
κ	Rate of crack growth $= \mathrm{d}l/\mathrm{d}t$.
λ, p	Exponents in strip fracture equation.
μ	$1/m$.
Σ	Failure front.
σ_B	Ultimate stress.
σ_S	Yield stress.
ψ	Continuity.

Subscript

0 Initial value.

VISCOUS RUPTURE THEORY

The fracture time according to Hoff's theory [1] is determined by unbounded quasiviscous flow of the body, and here, in the final flow stage, stresses grow boundlessly. As in the case of the viscous fracture of a rod subject to a constant force P, we have

$$t_1 = \frac{1}{mB_1\sigma_0{}^m} \qquad \left(\sigma_0 = \frac{P}{F_0}\right) \tag{1}$$

Here the stress $\sigma \to \infty$ as the current cross-sectional area of the rod $F \to 0$. According to this solution, very great stresses correspond to short time fractures. There is another drawback in this theory: its final results depend on the form of the creep law (in consequence of limit transition $\sigma \to \infty$). However, this theory is widely used because its drawbacks are not always very significant.

Several years ago Rosenblium [2] suggested a simple way of improving Hoff's theory, which takes into consideration plastic deformations. According to this new theory, the flow process develops as follows: first, when stresses are insignificant the body is subject to creep; then as stresses grow and reach the yield stress, plastic flow and then 'plastic fracture' begin to take place. In this theory fracture takes place somewhat earlier than in that of Hoff and the discrepancy between theoretical and experimental data becomes smaller. In the case of stretching a rod, instead of eqn. (1) we obtain

$$t_{11} = t_1\left[1 - \left(\frac{\sigma_0}{\sigma_S}\right)^m\right], \qquad (\sigma_0 \leq \sigma_S) \tag{2}$$

The correction factor is important when the creep index m is small and when the initial stress σ_0 is large.

BRITTLE RUPTURE THEORY

Let us consider brittle fracture which is explained by gradual development of cracks. The state of partial cracking of a material is characterised by the scalar quantity $0 \leq \psi \leq 1$ which is defined by the following equation [3, 4]:

$$\frac{d\psi}{dt} = -A\left(\frac{\sigma_{max}}{\psi}\right)^n \tag{3}$$

It should be noted that σ_{max} may be replaced by the more general criterion, *i.e.* by an effective stress σ_{eff} [4–6].

In the absence of initial damage $\psi = 1$ for $t = 0$, then we have

$$1 - \psi^{n+1} = A(n+1)\int_0^t \sigma_{max}{}^n \, dt \tag{4}$$

It is generally accepted that, at the moment of failure t', the continuity $\psi = 0$. When the rod is stretched by a constant stress σ_0 we have

$$t' = 1/[(n+1)A\sigma_0{}^n] \tag{5}$$

This formula is acceptable in the case of long time fracture. In the case of short time fracture the stresses are too great to allow its acceptance. This defect can be eliminated in the following way. Let us treat the ratio σ_{max}/ψ as 'true stress'. It is understood that the true stress does not exceed the ultimate stress σ_B,

$$\frac{\sigma_{max}}{\psi} \leq \sigma_B \tag{6}$$

Then, at the moment of fracture t'', we have $\psi = \psi' = (\sigma_{max}/\sigma_B)$, and

$$A(n+1)\int_0^{t''} \sigma_{max}{}^n \, dt = 1 - \left(\frac{\sigma_{max}}{\sigma_B}\right)^{(n+1)} \tag{7}$$

Turning back to the former example of stretching a rod, we get

$$t'' = t'\left[1 - \left(\frac{\sigma_0}{\sigma_B}\right)^{(n+1)}\right] \tag{8}$$

The correction factor is insignificant when the stress σ_0 is not large and when the index n is large. When $\sigma_B \to \infty$ we get the former eqn (5).

FAILURE FRONT

In the non-homogeneous state of stress of the brittle body at the stage of latent failure, $0 \leq t \leq t_I$, we have $\psi > \psi'$ at each point of the body. The fracture determined by eqn (7) may not take place (even if $t'' \to \infty$) when the stress is diminishing fast enough. Here we shall consider only a simple, but important, case when the stress σ_{max} does not diminish.

At the instant $t = t_I$, at some point (or region) of the body local failure occurs; the front of failure Σ is formed. At the surface Σ, $\psi = \psi'$, *i.e.*

$$A(n + 1) \int_0^t \sigma_{max}{}^n(\tau)\, d\tau = 1 - \left[\frac{\sigma_{max}(t)}{\sigma_B}\right]^{(n+1)} \tag{9}$$

This equation may be represented in another form

$$\frac{du}{dt} = -\left(\frac{A\sigma_{max}{}^n + \psi'^n\,(\partial\psi'/\partial t)}{(\partial/\partial u)\,\{A \int_0^t \sigma_{max}{}^n(\tau)\, d\tau + [1/(n + 1)]\, \psi'^{n+1}\}}\right)_\Sigma \tag{10}$$

where u is the normal to the front. The eqns (9) and (10) are somewhat more complicated than the former equations [3], which are deduced from eqns (9) and (10) when $\psi' \to 0$.

EXAMPLE OF BRITTLE FRACTURE IN BENDING

We shall consider the problem of brittle fracture of a rod of rectangular cross-section ($2b$ is the width and $2h_0$ is the initial height) in pure bending. During the stage of partial failure, $0 \leq t < t_I$, the entire section of the beam resists the bending. At the instant $t = t_I$ the surface layer fails. In the subsequent period, the failure front Σ moves down into the rod. Let the height of the unfractured rod be $2h$ at time t; at some instant $t \geq t_I$, the stress will be

$$\sigma = \frac{M}{I_m}\, y^\mu \qquad \left(y > 0,\ \mu = \frac{1}{m}\right)$$

where M is the bending moment and the generalised moment of inertia

$$I_m = \frac{2b}{1 + (\mu/2)}\, h^{2+\mu}$$

Here y is measured from the current position of the neutral axis. It should be noted that $\sigma_{max} = (\sigma)_{y=h}$.

Passing over the transformations analogous to those made in earlier work [7] we obtain the differential equation

$$\frac{d^2h}{dt^2} + 2(m-1)\frac{1}{h}\left(\frac{dh}{dt}\right)^2 - (2m+3)\frac{C}{h^y}\left(\frac{dh}{dt}\right)^3 = 0 \qquad (\text{for } t > t_I) \tag{11}$$

where

$$C = \frac{M(1+\mu/2)}{AB\sigma_B{}^{m+1}}$$

For $t = t_I$, we have $h = h_0$. Integrating this equation, and determining the arbitrary constants from the initial conditions, we find

$$\frac{t}{t_I} = 1 + \frac{2}{2m-1}\left(1 + \frac{Ch_0{}^2}{(2m+1)t_I}\right)\left[1 - \left(\frac{h}{h_0}\right)^{(2m-1)}\right] + \frac{2m+3}{2m+1}\frac{Ch_0{}^2}{2t_I}\left(\frac{h}{h_0}\right)^2 \tag{12}$$

For $h = 0$, complete failure of the beam takes place. The failure time t'' is given by

$$t'' = t_I{}^0\left[1 + \frac{2}{2m-1} - \frac{4m^2-3}{4m^2-1}\left(\frac{\sigma_{max}{}^0}{\sigma_B}\right)^{(m+1)}\right] \tag{13}$$

where $t_I{}^0$ is the time t_I for $\sigma_B = \infty$, *i.e.* for the former determination of the failure moment ($\psi' = 0$). The last term in the square brackets characterises the influence of the new criterion. This term is always smaller than unity; it is much smaller than unity when the initial bending stress is relatively small and also when the creep index is large. For a very large index m, we have $t'' \to t_I{}^0$.

An analogous analysis can be made in other problems which have been solved by applying the former criterion [7].

GENERALISED FORM OF GRIFFITH–IRWIN THEORY

Notches (defects or cracks) localise the fracture in the body, thus shortening its 'life' and so the whole concept of fracture is changed. The problem of crack propagation is a very complicated one. It involves the analysis of the fine structure of the stress state at the crack tip, when the body is in a condition of creep or plastic flow. Also, physical models more complicated than those used in the Griffith–Irwin theory are required.

Here we suggest the following general theory. Let us consider the equation of energy balance at the crack tip in any continuum, which has been deduced by Cherepanov [8]:

$$R \int_0^{2\pi} (U \cos \theta - A) \, d\theta = 2\gamma \tag{14}$$

where R is the radius of the small circle C, including the tip, U is deformation work, A is stress work on the contour C and γ is surface energy. Equation (14) characterises singularity of the stress field and determines the intensity coefficient N. In the case of an elastic body, eqn (16) leads to the well-known Irwin formula $\pi N^2(1 - \nu^2) = E\gamma$.

When analysing the crack propagation in the inelastic continuum we must not consider the surface energy as constant. First, it is necessary to take into account the accumulation of very small damages (having different origins for different materials), scattered throughout the body. As long as the continuum mechanics method is used it is expedient, for the analysis of scattered microdamage, to introduce a certain value ψ to denote the extent of damage and a certain kinetic equation for it. The damage facilitates the propagation of the major crack.

The fine processes developing at the crack tip are another important factor. The processes become very complicated when real materials are to be dealt with. The kinetics of damage may, for instance, be expressed by eqn (3). The damage increases, and the surface energy γ diminishes. We assume that the surface energy is proportional to the relative proportion of the non-fractured bonds

$$\gamma = \gamma_0 \frac{\psi - \psi'}{1 - \psi'} \kappa \tag{15}$$

where γ_0 is a limit value of the surface energy, and κ depends on the processes at the crack tip. When the crack reaches the fractured zone where $\psi = \psi'$, we have $\gamma = 0$. It should be interpreted in the following way: in the fractured zone the crack runs into one of the cracks which has developed previously.

At the crack tip the stresses are very intensive and this causes relaxing of bonds (as a result of flow) and fracturing of bonds (as a result of heat fluctuations). The intensity of these processes depends on the period of time the material remains in the dangerous zone, that is on the rate of crack growth

$$\kappa = \kappa \left(\frac{dl}{dt} \right) \tag{16}$$

When cracks develop slowly it is the fluctuating mechanism that usually plays the main role. In this case it is possible to use the following approximation

$$\kappa = 1 - \alpha \exp\left(-\beta \frac{\mathrm{d}l}{\mathrm{d}t}\right)$$

where $1 > \alpha > 0$, $\beta > 0$ are constants.

The suggested theory involves distributed cracking and the growth of major cracks (according to the generalised Griffith–Irwin theory). Here eqn (18) determines the stress intensity N as the time function.

SIMPLIFIED ANALYSIS OF CRACK GROWTH

Application of this theory to non-linear creep conditions leads to great mathematical difficulty. So here we suggest a simplified analysis of crack growth in creep conditions. This makes it possible to estimate the kinetics of crack growth from notches in different stress fields (tension, bending, etc).

According to observations, the stress concentration at the crack tip leads to local plastic deformations. Therefore let us assume that the maximum normal stress is equal to the yield stress σ_S. We shall consider only the plane deformation or stress states and use the yield criterion of Tresca–St Venant. The stress concentration decreases rapidly. We assume that this decrease of maximum stress in front of the crack tip is described as follows:

$$\frac{\partial \sigma_{\max}}{\partial x} = k(\sigma_S - \bar{\sigma}) \tag{17}$$

where $k < 0$ is a coefficient determined by experiment, and $\bar{\sigma}$ is the characteristic field stress near the crack tip.

The time of latent failure before a sharp notch develops into a crack is

$$t_{\mathrm{I}} = \left[1 - \left(\frac{\sigma_S}{\sigma_B}\right)^{n+1}\right] \frac{1}{(n+1)A\sigma_S{}^n}$$

Let the crack propagation be described by eqn (10), and let u be the distance reached by the crack at the time t (Fig. 1). In this case, ψ' is a constant and eqn (10) becomes

$$\frac{\mathrm{d}u}{\mathrm{d}t} = -\frac{1}{nk}\left[\int_0^t (1 - \sigma_*)\,\mathrm{d}t\right]^{-1} \tag{18}$$

where $\sigma_* = \bar{\sigma}/\sigma_S$. From this we get the differential equation

$$\frac{d^2u}{dt^2} - nk(1 - \sigma_*)\left(\frac{du}{dt}\right)^2 = 0$$

Its solution when initial conditions are such as

$$u = 0, \qquad \frac{du}{dt} = v_0 \quad \text{for} \quad t = t_I$$

[where v_0 is calculated according to eqn (2)] is

$$\int_0^u \exp(-nku) \times \exp\left(-nk\int_0^u \sigma_* \, du\right) du = v_0(t - t_I)$$

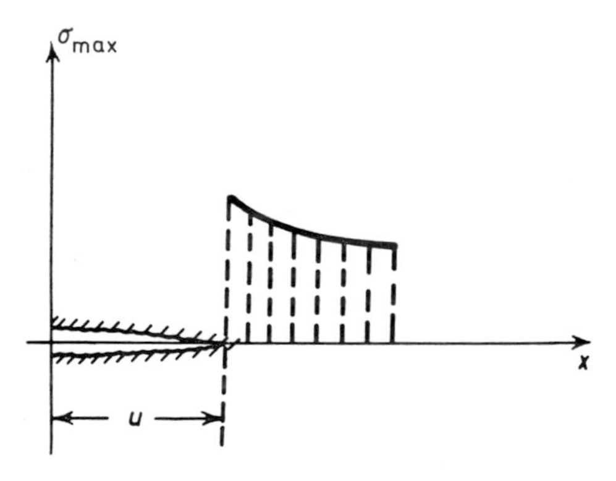

FIG. 1. Stress distribution at crack tip.

The final stage is often accompanied by viscoplastic flow. So we may assume that the fracture appears at $t = t''$ when stress $\bar{\sigma}$ reaches the yield stress.

EXAMPLE OF STRIP STRETCHING

Let us consider the example of stretching a strip $2b$ wide by applying forces $2P$. At the initial stage, the strip has two small symmetric sharp notches. Then,

$$\sigma_* = \frac{P}{b\sigma_S}\left(1 - \frac{u}{b}\right)^{-1}$$

and the fracture time is

$$\frac{t''}{t_I} = 1 - \lambda\, e^{-\lambda} \left(1 - \frac{P}{\lambda}\right) \int_{p/\lambda}^{1} e^{\lambda s} s^p \, ds$$

where

$$nkb = \lambda, \qquad \frac{P\lambda}{B\sigma_S} = p$$

The problem of bending a notched strip (initially $2h_0$ wide) is somewhat more complicated. Here the greatest bending stress in the strip $2h = 2h_0 - u$ wide is the stress $\bar{\sigma}$.

CONCLUSIONS

The theories of viscous and brittle fracture under creep conditions are made more exact by introducing the 'maximum' stress. A general rupture theory combining the scattered fractures and the growth of the major crack is proposed. A simplified analysis of crack growth under creep conditions is also proposed.

REFERENCES

1. HOFF, N. J. (1953). 'The necking and the rupture of rods subjected to constant tensile loads.' *J. Appl. Mech.*, **20,** 105–108.
2. ROSENBLIUM, V. I. (1963). *Creep and creep rupture* (in Russian). (Proc. of symposium) 1. Novosibirsk.
3. KACHANOV, L. M. (1958). 'On the time to failure under creep conditions' (in Russian). *Izv. Ak. Nauk SSSR Otdel. Tekh. Nauk* (8), 26–31.
4. JOHNSON, A. E., HENDERSON, J. and KHAN, B. (1962). *Complex-stress Creep, Relaxation, and Fracture of Metallic Alloys.* Edinburgh: HMSO.
5. ODQVIST, F. (1966). *Mathematical Theory of Creep and Creep Fracture.* Oxford: Clarendon Press.
6. RABOTNOV, YU. N. (1966). *Creep of Structural Elements* (in Russian). Moscow, Nauka. English Translation London: North-Holland Publishing Co. 1969.
7. KACHANOV, L. M. (1961). 'Rupture time under creep conditions.' In RADOK, J. R. M., *Problems of Continuum Mechanics; Contributions in Honour of the Seventieth Birthday of N. I. Muskhelishvili*, pp. 202–218. Philadelphia Society for Industrial and Applied Mathematics, 1961.
8. CHEREPANOV, G. P. (1967). 'Crack propagation in continuous media' (in Russian). *Pripl. Matem. Mech.*, **31** (3), 503.

Chapter 3

THEORIES OF CREEP RUPTURE UNDER MULTIAXIAL STATE OF STRESS

F. K. G. ODQVIST

SUMMARY

This chapter reviews the present stage of development of creep rupture, in particular relating to Kachanov's theory. Kachanov has found that the ratio of the time to ultimate creep rupture (t_{II}) *to the time to latent failure* (t_I) *(Dr Johnson's 'incubation period') usually stays between* 1 *and* 1·5 *for such structures as cylindrical tubes under internal pressure and beams of rectangular cross-section in bending. It is shown in this chapter that the ratio* t_{II}/t_I, *which is of great importance for designers, may well exceed* 3 *in certain cases.*

NOTATION

$\left.\begin{matrix} a \\ b \end{matrix}\right\}$ Defined in Fig. 4.

C Material constant.

N The unit vector of the normal of a failure front Σ.

n Exponent in Norton's law.

p Internal pressure in a tube.

q $dt/d\beta$.

r Defined in Fig. 4.

S_{ij} Stress deviation tensor.

t Time.

t_R Time to rupture.

t_I Time for continuity ψ to reach zero.

t_{II} Time to completion of creep rupture.
t' Variable of integration.
α a/b_0
β b/a.
$\dot{\varepsilon}$ Creep rate.
$\dot{\varepsilon}_c$ Arbitrarily chosen creep rate (often 10^{-7}/hr).
$\dot{\varepsilon}_{ij}$ Creep rate tensor.
ν Material constant.
ρ r/a.
σ Uniaxial tensile stress.
σ_c Material constant.
σ_e Effective stress.
σ_θ Circumferential stress.
ψ Continuity.

Subscripts
K Constant.
0 Initial state.
Σ At the failure front.

EXPERIMENTAL INVESTIGATIONS

In his excellent survey article, Finnie [1] states that 'The great majority of the experimental information in the literature on creep-rupture under multiaxial stresses has come from the work of Dr Johnson at the British National Engineering Laboratory', with which statement the present author wholeheartedly agrees. Dr Johnson's work on the subject has resulted in a series of publications which began in the late 1940s and two extensive papers [2, 3] sum up the theoretical and experimental work of him and his collaborators.

Dr Johnson [2] stated 'General comment on creep fracture work would seem to be that, whatsoever the nature of fracture, where it is accompanied by gradual and general cracking during creep, the stress criterion of tertiary creep and fracture is the maximum principal stress. On the other hand, where no general cracking occurs during creep, the criterion of tertiary creep and fracture would seem to be octahedral stress. Physically, the maximum-principal-stress criterion may correspond to internal grain rupture due to distortion, or rupture at boundaries due to stress concentrations arising from grain distortion. It is possible that the octahedral-stress criterion corresponds to cracks and voids formed by migration and agglomeration of lattice vacancies either at boundaries or within grains.'

Johnson *et al.* [3] said 'Where material showed general or continuous cracking in tertiary creep, both tertiary creep strain and fracture periods were functions of maximum principal stress, this stress possibly being a measure of the so-called incubation period of propagating cracks. Where no cracking appeared until after fracture, tertiary creep strain and fracture periods were functions of the Mises second order invariant or octahedral stress.'

Dr Johnson's experiments with multiaxial creep tests were carried out very carefully on about half a dozen different materials, including carbon steels, light alloys and copper. He found from experiments with 0·5 per cent Mo steel at 550°C, copper at 250°C and Nimonic 75 at 650°C that the time to rupture, t_R, was determined by σ_{max}, the maximum principal stress. In contrast with this, he also found, for a 0·2 per cent C steel at 450°C and an aluminium alloy RR 59 at 200°C, that t_R was determined by the effective stress σ_e, which is equivalent to the octahedral stress or the von Mises' second order invariant, and is defined through

$$\sigma_e^2 = \frac{3S_{ij}S_{ij}}{2} \tag{1}$$

where S_{ij} = stress deviation tensor, and the summation convention adopted here and later.

Dr Johnson also observed that, when σ_e determined the life, the material showed no evidence of cracking except at the actual site of fracture. Again, when lifetime was determined by the maximum tensile stress, the material showed continuously propagating general cracking, well in advance of final separation. For details of the experiments, the reader is referred to the original publications. For a recent review of Dr Johnson's work, *see* Dr Johnson [4].

Commenting upon this, Finnie [1] calls attention to the alleged fact that there is, unfortunately, one very important and general observation which cannot be explained if the maximum tensile stress is indeed the condition controlling rupture in certain materials, but is readily explained if the effective stress σ_e controls the life. This is, that a circumferentially notched bar will very often have a longer life than that of a smooth bar of the same cross-sectional area. A tentative explanation of this contradiction has been offered by Voorhees and Freeman [5], who suggested that σ_{max} controls crack initiation while σ_e governs crack propagation; this explanation may have some relevance.

The author, however, believes that this kind of observation could be preferably explained as a volume effect (*see* W. Weibull [6]). It must be remembered that the stress concentration due to the notch has a local character in contrast to the uniform stress distributed over a large portion of the smooth bar. In the immediate vicinity of a notch where high stresses are likely to prevail, in spite of some relaxation during the first stages of a creep test, the vacancies and voids from which continuous cracking develops may hardly be considered as uniformly distributed throughout this small region. In addition to this drawback, from the point of view of applicability of Kachanov's theory, as explained in the following sections, inhomogeneities of the material exist in this region arising from the production of the notch.

Dr Johnson, Henderson and Khan [3] stated 'The materials appear to fall into two categories. One group of these exhibit no cracking visible above the limit of an electron microscope range until a relatively few hours before fracture; while the second group shows generally distributed and continuously propagated cracking during tertiary creep. These characteristics are valid for each material throughout the whole working temperature range. Within both groups, the final fracture may be transcrystalline, intercrystalline or mixed, but this does not bear any significant relation to the stress criteria of tertiary creep or fracture.'

I consider this statement as absolutely fundamental for future research in the field, supported as it is by an experience as rich as Dr Johnson's.

The deterioration prior to rupture has been studied in detail by an increasing number of investigators using the refined methods of electron microscopy. I shall refrain from entering upon a closer study of their findings, since I consider these are not yet ripe for interpretation within the frame of applied mechanics. Certain observations by Söderquist [7], which are reviewed in the following paragraph, give further support to Dr Johnson's statement.

Söderquist has studied the creep properties of a cast magnesium alloy (ZRE 1:0·6 per cent Zr, 2·95 per cent rare earths, remainder Mg) at 240° and 260°C and also under states of multiaxial stress (*see* p. 33 of ref. 8). Test pieces, machined to thin plates with or without a circular hole, could be subjected independently to arbitrary tension in two perpendicular directions, in very much the same way as described by Dr Johnson [9]. Figure 1 shows deterioration in the neighbourhood of the circular hole in a plate specimen crept for 435 hr under uniform axisymmetric radial tension of 3·92 hbar (4 kg/mm^2). In this case, σ_{max} refers to the hoop stress and its maximum occurs at or near the hole and always on planes through the axis of symmetry. The deterioration is reasonably uniform round the hole. At higher magnification, Fig. 2, it is seen that small cracks are generally radially directed, *i.e.* at right angles to the direction of σ_{max}. These results seem to confirm, on the whole, Dr Johnson's conclusions [3].

FIG. 1. Creep damage around circular hole in plate under uniform radial tension (Söderquist) (5×).

Similar confirmatory observations have been made by Kachanov [10] who conducted experiments with thick walled steel tubes under combinations of axial load and internal pressure so as to have σ_{max} act upon planes either through the tube axis or perpendicular to it. Further confirmation is given later at the end of this chapter.

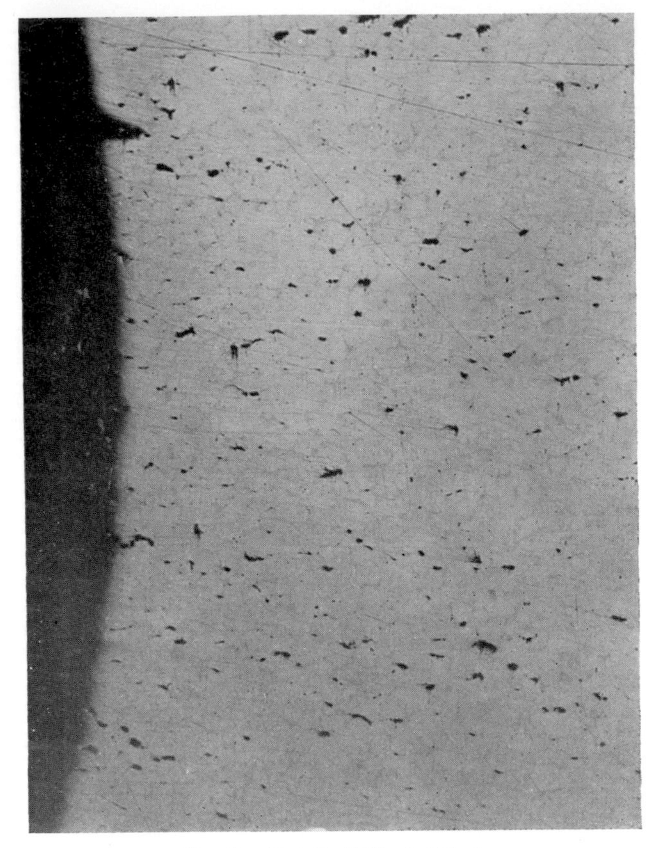

FIG. 2. Detail of Fig. 1 (60×).

KACHANOV'S THEORY OF CREEP RUPTURE UNDER UNIAXIAL STRESS

This section is concerned with the phenomenological theory of creep rupture originating from Kachanov [10] and Odqvist [8]. The fundamental assumption of this theory is that solid material exposed to stress suffers deterioration, thus reducing its 'continuity' ψ from unity to zero during its lifetime. In the special case of uniaxial tensile stress σ, Kachanov postulated for ψ the differential equation

$$\frac{\mathrm{d}\psi}{\mathrm{d}t} = -C\left(\frac{\sigma}{\psi}\right)^{\nu} \tag{2}$$

where C and ν are material constants and σ/ψ may be interpreted as a certain 'effective' stress. If σ is known as a function of the time t, eqn (2) may be integrated with the initial condition

$$\psi = 1 \quad \text{for} \quad t = 0 \text{ (virgin state)} \tag{3}$$

yielding

$$1 - \psi^{\nu+1} = C(1 + \nu) \int_0^t \sigma^\nu \, \mathrm{d}t' \tag{4}$$

where the variable of integration has been denoted by t'.

According to eqn (4), the function ψ will decrease with t. For $t = t_R$, the time to rupture, ψ will reach zero, namely

$$1 = C(1 + \nu) \int_0^{t_R} \sigma^\nu \, \mathrm{d}t' \tag{5}$$

If $\sigma = \sigma_K =$ constant, we may put $t_R = t_K$ and so obtain

$$1 = C(1 + \nu) t_K \sigma_K^{\ \nu} \tag{6}$$

Hence, eqns (4) and (5) may also be written in the form

$$(1 - \psi^{\nu+1}) t_K \sigma_K^{\ \nu} = \int_0^t \sigma^\nu \, \mathrm{d}t' \tag{7}$$

$$t_K \sigma_K^{\ \nu} = \int_0^{t_R} \sigma^\nu \, \mathrm{d}t' \tag{8}$$

where C has been eliminated and either σ_K or t_K may be chosen arbitrarily. Often t_K is taken as 10 000 hr, in which case σ_K is the stress giving creep rupture in 10 000 hr if kept constant during that time.

The considerations so far are independent of the phenomenon of creep and may be imagined to apply to the limiting case of perfectly brittle solids.

Kachanov now makes the important assumption that *deterioration and creep are independent phenomena*. As a special case σ may be determined by creep action.

Neglecting primary creep and considering secondary creep only, Kachanov, combining with Hoff's [11] stability theory of ductile creep rupture, succeeds in representing, at least qualitatively, the entire creep rupture curve by choosing rupture time on the conservative side for a given load; here by 'creep rupture curve' is meant a plot of initial stress σ_0 against rupture time t_R. Creep rupture curves have been published by many investigators; among the earlier, extended over nearly 100 000 hr, are those by Richard [12], Fig. 3, in which, in the log–log plot, the slope of the left part of the creep rupture curve according to Kachanov should be $1/n$ where n is the exponent in Norton's law

$$\frac{\dot{\varepsilon}}{\dot{\varepsilon}_c} = \left(\frac{\sigma}{\sigma_c}\right)^n \tag{9}$$

which connects the creep rate $\dot{\varepsilon}$ in secondary creep with the stress σ and σ_c is a material constant. The quantity $\dot{\varepsilon}_c$ may be chosen arbitrarily and is often taken as $\dot{\varepsilon}_c = 10^{-7}$/hr. The right part of the creep rupture curve has

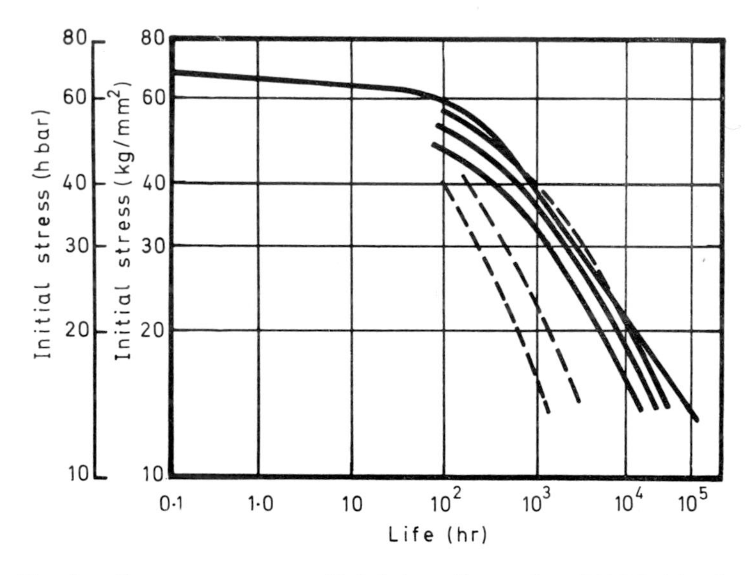

FIG. 3. Creep rupture curve of brittle low alloy steel at 500°C (Richard).

an asymptote with the slope $1/v$, according to eqn (6). Figure 3 represents the creep rupture curve of a brittle low alloy steel at 500°C. In the figure, the left part of the creep rupture curve corresponds mainly to ductile and the right, steeper part to more brittle type of fracture. Full details are given by Odqvist [8], as well as an improved theory which takes into account the influence of primary creep.

Creep rupture curves generally appear more or less concave from below, as in Fig. 3. This implies a relation of the form

$$\nu \leq n \tag{10}$$

for the exponents of eqns (6) and (9).

Tertiary creep is characterised by a gradually increasing slope of the curves connecting creep strain ε and time t in a test with constant tensile load, and is the kind of creep curve from which equations of the type of eqn (9) may be derived. Part of the tertiary creep may be explained from the fact that the cross-section of the test specimen is diminished during the test, while the load remains constant; this effect is taken into account in Kachanov's theory. Other causes of tertiary creep also exist; for example, in carbon steels spheroidisation of cementite may have such consequences. The formation of voids and 'general cracking' too, as suggested by Dr Johnson, will have a similar effect. In order to take into consideration the effects of coupling between deterioration and creep, Rabotnov [13] has generalised Kachanov's theory. On account of the mathematical complications involved, however, the new theory is still in its infancy.

GENERALISATION ON MULTIAXIAL STATE OF STRESS

Generalising on multiaxial state of stress, Kachanov merely replaces σ in eqns (2)–(8) by σ_{max}, the largest principal stress, and adds the assumption that the magnitude of the tensile stress is not small compared with the magnitude of the compression stresses. He also noted that the ratio σ_{max}/ψ may be interpreted as a certain effective stress. In the following, strengthened by the arguments put forward by Dr Johnson, we shall simply assume that

$$\sigma_{max} > 0 \tag{11}$$

is a necessary and sufficient condition for deterioration to start.

Let us consider a body or structure to which a system of loads in equilibrium be applied at $t = 0$, this being the stressless virgin state of the material. Then, in general, an elastic stress distribution will be formed at the first moment of loading, if we neglect inertia effects. As t increases, a redistribution of stress will occur as a consequence of creep. At any value of t, the state of stress and in particular σ_{max} will be a well defined function of the coordinates of the point and of the time. Reasoning and considering eqn (7), we have

$$t_K \sigma_K^{\nu}(1 - \psi^{\nu+1}) = \int_0^t (\sigma_{\max})^{\nu} \, dt' \tag{12}$$

which shows that deterioration will start at all points where eqn (11) holds and continuity ψ will decrease monotonically from its initial value of $\psi = 1$. If eqn (11) ceases to hold, then ψ will remain constant from that very moment. If we exclude this case, then finally, at a certain time $t = t_{\mathrm{I}}$, determined from

$$t_K \sigma_K^{\nu} = \int_0^{t_{\mathrm{I}}} (\sigma_{\max})^{\nu} \, dt' \tag{13}$$

ψ reaches zero at one or several points of the body or structure, supposing that the stability has not been impaired for $t < t_{\mathrm{I}}$. Kachanov calls the time period $t' = 0 - t_{\mathrm{I}}$, *the stage of latent failure*. Obviously this coincides with Dr Johnson's 'incubation period' as quoted earlier.

For $t > t_{\mathrm{I}}$ a failure front Σ, characterised by the condition

$$\psi = 0 \tag{14}$$

will travel through the body.

Under given external loads, the redistribution of stress will generally accelerate as soon as $t > t_{\mathrm{I}}$. This time period is called *the stage of propagation of failure*. When ψ has reached the value $\psi = 0$ all through the body, its carrying capacity is expended and *creep rupture is completed* at $t = t_{\mathrm{II}}$. The simple example of an ordinary tensile test specimen with reduced sectional area along the gauge length shows that it is by no means necessary that $\psi \equiv 0$ all over the entire body. It is sufficient that $\psi \equiv 0$ throughout a cross-section in order that separation shall take place and $t = t_{\mathrm{II}}$ shall be reached.

In most applications the influence of transient and primary creep is neglected. It is simply assumed that the state of stress in the body be determined from a state of secondary (stationary) creep, the constitutive equations of which have the form

$$\dot{\varepsilon}_{ij} = \tfrac{3}{2}\dot{\varepsilon}_c \left(\frac{\sigma_e}{\sigma_c}\right)^{n-1} \frac{S_{ij}}{\sigma_c} \tag{15}$$

where $\dot{\varepsilon}_{ij}$ is the creep rate tensor, S_{ij} the stress deviation tensor, σ_c and n are the material constants already introduced and $\dot{\varepsilon}_c$ may be arbitrarily chosen, say $=10^{-7}$/hr. The quantity σ_e is defined by eqn (1) as before.

As already stated, eqn (14) represents geometrically a surface Σ travelling through the body into that part for which $\psi > 0$, the normal N of Σ being directed towards this region.

We may then, differentiating eqns (12) and (14), compute the velocity $\mathrm{d}N/\mathrm{d}t$ of the failure front Σ. Once we have

$$\frac{\partial \psi}{\partial t} + \frac{\partial \psi}{\partial n}\frac{\mathrm{d}N}{\mathrm{d}t} = 0 \tag{16}$$

but from eqn (12)

$$-(\nu + 1)t_K \sigma_K{}^{\nu} \psi^{\nu} \frac{\partial \psi}{\partial t} = \sigma_{\max}{}^{\nu}$$

$$-(\nu + 1)t_K \sigma_K{}^{\nu} \psi^{\nu} \frac{\partial \psi}{\partial N} = \frac{\partial}{\partial N}\int_0^t (\sigma_{\max})^{\nu}\, \mathrm{d}t'$$

and these expressions for the derivatives of ψ are inserted in eqn (16) to yield Kachanov's equation

$$\frac{\mathrm{d}N}{\mathrm{d}t} = -\frac{(\sigma_{\max}{}^{\nu})_{\Sigma}}{\left\{\dfrac{\partial}{\partial N}\displaystyle\int_0^t (\sigma_{\max})^{\nu}\, \mathrm{d}t\right\}_{\Sigma}} \tag{17}$$

the terms within brackets being evaluated at the surface Σ as limit values from the side where $\psi > 0$. Equation (17) proves that, in general, $\mathrm{d}N/\mathrm{d}t$ is a finite quantity. This equation is an integro-differential equation for the determination of Σ as a function of time. In refs. 8 and 10 a number of examples of practical interest are presented.

Where the ratio $t_{\mathrm{II}}/t_{\mathrm{I}}$ depends on the exponents ν and n, knowledge of this ratio is of importance to designers or computing engineers. If a structure, as often happens in practice, is dimensioned with respect to t_{I}, the fact that $t_{\mathrm{II}}/t_{\mathrm{I}} > 1$ adds an extra safety factor, which would be useful to know. In all cases considered so far, the ratio $t_{\mathrm{II}}/t_{\mathrm{I}}$ remains below about 1·5. The following section indicates that much higher values are possible.

CREEP RUPTURE OF THICK WALLED TUBES WITH INTERNAL PRESSURE

This problem has been treated by Kachanov, who integrated numerically eqn (17).

For a thick walled cylindrical tube, with sectional dimensions as in Fig. 4, the circumferential stress σ_θ in points far from the ends of the tube where plane strain may be assumed, will be [8]:

$$\sigma_\theta = p\,\frac{b^{-2/n} - [1 - (2/n)]r^{-2/n}}{a^{-2/n} - b^{-2/n}} \tag{18}$$

Here p is the internal pressure, a and b the inner and outer radii and r the centre distance for an arbitrary point of the tube. In this case $\sigma_{\max}$ everywhere coincides with σ_θ. It is seen from eqn (18) that $\sigma_{\max}$ reaches

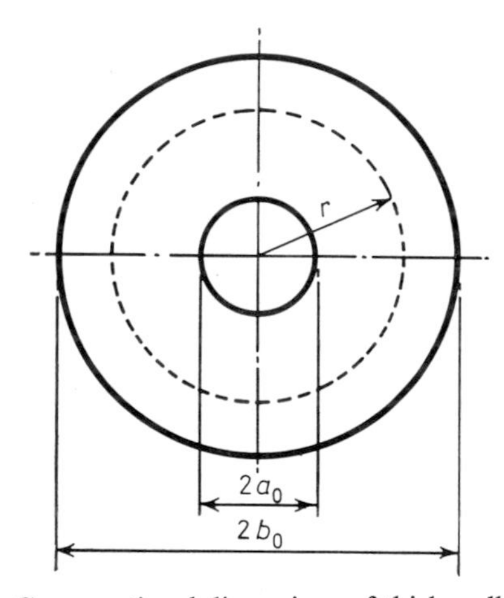

FIG. 4. Cross-sectional dimensions of thick-walled tube.

its largest value for $r = b$, if $n > 2$, and for $r = a$, if $n < 2$. For $n = 2$, σ_θ remains a constant. In this latter case we have always $t_{\mathrm{II}} = t_{\mathrm{I}}$. We are interested in exploring the values of $t_{\mathrm{II}}/t_{\mathrm{I}}$ as a function of the exponents n and ν for different values of the ratio $\beta = b/a$. In order to achieve this, we have to distinguish the two cases $n > 2$ and $n < 2$,

(1) $n > 2$

Here we put $b/a = \beta$, $r/a = \rho$. Failure will start from the outer radius $\rho = \beta$.

Equation (13) yields

$$t_K \left(\frac{\sigma_K}{p}\right)^{\nu} = \int_0^t \left[(\beta')^{-2/n} - \left(1 - \frac{2}{n}\right)\beta^{-2/n}\right]^{\nu} \left[1 - (\beta')^{-2/n}\right]^{-\nu} \mathrm{d}t' \tag{19}$$

where $\beta = \beta'$ corresponds to $t = t'$. If eqn (19) is differentiated with respect to t and we remember $\mathrm{d}/\mathrm{d}t = \mathrm{d}\beta/\mathrm{d}t \cdot \mathrm{d}/\mathrm{d}\beta$, we obtain

$$0 = \frac{n}{n-2} \frac{(2/n)^{\nu-1}\beta^{1+2(1-\nu)/n}}{\nu(1-\beta^{-2/n})^{\nu}}$$

$$+ \frac{\mathrm{d}\beta}{\mathrm{d}t} \int_0^t \left[(\beta')^{-2/n} - \left(1 - \frac{2}{n}\right)\beta^{-2/n}\right]^{\nu-1} \left[1 - (\beta')^{-2/n}\right]^{-\nu} \mathrm{d}t' \tag{20}$$

this being a special case of the general integro-differential eqn (17) to determine the function $\beta = \beta(t)$.

Initial values are for $t = 0$, $b = b_0$, $\beta = \beta_0 = b_0/a$. Moreover we have the corresponding values of the variables

$$\begin{aligned} t' &= 0 \ldots t_{\mathrm{I}} \ldots t \ldots t_{\mathrm{II}} \\ \beta' &= \beta_0 \ldots \beta_0 \ldots \beta \ldots 1 \end{aligned} \tag{21}$$

In the interval $0 < t' < t_{\mathrm{I}}$ we have $\beta = \beta_0 =$ constant. The value of $\mathrm{d}\beta/\mathrm{d}t$ for $t = t_{\mathrm{I}}$ may then be easily obtained from eqn (20):

$$\left(\frac{\mathrm{d}\beta}{\mathrm{d}t}\right)_{\mathrm{I}} = -\frac{n}{\nu(n-2)} \frac{\beta_0}{t_{\mathrm{I}}} \tag{22}$$

Numerical integration has been carried out for $\beta_0 = 2$ by Rosenblium and by Kachanov [10] in the two cases $\nu = n = 4$ and 6. The result is reproduced in Fig. 5.

A great simplification is attained in the special case $\nu = 1$. Introducing β as a new independent variable, differentiating with respect to t, and putting $\mathrm{d}t/\mathrm{d}\beta = q$, we may then write eqn (20) in the form

$$0 = \frac{n}{n-2} \frac{\mathrm{d}}{\mathrm{d}\beta} [\beta(1 - \beta^{-2/n})^{-1} q] + q(1 - \beta^{-2/n})^{-1} \tag{23}$$

an ordinary differential equation of the second order for t as a function of β, to be integrated with the conditions.

$$\begin{aligned} \beta &= \beta_0 \\ t &= t_{\mathrm{I}} \\ q &= \frac{\mathrm{d}t}{\mathrm{d}\beta} = -\frac{n-2}{n} \frac{t_{\mathrm{I}}}{\beta_0} \end{aligned} \tag{24}$$

Elementary integration and satisfaction of eqn (24) yields

$$t = -\frac{n-2}{n}\frac{\beta_0 t_{\mathrm{I}}}{\beta_0^{2/n}-1}\left[\frac{1}{\beta}-\frac{n\beta^{2/n}}{(n-2)\beta}\right]-\frac{2}{n}\frac{t_{\mathrm{I}}}{\beta_0^{2/n}-1} \tag{25}$$

For $t = t_{\mathrm{II}}$ the failure front has reached the inner surface $\beta = 1$. Thus we obtain the simple final result:

$$\frac{t_{\mathrm{II}}}{t_{\mathrm{I}}} = \frac{2}{n}\frac{\beta_0 - 1}{\beta_0^{2/n} - 1} \tag{26}$$

(2) $1 < n < 2$

Although most structural metals belong to the group $n > 2$, the case $1 < n < 2$ may also claim some interest in view of applications to certain polymers and to concrete. Mathematically this case turns out to be simpler

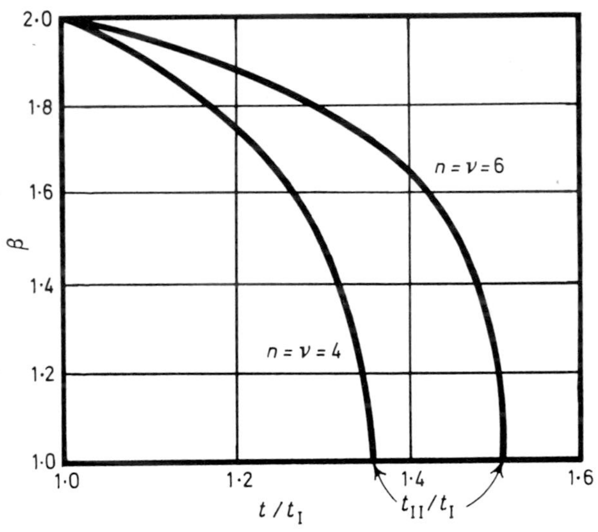

FIG. 5. Propagation of failure front, β as a function of t/t_{I} for $\beta_0 = 2$ and $\nu = n = 4$ and 6 (Kachanov).

than case (1) and in fact permits a solution in closed form. Here we may put in eqn (18) $b = b_0 =$ constant, $r/b_0 = \rho$, $a/b_0 = \alpha$, $a_0/b_0 = \alpha_0$ and so α becomes a function of t to be determined. We have

$$\sigma_{\max} = p\frac{1 + [(2/n) - 1]\rho^{-2/n}}{\alpha^{-2/n} - 1} \tag{27}$$

the maximum of which occurs at the inner surface $\rho = \alpha_0$ of the tube. Fixing our attention to the point $\rho = \alpha$, eqn (13) now has the form

$$t_K \left(\frac{\sigma_K}{p}\right)^\nu = \left[1 + \left(\frac{2}{n} - 1\right)\alpha^{-2/n}\right]^\nu \int_0^t \frac{dt'}{[(\alpha')^{-2/n} - 1]^\nu} \tag{28}$$

with the following correspondence of the variables

$$t' = 0 \ldots t_{\mathrm{I}} \ldots t \ldots t_{\mathrm{II}}$$
$$\alpha' = \alpha_0 \ldots \alpha_0 \ldots \alpha \ldots 1$$

For $t = t_{\mathrm{I}}$ we obtain from eqn (28)

$$t_K \left[\frac{\sigma_K}{p}\right]^\nu = t_{\mathrm{I}} \left[\frac{1 + [(2/n) - 1]\alpha_0^{-2/n}}{\alpha_0^{-2/n} - 1}\right]^\nu \tag{29}$$

If the factor outside the integral of the right member in eqn (28) is moved to the left and the equation is then differentiated with respect to t, we obtain an ordinary differential equation of the first order for α to be integrated with the condition $\alpha = \alpha_0$ for $t = t_{\mathrm{I}}$. For $t = t_{\mathrm{II}}$ we have $\alpha = 1$. The final result will be

$$\frac{t_{\mathrm{II}}}{t_{\mathrm{I}}} = 1 + \left[\frac{1 + [(2/n) - 1]\alpha_0^{-2/n}}{\alpha_0^{-2/n} - 1}\right]^\nu \int_{A_{\mathrm{I}}}^{A_{\mathrm{II}}} \left[\frac{n}{2 - n}(A^{-1/\nu} - 1) - 1\right]^\nu \mathrm{d}A$$

$$A_{\mathrm{I}} = \left[1 + \left(\frac{2}{n} - 1\right)\alpha_0^{-2/n}\right]^{-\nu} \tag{30}$$

$$A_{\mathrm{II}} = \left(\frac{n}{2}\right)^\nu$$

Also in this case a great simplification is obtained in the special case $\nu = 1$. We shall justify ourselves here by working out the final result in this case only. We obtain, putting $\alpha_0 = 1/\beta_0$

$$\frac{t_{\mathrm{II}}}{t_{\mathrm{I}}} = 1 + \frac{1 + [(2/n) - 1]\beta_0^{2/n}}{(\beta_0^{2/n} - 1)(2 - n)} \left\{n \ln \frac{n}{2}\left[1 + \left(\frac{2}{n} - 1\right)\beta_0^{2/n}\right] - n + \frac{2}{1 + [(2/n) - 1]\beta_0^{2/n}}\right\} \tag{31}$$

Figures 6 and 7 show the results of numerical calculations carried out so far. They give the ratio t_{II}/t_I in some cases of interest. Figure 6 shows t_{II}/t_I as a function of $\beta = \beta_0$ in the special case $\nu = 1$. Figure 7 shows t_{II}/t_I as a function of n for different values of ν in the special case $\beta_0 = 2$.

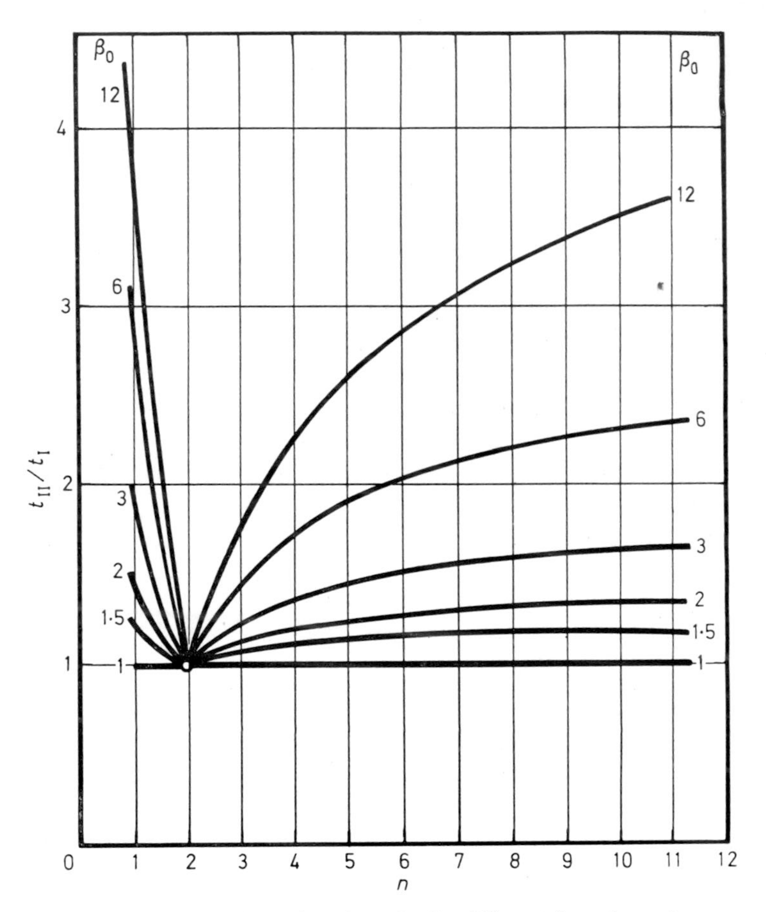

FIG. 6. Ratio t_{II}/t_I as function of n for different β_0 and $\nu = 1$.

It is based upon our Fig. 6 and the two points of Kachanov, taken from Fig. 5; the dotted curves are extrapolations and must be taken with caution but the general trend may be seen from the figures. In any case, values $t_{II}/t_I = 2$ to 3, *i.e.* greater than the previously found values, seem to be quite realistic.

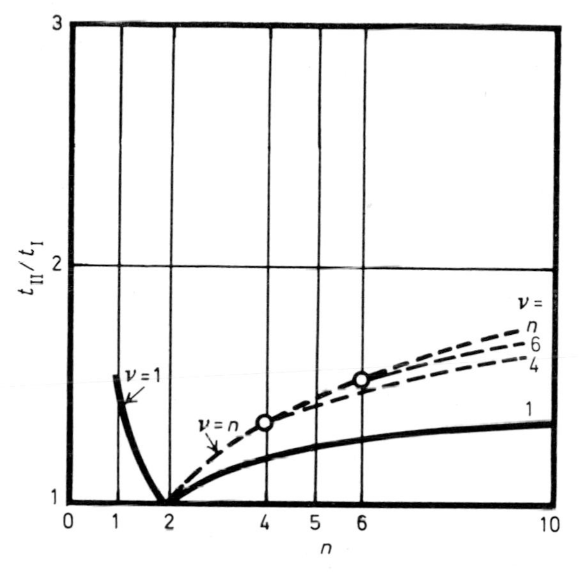

FIG. 7. Ratio t_{II}/t_I as function of n for different ν and $\beta_0 = 2$. Marked points by Kachanov for $\nu = n = 4$ and 6.

CONCLUSIONS

The foundation of Kachanov's theory of creep rupture under multiaxial states of stress was laid by Dr Johnson.

It is noticed that for values of Norton's exponent $n > 2$, *i.e.* for most structural metals in their ordinary working range, creep rupture starts from the outer surface of a tube under internal pressure. Observations in support of this behaviour were made already by Kachanov [10]. Further investigations by Tucker, Coulter and Kooistra [14] carried out on tubes of different steels under internal pressure point in the same direction. These observations constitute, in the writer's opinion, an additional strong support for Kachanov's theory of creep rupture.

A new result here is that the ratio t_{II}/t_I of the time of final separation to the time of incubation may easily reach values of more than 2 for tubes where the outer radius does not exceed four or five times the inner one.

All these results relate to isothermal conditions. The corresponding problem for tubes with added thermal gradient through the tube wall should be the subject of further research, and here Dr Johnson had already made preliminary investigations [4].

REFERENCES

1. Finnie, I. (1966). 'Stress analysis for creep and creep rupture.' Abramson, N. (Ed.), *Mechanics Surveys*. New York: Spartan.
2. Johnson, A. E. (1960). 'Complex stress creep of metals.' *Metall. Rev.*, **5**(20), 447–506.
3. Johnson, A. E., Henderson, J. and Khan, B. (1963–64). Multiaxial creep strain/complex-stress/time relations for metallic alloys with some applications to structures.' *Proc. Jnt Int. Conf. on Creep. Proc. Instn Mech. Engrs*, **178**(3A), 2-27–2-43.
4. Johnson, A. E. (1967). 'Some progress in creep mechanics.' *Folke Odqvist* 289–327. London: John Wiley & Sons.
5. Voorhees, H. R. and Freeman, J. M. (1956). 'Notch sensitivity of heat resistant alloys at elevated temperatures.' WADC, T.R.54-175/2, and WADC, T.R.59-470. Ohio: Wright Air Development Center.
6. Weibull, W. (a) (1939). 'A statistical theory of the strength of materials.' *Proc. R. Swedish Inst. Engng. Res.*, **151.** (b) (1939). 'The phenomenon of rupture in solids.' *Proc. R. Swedish Inst. Engng. Res.*, **153.**
7. Söderquist, B. (1968). 'Creep rupture under uniform radial tension of a disk with a circular hole.' *Acta. Polytech. Scand.* (ser. f), No. 53, 1–36.
8. Odqvist, F. K. G. (1966). *Mathematical Theory of Creep and Creep Rupture*. Oxford: Clarendon Press.
9. Johnson, A. E. (1951). 'Creep under complex stress systems at elevated' temperatures.' *Proc. Instn. Mech. Engrs.*, **164**(4), 432–447.
10. Kachanov, L. M. (a) (1958). 'On the time to failure under creep conditions (in Russian). *Izv. A.N. USSR, Otd. Tekhn. N.*, **8,** 26–31. (b) (1961). *Problems in Continuum Mechanics Contributions in Honour of Seventieth Birthday of N. I. Muskhelishvili* (Ed. J. R. M. Radok), Philadelphia, pp. 202–218. (c) (1960). *Theory of Creep*. Moscow.
11. Hoff, N. J. (1953). 'The necking and the rupture of rods subjected to constant tensile loads.' *J. Appl. Mech.*, **20,** 105–108.
12. Richard, K. (1955). 'Results obtained by the co-operative group on long time tests' (in German). *Mitt. Ver. Grossdampfk. Bes.* (39), 836–842.
13. Rabotnov, Yu. N. (a) (1963). 'On creep rupture' (in Russian). *Prikl. Mekh. Tekh. Fys. A.N. USSR*, **1,** 113–123. (b) (1963). 'On the equation of state of creep.' Jnt. Int. Conf. on creep, ASME, ASTM, *Instn. Mech. Engrs.* New York, London, paper 68. (c) (1966). *Creep of Structural Elements* (in Russian). Moscow.
14. Tucker, J. T., Coulter, E. E. and Kooistra, L. F. (1960). 'Effect of wall thickness on stress-rupture life of tubular specimens.' *J. Basic Engng.*, **82D**, 465–476.

Chapter 4

SOME STRUCTURAL THEOREMS OF CREEP AND THEIR IMPLICATIONS

F. A. LECKIE

SUMMARY

The implications of some energy theorems which have recently been discovered are discussed in relation to the performance of structures and materials which suffer creep and plastic deformations.

NOTATION

A	Surface.
A_T	Loaded surface.
A_u	Surface with zero displacement.
b, d	Breadth and depth of beam.
C_{ijkl}	Elastic constants.
$\dot{D}_c$	Rate of dissipation of creep energy.
E	Modulus of elasticity.
$\dot{e}_{ij}$	Elastic strain rate.
$H(t)$	Heaviside step function.
$k(t)$	Function of time.
M	Bending moment.
P_i	Applied loading.
P_i^L	Limit load.
P_i^s	Shakedown load.
P_i^*	Spatial loading function.
$\dot{p}_{ij}$	Plastic strain rate.
t	Time.

U^0	Elastic energy corresponding to initial elastic solution.
U^s	Elastic energy corresponding to stationary stress solution.
$\dot{u}_i$	Displacement rate.
V	Material volume.
$\dot{v}_{ij}$	Creep strain rate.
$\dot{v}_0$	Material constant.
x	Spatial ordinate.
α, β	Defined by eqn (14).
ε	Total strain.
$\dot{\varepsilon}_{ij}$	Total strain rate.
$\dot{\varepsilon}_0$	Material constant.
$\bar{\varepsilon}$	Strain due to $\bar{\sigma}$.
κ	Curvature.
κ_e	Elastic curvature.
κ_s	Curvature due to stationary stresses.
λ	Proportionality constant.
λ_1	Value of λ at first yield.
σ	Applied uniaxial stress.
$\bar{\sigma}$	Stress defined by eqn (13).
σ_{ij}	Stress tensor.
$\bar{\sigma}_{ij}$	Stress distribution.
σ_0	Material constant.
σ_y	Uniaxial yield stress.
$\sigma_1, \sigma_2, \sigma_3$	Principal stresses.
$\psi(t)$	Time function.

Superscripts

a, b, c	Arbitrary indices.
n	Material constant.

INTRODUCTION

It is well known that it is possible to determine the stress distribution and deformations of a structure in a state of creep, if an iteration technique is used in conjunction with the elastic solution and the constitutive relations of the material. On this basis, procedures have already been developed for the solution of creep problems by computers [1], but it would appear that there is as yet insufficient creep information to justify the computational effort. Moreover, it is unlikely that sufficient creep information will be available in the immediate future, because creep testing of materials under non-proportional loading and temperature variation is extremely difficult and time consuming; and tests [2] which have already been carried out indicate an unpromising future. Further complications arise when plastic deformations are included in regions of high stress concentration. The neglect of the plastic contribution to strain could possibly result in predictions of deformation considerably less than those actually observed.

The addition of plastic strains will make computational work even more complicated, and while computation is still conceptually possible it is not difficult to imagine problems completely outside the potential of existing computers. Furthermore computer solutions are unlikely to yield the type of general result which appear in the theory of plasticity and which prove to be of such power in design.

In this chapter, some general results are presented on the basis of certain material idealisations. These idealisations provide a reasonable approximation to material behaviour and yet remain sufficiently simple to be dealt with by conventional small deflection mechanics. The significance of some of the general results is discussed in relation to laboratory testing and design.

MATERIAL IDEALISATION

It is assumed that the total strain rate can be separated into an elastic, a time hardening creep and a plastic component in the following way

$$\dot{\varepsilon}_{ij} = \dot{e}_{ij} + \dot{v}_{ij} + \dot{p}_{ij} \tag{1}$$

where $\dot{\varepsilon}_{ij}$ is the total strain rate, $\dot{e}_{ij}$ is the elastic strain rate, $\dot{v}_{ij}$ is the creep strain rate and $\dot{p}_{ij}$ is the plastic strain rate. The elastic strain is assumed to be a linear function of stress so that

$$e_{ij} = C_{ijkl}\sigma_{kl} \tag{2}$$

the constant C_{ijkl} being independent of time and stress history. The creep strain rate is assumed to obey the general time hardening law of the form

$$\frac{\dot{v}_{ij}}{\dot{v}_0} = \varphi^n \frac{\partial\varphi}{\partial(\sigma_{ij}/\sigma_0)} k(t) \tag{3}$$

where φ is homogeneous of degree one in σ_{ij}/σ_0 and has the value unity when σ_{ij} is the uniaxial tension σ_0. The material constants are $\sigma_0 \dot{v}_0$ and n, and $k(t)$ is a positive function of time. It can be argued that time hardening laws are physically unsound, but it has been shown [3] that provided $k(t)$ is properly chosen, the time hardening law is satisfactory.

The rate of dissipation of creep energy is given by

$$\dot{D}_c(\sigma_{ij}) = \sigma_{ij}\dot{v}_{ij} = \sigma_0\dot{v}_0\varphi^{n+1}k(t)$$

where $\sigma_0\dot{v}_0$ and $k(t)$ are positive and so the term φ^{n+1} is always positive.

The assumption of homogeneity implies that reversal of stress reverses the sign of the creep strain rate, but does not imply material isotropy.

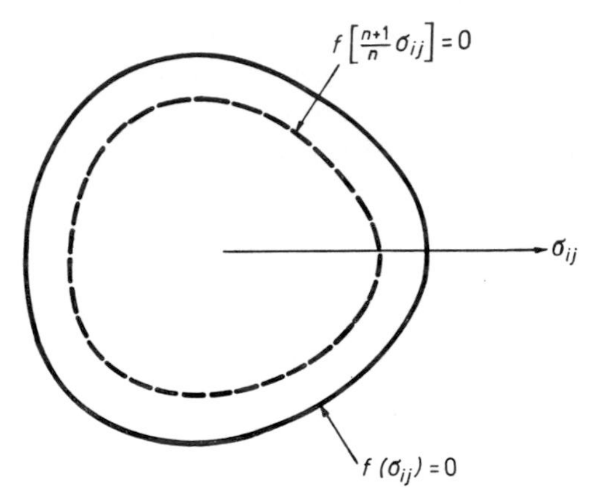

FIG. 1. Yield surface.

At any particular instant, the inequality discussed by Martin [4] can be applied to two states of stress and creep strain rate $\sigma_{ij}{}^a$, $\dot{v}_{ij}{}^a$ and $\sigma_{ij}{}^b$, $\dot{v}_{ij}{}^b$ which satisfy eqn (3) at that particular instant. Hence

$$\frac{1}{n+1}\sigma_{ij}{}^a\dot{v}_{ij}{}^a + \frac{n}{n+1}\sigma_{ij}{}^b\dot{v}_{ij}{}^b \geq \sigma_{ij}{}^a\dot{v}_{ij}{}^b \tag{4}$$

Plastic deformation can occur when the stress σ_{ij} is on the convex yield surface defined by $f(\sigma_{ij}) = 0$ (Fig. 1). In the region interior to $f(\sigma_{ij}) = 0$ then $f(\sigma_{ij}) < 0$ and in the region exterior to $f(\sigma_{ij}) = 0$ then $f(\sigma_{ij}) > 0$. Plastic deformations are normal to the yield surface at σ_{ij}, and, on account of the convexity of the yield surface,

$$(\sigma_{ij} - \sigma_{ij}{}^c)\dot{p}_{ij} > 0 \tag{5}$$

where $\sigma_{ij}{}^c$ is any stress within or on the yield surface.

If it is assumed that the material is isotropic, incompressible and that strain rates are independent of hydrostatic pressure, then in principal stress space, surfaces of constant creep energy dissipation rate and the yield surfaces form cylinders with common axis $\sigma_1 = \sigma_2 = \sigma_3$. On a surface $\sigma_1 + \sigma_2 + \sigma_3 =$ constant, the yield surface and a surface of constant creep energy dissipation rate lie between the Tresca hexagon and the maximum reduced stress hexagon, Fig. 2. The yield surface usually lies near the von Mises circle, and therefore cannot deviate very much from the creep dissipation surface. As both creep and plastic strains are curbed

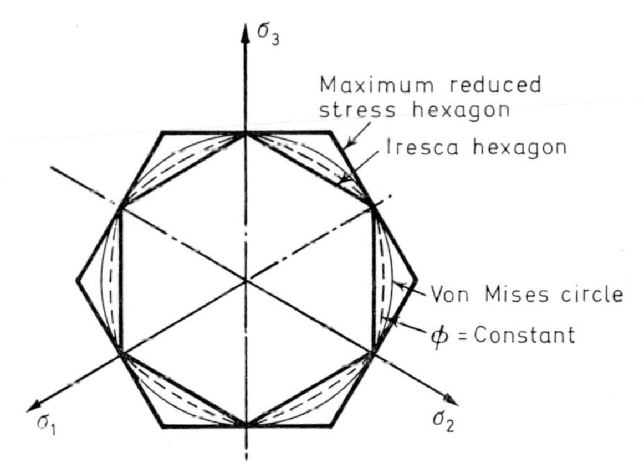

FIG. 2. Yield surfaces and surfaces of constant energy dissipation.

by similar mechanisms, it appears reasonable, considering the lack of experimental evidence, to assume the two surfaces coincide. Thus, constant creep energy dissipation rate surfaces are sometimes assumed to be similar to the yield surface. Although this is not a necessary restriction in the following work, it can provide considerable simplification.

STRUCTURAL THEOREMS

On the basis of the constitutive eqns (1), (2) and (3) the resulting inequalities (4) and (5) and the virtual work principle certain structural theorems have been obtained [3, 5]. Those theorems which appear to be of greatest significance arise (a) when elastic and creep deformations are considered to be of significance and plastic deformations are non-existent, and (b) when elastic deformations may be neglected by comparison with creep and plastic deformations.

Elastic and creep deformations

The problem considered is that of a body consisting of an elastic creeping material which is subjected to proportional loading on the surface A_T of the form

$$P_i = P_i^*(x)\psi(t)$$

and the displacements on the remaining surface A_u remain zero. The stress and strain distributions vary with time because of the variation in load and the interaction between the elastic and creep strains. As far as the accumulation of strain is concerned it is convenient to consider the strains in terms of the elastic solution, the stationary solution and the effect of stress redistribution.

For the case of an initially stress free body subjected to a step loading of the form

$$P_i = P_i^*(x)H(t)$$

where $H(t)$ is the Heaviside step function, it has been shown by Leckie and Martin [3] that the dissipation of energy due to stress redistribution lies between the limits

$$(U^s - U^0) \quad \text{and} \quad (n + 2)(U^s - U^0) \tag{6}$$

In these expressions U^s is the elastic energy corresponding to the stationary stress solution and U^0 is the initial elastic solution. Marriott [6] has shown in a less rigorous way that the energy dissipation is approximately $(n + 3)/2$ $(U^s - U^0)$. On account of the theorem of minimum complementary energy it is to be expected that $(U^s - U^0)$ is small by comparison with U^0 and provided n is not greater than about 5 the effect of stress redistribution is small.

So far it appears that equivalent results have not been determined for variable loading, although Marriott on the basis of calculations made on simple structures concludes that the effect of stress redistribution is unlikely to be of great significance.

Plastic and creep deformations

A theorem with design significance [5] can be obtained for situations in which plastic and creep deformations predominate and elastic deformations can be neglected. With this assumption, no stress redistribution takes place and the stress distribution remains constant. Under variable loading the stress distribution remains identical but the magnitude varies in direct proportion to the value of the load. This stress distribution and the corresponding deformations are referred to as the stationary plastic creep solution [7].

The theorem gives an upper bound to the rate of work done by the external forces in terms of stress distributions $\bar{\sigma}_{ij}$ in equilibrium with the applied load P_i.

Then

$$\dot{D} = \int_A P_i \dot{u}_i \, \mathrm{d}A < \int_V \bar{\sigma}_{ij} \bar{\dot{v}}_{ij} \, \mathrm{d}V = \int_V \dot{D}_c(\bar{\sigma}_{ij}) \, \mathrm{d}V \tag{7a}$$

provided

$$f\left(\frac{n+1}{n} \bar{\sigma}_{ij}\right) < 0 \tag{7b}$$

It is rather surprising that no plastic strains are involved in the inequality, but it is noted that the stress $\bar{\sigma}_{ij}$ must lie within a surface which is $n/(n+1)$ of the yield surface and consequently is only applicable to loads less than $n/(n+1)$ of the limit load of the structure.

The result of eqn (7) can be simplified if the assumption of material behaviour described under 'Material Idealisation' is applied to eqn (7).

For a structure in which all points of the body are at a state of plastic yield (σ_{ij}^L) at collapse then the stress distribution $\lambda\sigma_{ij}^L$ is in equilibrium with the load λP^L and remains on a surface of constant $\dot{D}_c$.

Then the result of eqn (7) becomes

$$\int_A P_i \dot{u}_i \, \mathrm{d}A < V\dot{D}_c(\lambda\sigma_y) \quad \text{provided} \quad \lambda < \frac{n}{n+1} \tag{8}$$

where σ_y is the uniaxial yield stress and V is the material volume. This result provides a bound on the total energy dissipation rate in terms of the limit load and data from a single uniaxial test.

For a minimum weight design structure, the plastic energy dissipation rate remains constant throughout the body at collapse. The stresses $\lambda\sigma_{ij}^L$ will also be on a surface of constant creep energy dissipation, and it is evident that $\lambda\sigma_{ij}^L$ provides the creep solution for $P_i = \lambda P_i^L$, since the plastic strain rates at collapse and the creep strain rates given by this distribution of stress differ only by a constant multiple throughout the body. Thus, plastic deformation does not commence until $\lambda = 1$ and equality holds in eqn (7) when $\bar{\sigma}_{ij} = \lambda\sigma_{ij}^L$, for all $\lambda < 1$. Further, as $\dot{D}_c$ is constant throughout the body and equal to the value associated with a uniaxial stress of $\lambda\sigma_y$, then [8]:

$$\int_A P_i \dot{u}_i \, \mathrm{d}A = V\dot{D}_c(\lambda\sigma_y) \tag{9}$$

SIGNIFICANCE OF THEOREMS

Loading levels on structures

The limit load P_i^L of the structure remains unaffected by the presence of creep, and for proportional type loading it is convenient to consider states of loading λP_i^L where $\lambda < 1$.

For $\lambda < \lambda_1$ no plastic deformations occur, and at $\lambda = \lambda_1$ the maximum stresses reach the yield stress. Within the range $\lambda_1 < \lambda < 1$, both plastic and creep strain occur and the deformation rate of the body increases rapidly as λ approaches unity, achieving infinite values at $\lambda = 1$. This behaviour is demonstrated by the analytic solution for beams under pure flexure in Fig. 3, and a thick cylinder under internal pressure in Fig. 4.

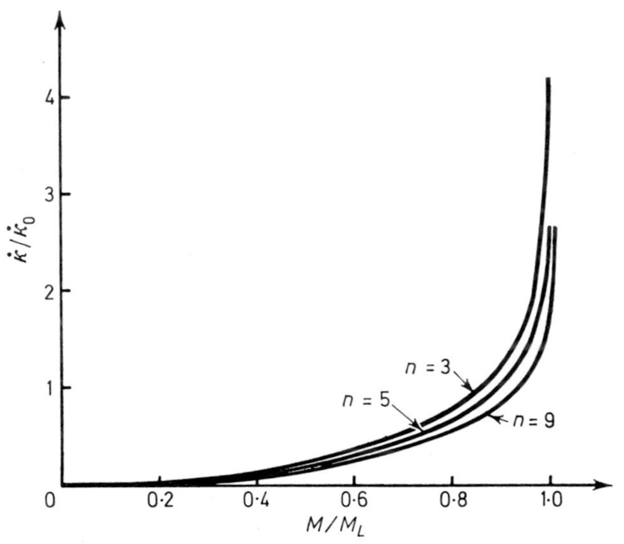

FIG. 3. Curvature rate/moment relations for beam in pure flexure.

As the upper bound in eqn (7) involves only the creep energy dissipation rate, it implies that for $\lambda < [n/(n + 1)]$, the plastic energy dissipation rate remains small compared with the creep energy dissipation rate. Thus, the effects of severe stress concentrations have little effect on the total deformation rates.

Since it has been shown that for a minimum weight design plastic deformation does not commence until $\lambda = 1$, it is clear that the previous statement is conservative and that for most structures which are neither wholly 'good' nor wholly 'bad' the value of λ at which plastic strains become significant will lie between $n/(n + 1)$ and unity.

The few analytic solutions [8] which have been obtained do show that once plastic deformations become significant then the deformation rates

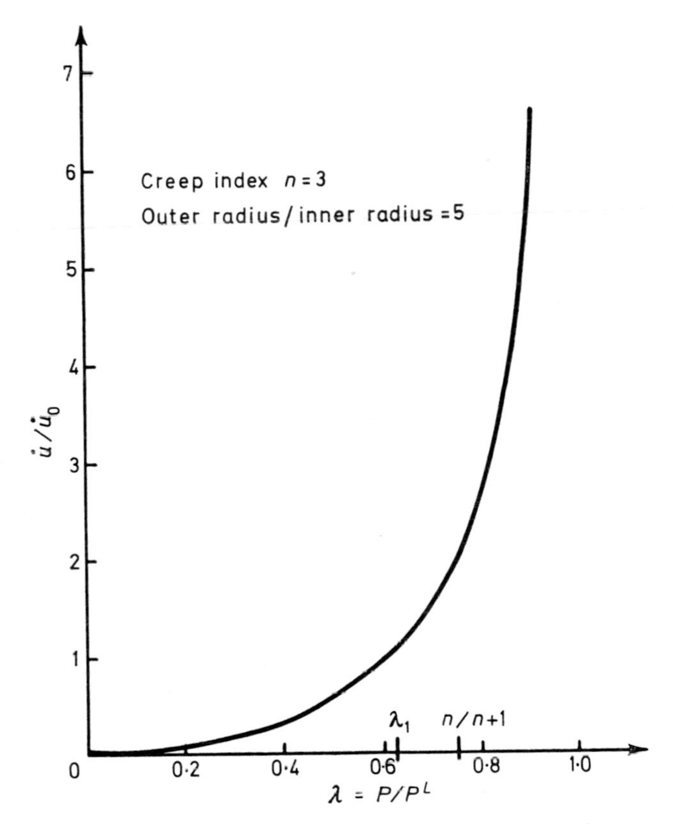

FIG. 4. Creep rate at bore for thick cylinder under internal pressure.

increase very rapidly, and the inference seems to be that the loading parameter for structures in the creeping range should not exceed a value of $\lambda = [n/(n + 1)]$.

In a private communication to the author, Dr Ponter reported that a similar result can be obtained for structures subjected to variable loading. His result implies that if the shakedown load is P_i^s then incremental plastic deformations can be neglected if the maximum value of the variable loading is less than $n/(n + 1)\ P_i^s$.

Determination of test stress level

For structures subjected to loads less than $n/(n+1)\,P_i^L$ it is possible using the result of eqn (6) to isolate a single uniaxial test which can be identified with the creep performance of the structure. This may be represented symbolically as shown in Fig. 5.

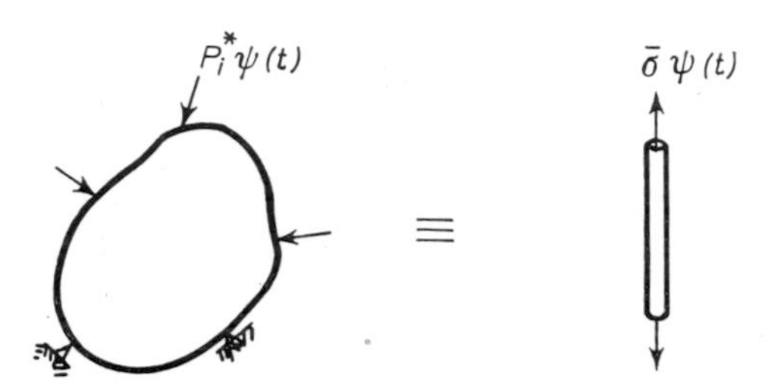

FIG. 5. Equivalent uniaxial stress.

The process can be most readily illustrated in terms of a beam under pure bending. The effect of stress redistribution is shown in Fig. 6 for a rectangular beam subjected to constant bending moment. It can be seen that the effect in this case is about 10 per cent of the initial elastic response, a result in keeping with the result of eqn (6).

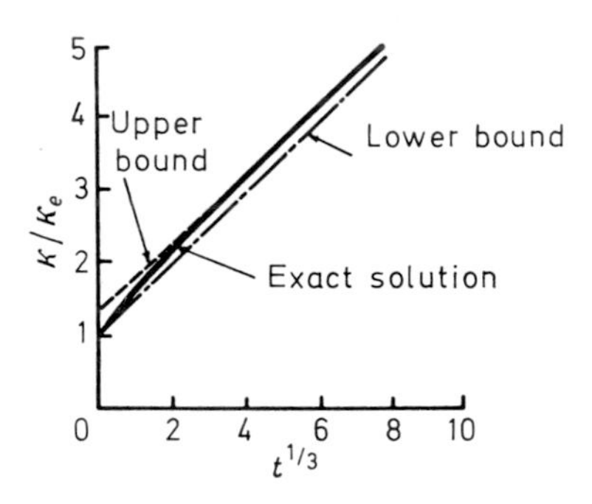

FIG. 6. Transient solution for rectangular beam under pure bending.

If the constitutive law for applied uniaxial stress is

$$\varepsilon = \frac{\sigma}{E} + \dot{\varepsilon}_0 \left(\frac{\sigma}{\sigma_0}\right)^n f(t)$$

then the elastic curvature is $\kappa_e = 12M/Ebd^3$ and the curvature due to the stationary stresses is

$$\kappa_s = \left(\frac{2n+1}{n}\right)^n 2^{n+1} \left(\frac{\dot{\varepsilon}_0}{\sigma_0{}^n b^n d^{2n+1}}\right) M^n f(t)$$

From the previous work

$$M\kappa = M\kappa_e + \left(\frac{2n+1}{n}\right)^n 2^{n+1} \left(\frac{\dot{\varepsilon}_0}{\sigma_0{}^n b^n d^{2n+1}}\right) M^{n+1} f(t)$$

$$+ \text{ Terms due to stress redistribution} \qquad (10)$$

and it is assumed that the terms due to stress redistribution can be neglected.

Now assume a stress exists for which

$$\bar{\sigma} = \frac{\alpha M}{bd^2} \qquad (11)$$

then the corresponding strain is

$$\bar{\varepsilon} = \frac{\alpha M}{Ebd^2} + \dot{\varepsilon}_0 \left(\frac{\alpha M}{\sigma_0 bd^2}\right)^n f(t) \qquad (12)$$

An energy balance per unit length of the beam is given by

$$\beta b \, \mathrm{d}\bar{\sigma}\bar{\varepsilon} = M\kappa \qquad (13)$$

If eqns (10), (11) and (12) are substituted in eqn (13), then the following expressions are determined for α and β

$$\alpha = \left(\frac{2n+1}{n}\right)^{n/(n-1)} \frac{2}{3^{1/(n-1)}}; \qquad \beta = \frac{12}{\alpha^2} \qquad (14)$$

From these results, the expression for κ becomes

$$\kappa = \frac{12}{\alpha d} \bar{\varepsilon} \qquad (15)$$

The curvature κ is obtained from the above formula by finding the strain $\bar{\varepsilon}$ associated with the stress $\bar{\sigma}$. For a given value of n, α can be calculated from eqn (14) and the corresponding tensile test is carried out at the stress given by eqn (11). In the case of the beam, the value of α is almost independent of n (Table 1) so that a knowledge of the exact value of n is unnecessary if α is assumed to be 4·10.

TABLE 1

n	1	2	3	4	5	7
α	6	4·16	4·12	4·10	4·08	4·04

This method describes the prescription for determining the deformation of a beam on the basis of one appropriate tensile test. The method, which is similar in spirit to that proposed by MacKenzie [9], can be extended further and will be the subject of later work.

If this procedure is followed then a lower bound will be obtained on the deformation. The procedure is easily extended to give an upper bound by using the upper bound of eqn (6) instead of neglecting the stress redistribution terms. Or, if a computer solution is available, then the appropriate stress can be found by using the approximation illustrated in Fig. 6.

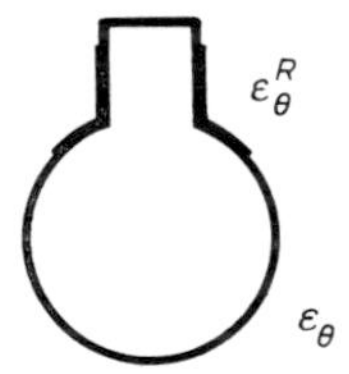

FIG. 7. Reinforced nozzle in pressure vessel.

Calculation of conditions of rupture

The now standard calculations for creep rupture due to Hoff [10] and Odqvist [11] are based on the effects of geometry change on the creep rates. The time to rupture for a tensile specimen, for example, is often based on the time the cross-sectional area takes to become zero. In the book by Hult [12], it is mentioned that the limit would be more reasonably defined when the effective stress becomes equal to the yield stress. The point is

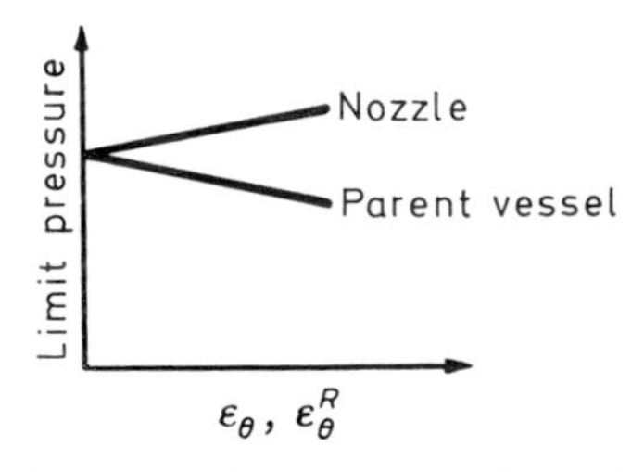

FIG. 8. Change of limit pressure with deformation.

made that thc times to rupture calculated either way are little different and consequently the argument is rather academic. This may be so in the simpler situations but the equivalent of the stress criterion can be readily applied to structures which are more complex.

On the basic of eqn (7) it seems reasonable to define the onset of rupture in a structure when the applied loads P_i become $n/(n + 1)$ of the limit

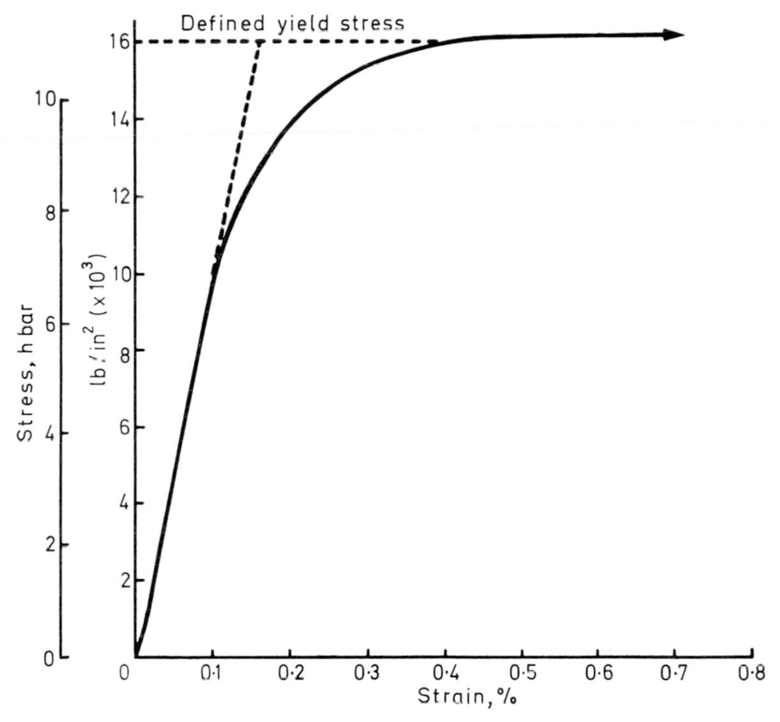

FIG. 9. Stress–strain diagram for aluminium.

load P_i^L. This is a particularly useful criterion since the effect of geometry changes on limit load [13] can readily be calculated. For example, in a pressure vessel of the type shown in Fig. 7 subjected to internal pressure, the nozzle is designed to have the same limit pressure as the parent vessel. However the limit pressure of the nozzle increases with deformation [14] while that of the parent vessel decreases, Fig. 8. Consequently, the possibility of creep rupture occurring in the nozzle can be dismissed. This does not exclude the possibility of the nozzle cracking in areas of high strain, which however is another problem!

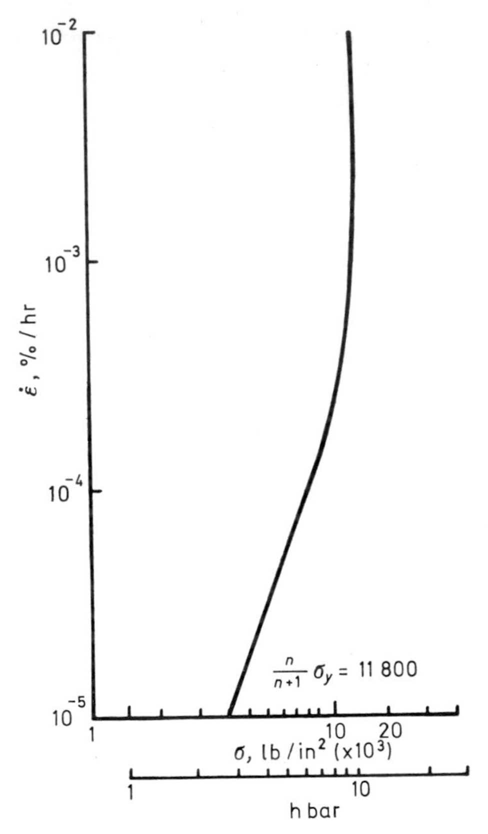

FIG. 10. Creep rate/stress diagram for aluminium.

Behaviour of metals under creep

It has been suggested before [15], that a metal with its agglomeration of crystals can reasonably be regarded as a structure with the crystals as the component parts. In a uniaxial test the limit load can then be identified as the yield stress (Fig. 9) and if the result of eqn (7) were applicable then it follows that creep rates should increase rapidly when the applied stress is in excess of $n/(n + 1)$ of the yield stress. The stress–strain curve and the steady state creep rate–stress curve for an aluminium which creeps at room temperature are shown in Figs. 9 and 10. The curves appear to support the postulate, and similar agreement has been reported verbally by Dr Henderson of the National Engineering Laboratory and Dr MacKenzie of the University of Glasgow for copper at 250°C and the aluminium RR 58 at 200°C.

REFERENCES

1. MENDELSOHN, A., HIRSCHBERG, M. H. and MANSON, S. S. (1959). 'A general approach to the practical solution of creep problems.' *J. Bas. Engng*, **81,** 585–598.
2. JOHNSON, A. E., HENDERSON, J. and KHAN, B. (1963). 'Multiaxial creep strain/complex stress/time relations for metallic alloys with same applications to structures.' *Proc. Instn Mech. Engrs*, **178**(3A), 2-27 2-43.
3. LECKIE, F. A. and MARTIN, J. B. (1967). 'Deformation bounds for bodies in a state of creep.' *J. Appl. Mech.*, **34**(2), 411–417.
4. MARTIN, J. B. (1966). 'A note on the determination of an upper bound on displacement rates for steady creep problems.' *J. Appl. Mech.*, **33**(1), 216–217.
5. LECKIE, F. A. and PONTER, A. R. S. (1968). 'Deformation bounds for bodies which creep in the plastic range.' *Engineering Dept. Report*, 68–10. University of Leicester.
6. MARRIOTT, D. L. (1968). 'Approximate analysis of transient creep deformation.' *J. Strain Anal.*, **3**(4), 288–296.
7. HULT, J. A. H. (1962). 'On the stationarity of stress and strain distribution in creep.' REINER, M. and ABIR, D. (Eds.), *Proc. Symp. Second-order Effects in Elasticity, Plasticity and Fluid Dynamics*. Oxford: Pergamon Press.
8. PONTER, A. R. S. and LECKIE, F. A. (1968). 'The application of energy theorems to bodies which creep in the plastic range.' *Engineering Dept. Report*, 68–11. University of Leicester.
9. MACKENZIE, A. C. (1968). 'On the use of a single uniaxial test to estimate deformation rates in some structures undergoing creep.' *Int. J. Mech. Sci.*, **10**(5), 441–455.
10. HOFF, N. J. (1953). The necking and rupture of rods subjected to constant tensile loads. *J. Appl. Mech.*, **20,** 105–108.
11. ODQVIST, F. K. G. (1966). *Mathematical Theory of Creep and Creep Rupture*. Oxford: Clarendon Press.
12. HULT, J. A. H. (1966). *Creep in Engineering Structures*. Waltham, Mass.: Blaisdell.
13. ONAT, E. T. (1960). 'Influence of geometry changes on the load-deformation behaviour of plastic solids.' LEE, E. H. and SYMONDS, P. S. (Eds.), *Plasticity: Proc. Symp. on Naval Structural Mechanics*, 2nd edn. New York: Pergamon.
14. ALLMAN, D. J. and GILL, S. S. (1968). 'The effect of change of geometry on the limit pressure of a flush nozzle in a spherical pressure vessel.' JEYMAN, J. and LECKIE, F. A. (Eds.), *Engineering Plasticity*, London, New York: Cambridge University Press.
15. LECKIE, F. A., PONTER, A. R. S. and SIM, R. S. (1968). 'On creep rates in metal.' *Engineering Dept. Report*, 68–12. University of Leicester.

Chapter 5

A RATIONAL APPROACH TO CREEP DESIGN

J. HULT AND U. EDSTAM

SUMMARY

Structural creep problems usually involve a non-linear creep law, which gives rise to mathematical difficulties and very often numerical calculations have to be performed at an early stage. The results of such calculations may preferably be presented in a dimensionless form, relating the creep problem to the corresponding problem of linear elasticity. A type of diagram is presented which brings together elasticity, creep and plasticity solutions for various types of problems.

NOTATION

A	Area of cross-section.
C	Constant of integration (eqn 2).
d, e_1, e_2	Defined in Fig. 1.
G	Defined in eqn (4).
H	Defined in Figs. 1, 2 and 3.
I	Moment of inertia.
K	Constant dependent on temperature.
$L_1, L_2, \ldots$	Loads.
M	Bending moment (Fig. 1).
N	Axial force (Fig. 1).
n	Defined by eqn (6) $= p/(1 + q)$.
p, q	Constants dependent on temperature.
S	Moment of area

t	Time.
w, x	Defined in Figs. 4 and 5.
z	Defined in eqn 23.
α	Defined in eqn 23.
β	Defined in eqn 23.
δ	Defined in Fig. 5.
ζ	Defined in Fig. 1.
θ	Dimensionless quantity (eqn 12).
ξ	Dimensionless coordinate, defined in eqn (35).
ρ	Defined in Fig. 1.
ε	Creep strain.
σ	Stress.
σ_e	Largest effective stress.

Superscripts

$'$	Indicates the dimensionless form.
$*$	Applies to the corresponding linearly elastic structure.

INTRODUCTION

In the design of engineering structures, three aspects invariably have to be considered, namely material, shape and loading. Material, shape and loading, however, may appear in unlimited variations, prohibiting any systematic study. Strength of materials, therefore, deals with a limited number of hypothetical, standardised classes of material, shape and loading amenable to analysis. Stresses and displacements have been determined for a number of relevant cases, and ready made formulae, diagrams, charts and tables are now available for the benefit of designers. An increasing number of shapes and loadings is continuously being added to this list, but materials other than the Hookean are very seldom examined. Two reasons have been quoted for this lack of coverage: most engineering materials may be considered as approximately Hookean and most structural problems become extremely complicated for non-Hookean materials. In this age of high temperature technology and high speed computers, neither of these reasons seems particularly valid. Increasing effort has, in fact, been exerted in recent years to develop strength of materials into this third dimension: non-Hookean materials. Among these materials, those displaying creep have dominated. The work of Dr

Johnson was largely devoted to this worthwhile task. A comprehensive survey of studies made between 1940 and 1959 was given by Dr Johnson [1].

Structural creep problems seldom may be analysed in closed form; rather extensive numerical work, usually requiring an electronic computer, must normally be done. This leads to the question of how the results should best be presented to allow maximum practical use. A designer often has to consider the effect of certain alterations in shape, loading or material properties. Computer data are therefore preferably presented in diagrams or charts. In this chapter, a graphical presentation will be demonstrated, which relates creep problems under combined loadings to similar Hookean problems already treated at length in existing literature. Since the theories of creep and linear elasticity are based on the same equations of equilibrium and compatibility, only the constitutive equations being different, much work might be saved by bringing out, in this way, common parts. The first account of the method was given previously [2]. A brief outline will be given here, and its application to a few structural creep problems will then be shown.

CREEP UNDER COMBINED LOADINGS

The constitutive equations describing creep deformation in structural materials are usually non-linear [3]. Two classes of creep behaviour have been established, relating to the development of creep strain, ε, under constant stress, σ:

(1) Primary creep, where the creep strain rate is decelerating.
(2) Secondary creep, where the creep strain rate is constant.

Both these may often, for engineering purposes, be described by a constitutive equation (creep law) of the form

$$\frac{\mathrm{d}\varepsilon}{\mathrm{d}t} = K\frac{\sigma^p}{\varepsilon^q} \tag{1}$$

where $q > 0$ for primary creep and $q = 0$ for secondary creep. The constants K, p and q all depend on temperature. If the stress σ is kept constant, eqn (1) may be integrated to give

$$\varepsilon^{(q+1)} = (q + 1)K\sigma^p t + C \tag{2}$$

The constant of integration, C, is determined from the conditions at

$t = 0$. Since creep strain requires time to develop, $\varepsilon = 0$ at $t = 0$, and hence $C = 0$. From eqn (2) it then follows

$$\varepsilon = G\sigma^{p/(1+q)} \tag{3}$$

where

$$G = [(1 + q)Kt]^{1/(1+q)} \tag{4}$$

From eqn (3) follows the elastic analogy, first stated by Hoff [4] in its general form:

In a structure subjected to a set of constant surface tractions and/or body forces and obeying the creep law of eqn (1), a constant state of stress will eventually develop as if the material obeyed the constitutive eqn (3). The analogy, of course, is also valid, and in fact of most use, for cases of multiaxial stress.

Hence each problem of stationary non-linear creep may be transformed to a problem of non-linear elasticity. For simplicity, eqn (3) will be written as:

$$\varepsilon = (\sigma/\sigma_n)^n \tag{5}$$

where

$$n = \frac{p}{1 + q} \tag{6}$$

$$\sigma_n = [(1 + q)Kt]^{-1/p} \tag{7}$$

After the displacements have been found for this associated problem of non-linear elasticity, the time dependent displacements for the original creep problem are obtained by reinserting σ_n according to eqn (7).

The dimensionless material parameter n, defined by eqn (6), is found, for most engineering materials, to be larger than unity. For many high temperature steels, n is in the range 2–5. For $n = 1$ the constitutive eqn (5) defines the linearly elastic (Hooke) material and for $n = \infty$ it defines the rigidly plastic (Saint Venant) material. Hence the stress distribution in the actual structure will, in a defined sense, be intermediate between those prevalent in the corresponding Hooke and Saint Venant structures. With the load consisting only of prescribed surface tractions and/or body forces, the stress distribution will depend uniquely on n. The quantity σ_n will enter only into expressions for strain and displacement.

If a non-linear elastic structure is subjected to r loads $L_1, L_2, \ldots, L_r$ acting simultaneously, the largest effective stress in the structure, σ_e, may be written as

$$\sigma_e = f(L_1, L_2, \ldots, L_r; n) \tag{8}$$

where f is linearly homogeneous in $L_1, L_2, \ldots, L_r$. Then eqn (8) may be rewritten as

$$\sigma_e = L_1 f(1, \alpha_2, \alpha_3, \ldots, \alpha_r; n) \tag{9}$$

where

$$\alpha_i = \frac{L_i}{L_1}, \qquad (i = 2, 3, \ldots, r) \tag{10}$$

For the corresponding linearly elastic structure ($n = 1$, indicated where necessary by an asterisk) the largest effective stress takes the form

$$\sigma_e^* = L_1 f(1, \alpha_2, \alpha_3, \ldots, \alpha_r; 1) \tag{11}$$

and hence

$$\frac{\sigma_e}{\sigma_e^*} = \frac{f(1, \alpha_2, \alpha_3, \ldots, \alpha_r; n)}{f(1, \alpha_2, \alpha_3, \ldots, \alpha_r; 1)} = \theta(\alpha_2, \alpha_3, \ldots, \alpha_r; n) \tag{12}$$

This relationship indicates a rational way of collecting and presenting data from analytical and numerical studies of structural creep problems. The quantity σ_e^* may usually be found in existing handbooks; the dimensionless quantity θ, expressing the influence of the non-linearity exponent n, depends on load ratios only. Hence by expressing the largest effective stress in the structure in the form

$$\sigma_e = \theta\, \sigma_e^* \tag{13}$$

a substantial amount of existing information, namely, σ_e^*, may be utilised. With $n > 1$, it may be shown that $\theta < 1$ for all possible load ratios α_i, $(i = 2, 3, \ldots, r)$.

Where only two load systems are involved ($r = 2$), the *stress reduction factor* θ depends only on the two quantities L_2/L_1 and n. This allows a simple graphical representation, which will be demonstrated below for the case of a T-bar subjected to simultaneously acting axial force and bending moment. In a later section, the creep deflection of beams under combined loadings will be analysed in a similar way.

PLANE BENDING OF MONOSYMMETRIC BAR

In a bar with monosymmetric cross-section, subjected to an axial force N and a bending moment M according to Fig. 1, let ρ denote the radius of curvature of the neutral fibre and other quantities be as defined directly in the figure. Since both positive and negative strains occur, the constitutive eqn (5) must be stated as

$$\varepsilon = |\sigma/\sigma_n|^n \mathrm{sgn}\sigma \tag{14}$$

with the inverse

$$\sigma = \sigma_n |\varepsilon|^{1/n} \mathrm{sgn}\varepsilon \tag{15}$$

Introducing here the strain of the fibre a–a,

$$\varepsilon = \frac{\zeta}{\rho} \tag{16}$$

the stress distribution takes the form

$$\sigma = \sigma_n |\zeta/\rho|^{1/n} \mathrm{sgn}\left(\frac{\zeta}{\rho}\right) \tag{17}$$

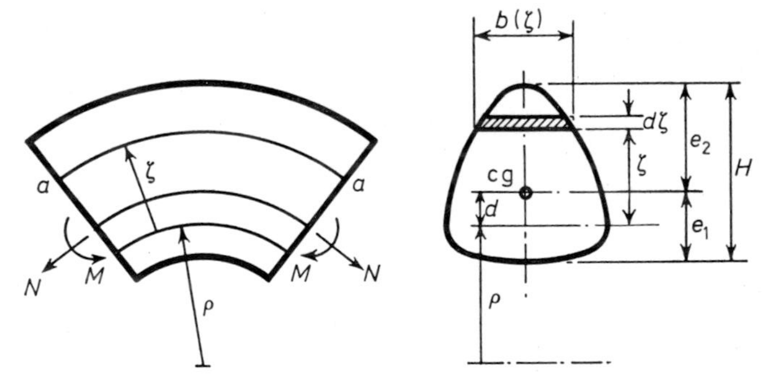

FIG. 1. Plane bending of monosymmetric bar.

From the equilibrium conditions

$$\begin{aligned} \int \sigma \, \mathrm{d}A &= N \\ \int \sigma(\zeta - d) \, \mathrm{d}A &= M \end{aligned} \tag{18}$$

it follows that

$$\begin{aligned} \sigma_n |\rho|^{-1/n} \mathrm{sgn}\rho \; S_n &= N \\ \sigma_n |\rho|^{-1/n} \mathrm{sgn}\rho \; I_n &= M + dN \end{aligned} \tag{19}$$

where

$$\begin{aligned} S_n &= \int |\zeta|^{1/n} \mathrm{sgn}\zeta \; b(\zeta) \, \mathrm{d}\zeta \\ I_n &= \int |\zeta|^{(1+(1/n))} b(\zeta) \, \mathrm{d}\zeta \end{aligned} \tag{20}$$

the integrals being taken from $-e_1 + d$ to $e_2 + d$. From eqn (19) follows

$$\frac{I_n}{S_n} + d = \frac{M}{N} \tag{21}$$

Since both I_n and S_n depend on d, this is an implicit relation for calculating d. It is brought to the dimensionless form

$$\frac{I_n'}{S_n'} + \beta = \frac{1}{\alpha} \tag{22}$$

by introducing the new variables

$$\alpha = \frac{NH}{2M}$$
$$\beta = \frac{2d}{H} \tag{23}$$
$$z = \frac{2\zeta}{H}$$

and

$$S_n' = \int |z|^{1/n} \operatorname{sgn} z \, b\left(\frac{zH}{2}\right) dz \qquad I_n' = \int |z|^{(1+(1/n))} b\left(\frac{zH}{2}\right) dz \tag{24}$$

the integrals now being taken from $-2e_1/H + \beta$ to $2e_2/H + \beta$. For any given load ratio α, the quantity β, *i.e.* the location of the neutral layer, may be calculated from eqn (22). The largest effective stress is then finally obtained from eqn (17) and the second eqn (19)

$$\sigma_e = \frac{|M + dN|}{I_n} |\zeta_{max}|^{1/n} \tag{25}$$

With $n = 1$ this transforms to the largest stress for the corresponding Hookean bar

$$\sigma^*{}_e = \frac{|M + d^*N|}{I_1} |\zeta^*{}_{max}| \tag{26}$$

and hence, from eqn (12)

$$\theta = \left|\frac{M + dN}{M + d^*N}\right| \frac{|\zeta_{max}|^{1/n}}{|\zeta^*{}_{max}|} \frac{I_1}{I_n} \tag{27}$$

or, with the notation of eqns (23) and (24)

$$\theta = \left|\frac{1 + \alpha\beta}{1 + \alpha\beta^*}\right| \left|\frac{z_{max}}{z^*_{max}}\right|^{1/n} \frac{I_1'}{I_n'} \tag{28}$$

After β has been determined from eqn (22), the stress reduction factor θ may be determined from eqn (28).

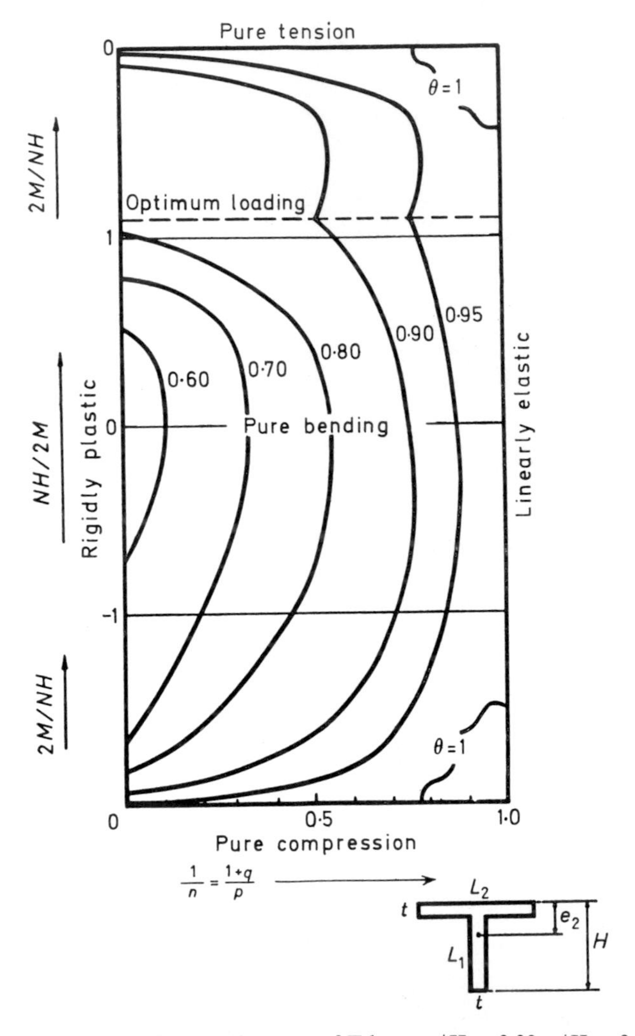

FIG. 2. Design chart for bending of T-bar; $e_2/H = 0{\cdot}30$; $t/H = 0{\cdot}12$.

Because of the monosymmetry of the cross-section and the non-linearity of the constitutive equation, a complete coverage of all loading cases requires α to vary from $-\infty$ (pure axial compression) through zero (pure bending) to $+\infty$ (pure axial tension). This may be presented in one diagram by using semi-invert representation [5]. A design chart of this kind for a T-shaped cross-section is shown in Fig. 2. The chart refers to an idealised cross-section without tapering but with overall shape similar to the standard type.

The discontinuity in slope of the θ-curves occurring at the load ratio $2M/NH = 0{\cdot}93$ indicates that the location of the most highly stressed fibre is switching from web to flange, when the load ratio passes this

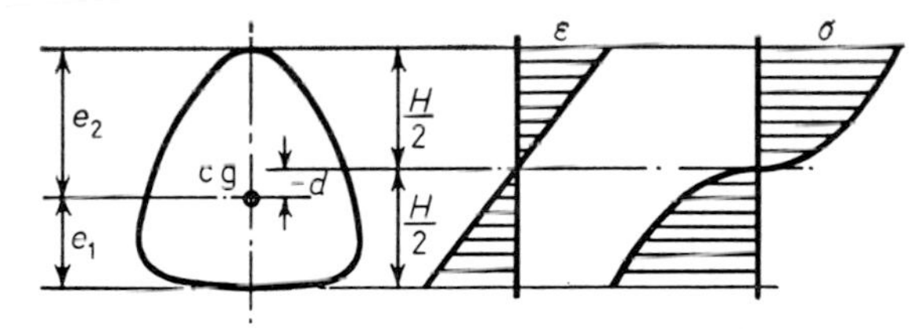

FIG. 3. Strain and stress distributions at optimum loading.

value. It is interesting to note that this *ratio of optimum loading* (causing maximum stress in both extreme fibres) is independent of $(1 + q)/p$. The same is found for the triangular cross-section, but is not generally true.

At optimum loading (denoted below by subscript o) the neutral layer is located at half the height of the cross-section, Fig. 3, and hence d may be determined directly from the cross-section geometry

$$d_o = e_1 - \frac{H}{2} \tag{29}$$

The corresponding optimum load ratio may then be calculated directly from eqn (21) or (22)

$$\left(\frac{2M}{HN}\right)_o = \frac{e_1 - e_2}{H} + \frac{\int_{-1}^{+1} |z|^{1+(1/n)}\, b\left(\frac{zH}{2}\right) \mathrm{d}z}{\int_{-1}^{+1} |z|^{1+(1/n)} \operatorname{sgn} z\, b\left(\frac{zH}{2}\right) \mathrm{d}z} \tag{30}$$

For a bisymmetrical cross-section, the optimum loading ratio corresponds to pure bending ($N = 0$), and the design chart is symmetric with respect to this line [2].

Since pure tension and compression are isostatic loading cases, the curve $\theta = 1$ will always consist of the top and bottom edges of the chart, connected by the right hand edge. The left hand edge may readily be graduated by using known solutions for rigidly plastic material. In this way the required computational work may be limited.

The chart gives a complete survey of all possible loading ratios and all possible creep parameters. It shows that the largest effective stress depends rather weakly on the creep parameters in the range of practical interest, namely:

$$0{\cdot}2 < \frac{1 + q}{p} < 0{\cdot}5$$

DEFLECTION DURING BENDING OF BISYMMETRIC BAR

In pure bending ($N = 0$) of a non-linearly elastic bar with bisymmetric cross-section, eqn (21) yields $d = 0$, *i.e.* the neutral layer will then contain the centre of gravity. Hence, from eqn (19)

$$\frac{1}{\rho} = \left|\frac{M}{\sigma_m I_n}\right|^n \operatorname{sgn} M \tag{31}$$

or, with the notation of Fig. 4, considering only cases of small slopes

$$\frac{\mathrm{d}^2 w}{\mathrm{d}x^2} = -c_n |M|^n \operatorname{sgn} M \tag{32}$$

where

$$c_n = (\sigma_n I_n)^{-n} \tag{33}$$

Together with relevant boundary conditions the differential eqn (32) determines the shape of the deflection curve. Owing to the non-linearity exponent n, deflections obtained for various load systems may not be superposed as is done for problems of linear elasticity. However, since M is a linearly homogeneous function of the acting loads, this problem may be treated similarly to the stress problem in the previous section. If, in particular, only two load systems are acting, design charts may be drawn on the same principles.

As an example, consider the cantilever beam shown in Fig. 5, loaded by a force P at the outer end and a uniformly distributed load Q acting along the whole beam length L. With x measured from the outer end, the bending moment is

$$M = Px + \frac{Qx^2}{2L} \tag{34}$$

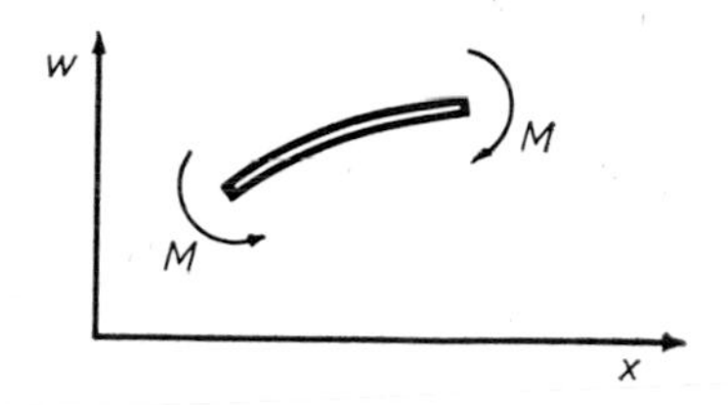

FIG. 4. Coordinates for beam deflection.

Introducing here the dimensionless coordinate

$$\xi = \frac{x}{L} \tag{35}$$

and the load ratios

$$p = \frac{P}{(|P| + Q)}$$
$$q = \frac{Q}{(|P| + Q)} \tag{36}$$

the bending moment takes the form

$$M = L(|P| + Q)\left(p\xi + \frac{q\xi^2}{2}\right) \tag{37}$$

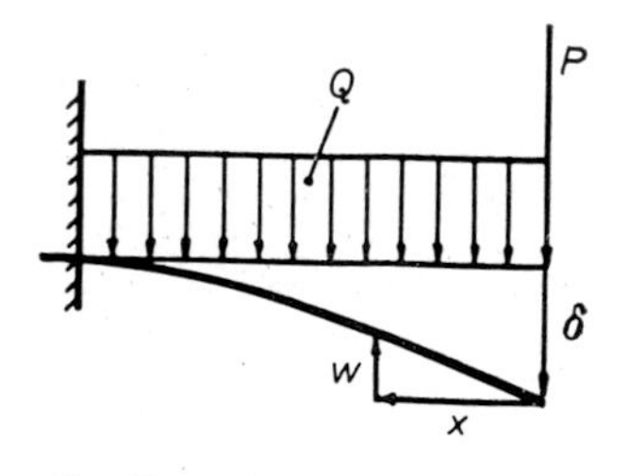

FIG. 5. Cantilever beam with two load systems.

The load Q is taken as positive, but P may be positive or negative, which motivates the load ratios to be defined as in eqn (36). Thus, p and q always remain finite quantities. An alternative definition might have been

$$p' = \frac{P}{(P^2 + Q^2)^{\frac{1}{2}}}, \qquad q' = \frac{Q}{(P^2 + Q^2)^{\frac{1}{2}}}$$

but that will not be used here.

The differential eqn (32) now transforms to

$$\frac{d^2w}{d\xi^2} = -c_n L^{(n+2)}(|P| + Q)^n \left(p\xi + \frac{q\xi^2}{2}\right)^n \operatorname{sgn}\left(p\xi + \frac{q\xi^2}{2}\right) \tag{38}$$

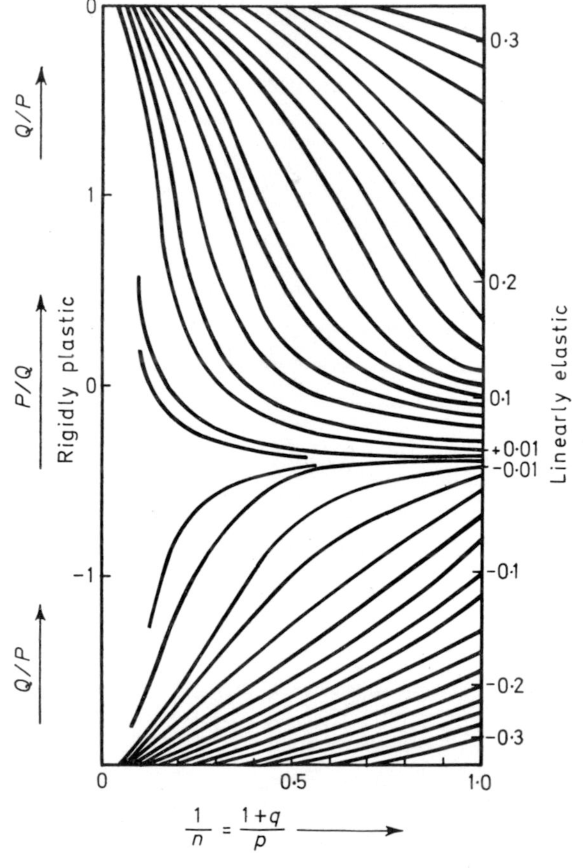

FIG. 6. Design chart for cantilever beam.

The boundary conditions take the form

$$w(0) = 0, \qquad w'(1) = 0 \tag{39}$$

Finally, the deflection δ of the outer end is obtained as

$$\delta = w(1) \tag{40}$$

It may be written in the form

$$\delta = \gamma c_n L^{(n+2)}(|P| + Q)^n \tag{41}$$

where γ is a function of the load ratio P/Q and the non-linearity exponent n. Integration of eqn (38) may be done in closed form when n is an integer, and is particularly simple if n is odd. Other n-values require an approximate numerical integration. A design chart for γ is shown in Fig. 6.

The design chart gives immediate information about the effect of changing the load ratio P/Q at any arbitrary values of the creep parameters p and q. It may also be used to determine the support reaction at a propped cantilever. With $\gamma = 0$, the chart gives $P/Q \approx 0{\cdot}4$ for all relevant creep parameters p and q.

Similar design charts have been drawn for other end conditions, and they all show the same general appearance. This may be utilised for rapid estimates of creep design charts for various similar problems.

CONCLUSIONS

The largest effective stress in a structure subject to stationary non-linear viscous creep under prescribed forces L_1, L_2, L_3, . . . may preferably be related to that in the corresponding linearly elastic structure. The ratio θ between these two design stresses depends only on the creep index n and the relative magnitudes of the load components L_1/L_2, L_1/L_3, ... If, in particular, only two loads are acting, the stress reduction factor θ may be presented in a simple type of diagram with n and L_1/L_2 as variables. Such a diagram, derived for a monosymmetric beam under simultaneous axial force N and bending moment M, shows that θ depends only slightly on the ratio N/M in a considerable range of interest. Figure 2 also shows the conditions for optimum loading, implying smallest maximum stress in the beam. Finally, the creep deflection of a cantilever under a two load system is studied.

REFERENCES

1. JOHNSON, A. E. (1960). 'Complex stress creep of metals.' *Metall. Rev.*, **5**(20), 447–506.
2. EDSTAM, U. and HULT, J. (1967). BROBERG, B. (Ed.), *Advances in Applied Mechanics* (Folke Odqvist volume), pp. 223–239. London: J. Wiley & Sons.
3. JOHNSON, A. E. (1967). 'Some progress in creep mechanics.' BROBERG, B. (Ed.), *Advances in Applied Mechanics* (Folke Odqvist volume), pp. 289–327. London: J. Wiley & Sons.
4. HOFF, N. J. (1954). 'Approximate analysis of structures in the presence of moderately large creep deformation.' *Quart. J. Appl. Math.*, **12,** 49–55.
5. JAKOBSSON, B. (1947). 'Semi-invert diagram' (in Swedish). *Tek. Tidskr., Stockholm*, **77,** 572–573.

SECTION II

CREEP PROPERTIES FOR DESIGN

Chapter 6

CREEP RECOVERY IN HEAT RESISTANT STEELS

R. M. GOLDHOFF

SUMMARY

This chapter deals with the phenomenology and engineering significance of creep recovery. A brief review of the subject is given. Some of the well known postulates regarding creep recovery are substantiated and extended by the introduction of new data. Analytical relations to describe the data are given, modulus effects are examined, and the 'plastic creep limit' is discussed. Particular attention is given to ductility or microstructural effects. The possible engineering significance of recovery effects is discussed with reference to plastic creep analysis, creep under varying stress, and damage.

NOTATION

A Constant.
A_0 'A' intercept on zero time axis.
B Constant.
E_L Elastic modulus.
E_R 'Relaxed' modulus.
K Constant (eqn (10)).
n Slope (*see* Fig. 5d).
T Temperature.
t Time.
x, y Constants at a given temperature: eqn (3).
ε_A Recovery strain.

ε_R Total recoverable strain.
σ Stress.
τ Constant from relaxation spectrum in eqn (1).

INTRODUCTION

The subject of creep recovery in metals is not one about which a great body of information exists. Phenomenological observations have been presented and analytical treatment of such results given, but mechanistic details and consideration of the practical significance of creep recovery are mainly speculative. The late Dr Johnson recognised the need for a better understanding of this phenomenon in his overall work on elevated temperature properties under complex loading, and he was responsible for much of what is known today. In this chapter, I want first to review briefly the subject of creep recovery, mainly by presenting data obtained in tests on the steels at present used in power generation equipment, and show how they confirm, at least in an engineering sense, some of the things we already know. Secondly, I want to discuss some of the possible practical significance of these results.

GENERAL BACKGROUND

The engineering significance that accompanies this subject, coupled with the general lack of information concerning its nature, prompts an attempt to describe the phenomenon briefly and in simple engineering terms. Mechanically speaking, the changes in the stress–strain relationship for a solid subjected to a small, external load are referred to as 'after-effects'. Some familiar examples are the phenomena of creep recovery, internal damping, the Bauschinger effect and stress relaxation.

After-effects are usually interpreted in terms of the distribution of such variables as stress and temperature within the solid. When, for example, a load is applied externally to a solid, any non-uniform distribution of the load, inhomogeneity or anisotropy in the material will give rise to a non-uniform distribution of the stress within the solid. Subsequently, so far as the material allows, these distributions tend to become uniform. Because of the relationship between the imposed strain and the stress, the tendency toward uniformity (relaxation) is manifested in the observable after-effects.

The mechanical description of this behaviour in elastic solids involves the relaxation of shear stress across particular regions of the solid. This reaction is described as 'viscous'. In metals, slip bands, twin boundaries and grain boundaries exhibit a behaviour that appears to be viscous in nature and gives rise to this phenomenon. The physical description of this

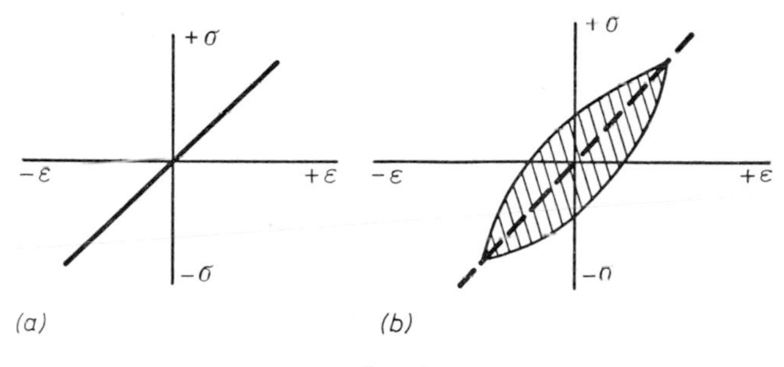

FIG. 1.

response is usually based on variable motion of atomic imperfections as a result of external stressing. In essence, then, the resulting redistribution of atomic arrangement or stress alters the strain in the region undergoing change and the sum of these rearrangements represents the external observation. After-effects may be either plastic or elastic in nature.

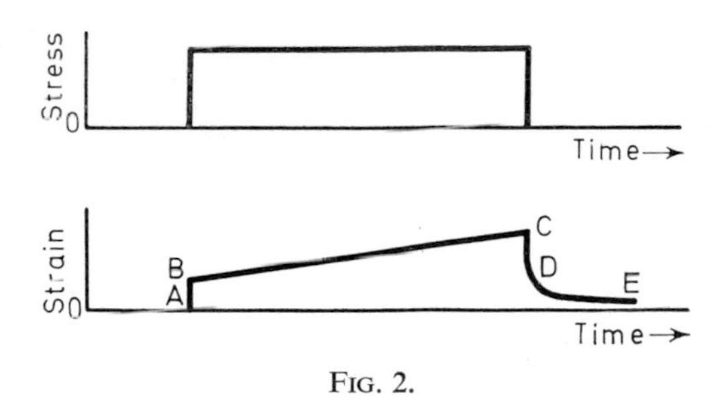

FIG. 2.

The concept of anelasticity arises from the comparison of the ideal stress–strain relations for a perfectly elastic body (shown schematically in Fig. 1a) and the real behaviour (shown schematically in Fig. 1b).

The loop shown in (b) is the so-called 'hysteresis loop' and is a measure of energy dissipation per cycle. Thus, if a material were perfectly elastic, it would vibrate indefinitely since no energy would be dissipated. Real metals, of course, cease to vibrate and the 'damping' characteristics responsible for this are a consequence of non-elastic behaviour. To distinguish this elastic after-effect from pure elastic or plastic deformation, the word 'anelasticity' has been used. By Zener's [1] definition, anelasticity denotes the property of solids by virtue of which stress and strain are not single valued functions of one another in the elastic range of deformation.

Creep recovery arises as an actual response of metals to applied stress in the elastic deformation range and occurs ideally as shown in Fig. 2.

Application of stress results in instantaneous extension (AB); with additional time, strain increases (BC). Upon removal of the stress there is an instantaneous contraction (CD), followed by a gradual contraction with time (DE). Such behaviour, of course, is dependent upon temperature and strain rate.

The portion of interest in this work is the time dependent, recoverable strain in the segment (DE). This description immediately suggests the well known classification of deformation types as follows:

Type of deformation	Time dependence	Stress proportionality	Recoverability
Elastic	Independent	Proportional	Recoverable
Anelastic	Dependent	Proportional	Recoverable in time
Plastic	Dependent	Not proportional	Not recovered in time

The suggestion is that creep recovery strain and anelastic strain are similar. The characteristics of creep recovery in metals can be summarised briefly (Johnson *et al.* [2]; Johnson *et al.* [3]; Lubahn and Felgar [4]; Henderson and Snedden [5]), as follows:

1 Creep recovery and anelastic strain exhibit the same characteristics. Both can be described mathematically by a large number of mechanical spring-dashpot combinations with a wide range of relaxation times.

2 At elevated temperatures the anelastic strain for a given time increases linearly with stress. Creep recovery follows a similar law but appears to be dependent on the prior duration of creep deformation.

3 The rate of recoverable creep under a given stress increases as the temperature is raised, but this has practical limitations.

4 When the stress is low enough, essentially all transient creep is linear with stress and recoverable. The converse is true at high stress, and the transition between the two types of behaviour is sensitive to temperature and prestrain.

The above suggests some possible engineering significance in the engineering use of alloys with regard to creep analysis, creep under varying stress and temperature, and potential damage during exposure.

MATERIALS AND PROCEDURE

Interest in this work centered about the transformable, heat resistant steels in common use in power generating equipment. The typical steels used are a 1 per cent Cr, 1 per cent Mo, 0·25 per cent V, 0·30 per cent C; a 1·25 per cent Cr, 1 per cent Mo, 0·2 per cent C; and an 11 per cent Cr, 1 per cent Mo, 0·5 per cent Ni, 0·2 per cent C steel. Earlier work with a Cr–Mo–V steel [6] was at much lower temperatures and with the steel in a heat treated condition long since abandoned. Some of the present work repeated the earlier work, but used current materials and conditions of interest. In general, the few experiments made were designed to show the effects of time, temperature, stress and stress history on creep recovery. All tests were made using cylindrical specimens stressed in uniaxial tension with strain measured by the best mechanical extensometry available. The sensitivity and accuracy of such tests were sufficient to illustrate the points made in this chapter.

RESULTS AND DISCUSSION

Calculating recovery strains

Normal creep-test procedure leads inherently to problems in accurate measurement of recoverable strain, which can best be described with reference to Fig. 3. In this work, a specimen was mounted in a creep furnace, stabilised at test temperature, and dead-weight loaded incrementally at a controlled rate to the intended stress as shown in Fig. 3a. The loading line is generally curved as shown because of the presence of inelastic strain, *i.e.* anelastic and plastic strain. Upon reaching the test stress, a zero point for strain and time is calculated as suggested in the figure and creep deformation proceeds under constant load giving the curve of Fig. 3b; both creep and anelastic strain accumulate during this phase. At a desired creep time, the load is removed rapidly by means of a jack while the temperature is maintained. In Fig. 3c, the elastic recovery followed by the creep recovery is shown.

An excellent measure of elastic strain is given by the interval of elastic recovery after load removal. However, the measurement is complicated by some creep recovery strain in this interval. The problem is to establish a zero point in time for the beginning of creep recovery. Several of the

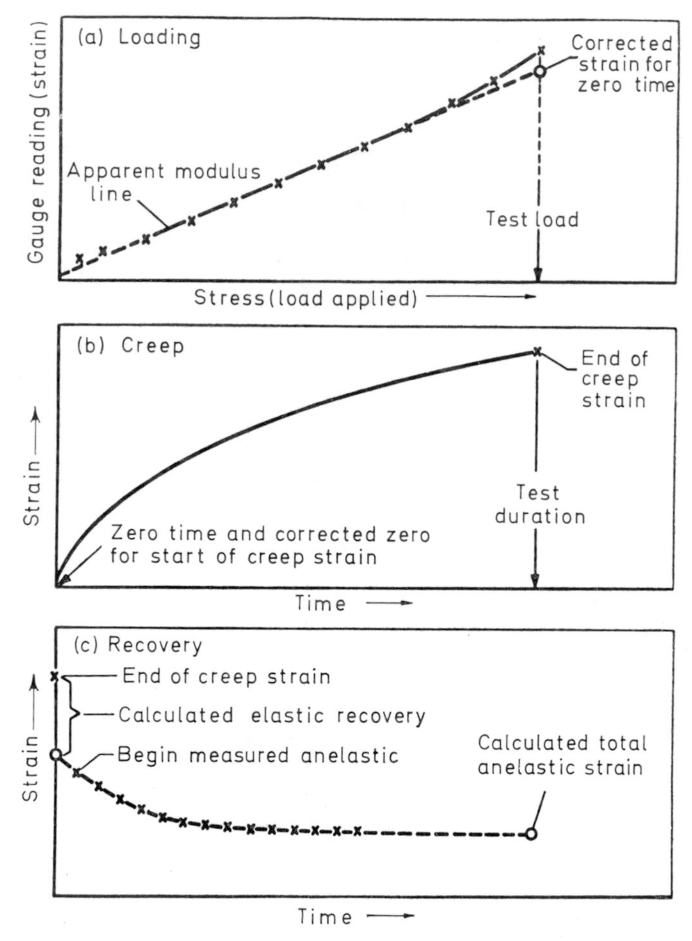

FIG. 3. Schematic creep recovery test.

first recovery data points were plotted [6] as a function of fractional powers of time until a straight line was obtained. This line was then extrapolated back to the zero recovery time axis to provide an intercept which defines the elastic strain and a zero point for creep recovery. The method used here was determined by noting that the recovery curve can be fitted by a function of the following type:

$$\varepsilon_A = \sum_i A_i(1 - e^{-t/\tau_i}) \qquad (1)$$

where ε_A = Recovery strain, t = Time, A = Constant, and τ_i = Constants from the 'relaxation spectrum'.

There is ample background for this equation [6, 7]. In its simplest form it can be shown to be a solution to the differential equation of motion of a mechanical spring-dashpot combination with τ the 'relaxation time'. The summation is introduced for the problem of time dependent deformation and τ_i is then the 'relaxation spectrum'. Graphical methods are available to determine the coefficients and τ series of the relaxation spectrum. In this instance, however, a trial-and-error program was used with a 12-term computerised solution to eqn (1), using the data from several Cr–Mo–V steel tests to determine statistically optimised values for the magnitude and interval of τ_i. The final equation used was:

$$\varepsilon_A = A_0 + A_1(1 - e^{-t/1000}) + A_2(1 - e^{-t/200}) + \ldots + A_{12}(1 - e^{-t/2 \cdot 05 \times 10^{-5}}) \qquad (2)$$

If the actual gauge readings during recovery are used in the computer input, the equation of best fit gives a value A_0, which, theoretically, is the intercept on the zero time axis. Also, at infinite time, ε_A is simply the sum of all the coefficients and thus an estimate of the total possible recovery strain is given. An approximate separation of the component strains on loading can also be made by assuming the apparent modulus shown in Fig. 3a. This is of interest in examining modulus effects. Equation (2) was applied to all the test data shown in this chapter.

Analytical relations in creep recovery

Henderson and Snedden [5] have recently reviewed much of the previous work on creep recovery. For their very comprehensive data on an aluminium alloy they concluded that no current theory describes recovery completely, but also found that recovery could be well represented by an equation of the form:

$$\varepsilon_A = C\sigma\tau^y t^x \qquad (3)$$

where σ = Stress, τ = Prior creep time, and t = Recovery time.

On the other hand, the case for the analogy between anelastic creep (the recoverable strain accumulated while straining under constant load)

TABLE 1

Test conditions and results

Steel type	Test No.	Stress lb/in² (hbar)	Temp. °F (°C)	Time of creep (hr)	Time of recovery (hr)	GL in (mm)	ε_A in (mm)	ε_{AX} in (mm)	ε_E in (mm)	ε_P in (mm)	ε_T in (mm)	E_R 10^3 lb/in² (10^3 hbar)	E_L 10^3 lb/in² (10^3 hbar)	E_D 10^3 lb/in² (10^3 hbar)	N	Regression coefficient	Standard error 10^{-3} in (μm)
Cr–Mo	1	28000 (19·3)	1050 (566)	166	287	7·0 (178)	0·0065 (0·165)	0·0097 (0·246)	0·0083 (0·211)	0·0525 (1·33)	0·0705 (1·79)	19900 (13·7)	23600 (16·3)	24500 (16·9)	10	0·99992	0·01981 (0·503)
Cr–Mo	2	17000 (11·7)	1050 (566)	8736	1968	7·0 (178)	0·0060 (0·152)	0·0061 (0·155)	0·0053 (0·135)	0·0532 (1·35)	0·0646 (1·64)	20000 (13·8)	22600 (15·6)	24500 (16·9)	8	0·99997	0·01748 (0·444)
Cr–Mo–V	3	25000 (17·2)	900 (482)	48063	4831	7·0 (178)	0·0042 (0·107)	0·0043 (0·109)	0·0071 (0·180)	0·0190 (0·48)	0·0304 (0·77)	23600 (16·3)	24700 (17·0)	25700 (17·7)	7	0·99920	0·06432 (1·634)
Cr–Mo–V	4	15000 (10·3)	900 (482)	31818	4246	7·0 (178)	0·0022 (0·056)	0·0023 (0·058)	0·0041 (0·104)	0·0040 (0·102)	0·0103 (0·262)	24900 (17·2)	25600 (17·6)	25700 (17·7)	7	0·99813	0·05144 (1·307)
Cr–Mo–V	5	23000 (15·9)	1000 (538)	2330	1007	7·0 (178)	0·0054 (0·137)	0·0058 (0·147)	0·0067 (0·170)	0·0226 (0·574)	0·0351 (0·89)	22100 (15·2)	24000 (16·5)	25000 (17·2)	9	0·99995	0·01555 (0·395)
Cr–Mo–V	6	10000 (6·9)	1000 (538)	37042	3393	5·0 (127)	0·0018 (0·046)	0·0019 (0·048)	0·0023 (0·058)	0·0069 (0·175)	0·0107 (0·272)	22700 (15·6)	24600 (17·0)	25000 (17·2)	6	0·99669	0·04751 (1·207)
Cr–Mo–V	7	7500 (5·15)	1000 (538)	2135	2233	7·1 (180)	0·0017 (0·043)	0·0018 (0·046)	0·0022 (0·056)	0·0014 (0·036)	0·0054 (0·137)	23100 (15·9)	24700 (17·0)	25000 (17·2)	8	0·99960	0·01578 (0·401)
Cr–Mo–V	8	5000 (3·45)	1000 (538)	2183	2209	7·1 (180)	0·0010 (0·025)	0·0011 (0·028)	0·0015 (0·038)	0·0006 (0·015)	0·0032 (0·081)	21600 (14·9)	24500 (16·9)	25000 (17·2)	8	0·99928	0·01282 (0·326)
Cr–Mo–V	9	30000 (20·7)	1050 (566)	790	864	7·0 (178)	0·0065 (0·165)	0·0069 (0·175)	–	–	–	–	–	–	8	0·99994	0·01954 (0·496)
(*Step-test*)								*Incremental loading*									
Cr–Mo–V	10	20000 (13·8)	1050 (566)	1130	598	7·0 (178)	0·0056 (0·142)	0·0060 (0·152)	0·0063 (0·160)	0·0398 (1·01)	0·0523 (1·33)	20800 (14·3)	22100 (15·2)	24500 (16·9)	8	0·99996	0·01459 (0·371)
Cr–Mo–V	11	10000 (6·9)	1050 (566)	1131	552	7·0 (178)	0·0021 (0·053)	0·0025 (0·064)	0·0032 (0·081)	0·0065 (0·165)	0·0117 (0·297)	20800 (14·3)	22100 (15·2)	24500 (16·9)	7	0·99959	0·02031 (0·516)
Cr–Mo–V	12	30000 (20·7)	1000 (538)	887	650	7·0 (178)	0·0028 (0·071)	0·0030 (0·076)	–	–	–	–	–	–	9	0·99968	0·01485 (0·377)

(Step-test)																	
							Incremental loading										
Cr–Mo–V (Low duct.)	13	20000 (13·8)	1000 (538)	1006	482	7·0 (178)	0·0021 (0·053)	0·0022 (0·056)	0·0060 (0·152)	0·0034 (0·086)	0·0115 (0·292)	22100 (15·2)	23400 (16·1)	25000 (17·2)	9	0·99961	0·01554 (0·395)
Cr–Mo–V (Low duct.)	14	10000 (6·9)	1000 (538)	1008	528	7·0 (178)	0·0010 (0·025)	0·0014 (0·036)	0·0028 (0·071)	0·0003 (0·008)	0·0045 (0·114)	22800 (15·7)	25000 (17·2)	25000 (17·2)	9	0·99773	0·01698 (0·431)
12 Cr (Relaxation)	15	26000 (17·9)	1000 (538)	1202	1098	6·8 (173)	0·0029 (0·074)	–	0·0074 (0·188)	–	–	21200 (14·6)	24600 (17·0)	25700 (17·7)	–	–	–
12 Cr (Relaxation)	16	25000 (17·2)	1000 (538)	1006	505	6·8 (173)	0·0024 (0·061)	–	–	–	–	–	–	–	–	–	–
12 Cr (Relaxation)	17	21400 (14·8)	1000 (538)	1103	1008	6·8 (173)	0·0019 (0·048)	–	0·0061 (0·155)	–	–	21200 (14·6)	24400 (16·8)	25700 (17·7)	–	–	–

Legend: GL, Gauge length of specimen; ε_A, Total measured anelastic (recovery) strain; ε_{AX}, Total calculated anelastic strain; ε_E, Total elastic strain; ε_P, Total plastic strain; ε_T, Total strain; E_R, Relaxed modulus (includes elastic and anelastic loading strain); E_L, Elastic modulus (excludes anelastic loading strain); E_D, Dynamic modulus (measured experimentally); N, Number of terms in recovery equation.

and creep recovery has been well documented [6]. Lubahn arrives at this conclusion by showing that recovery has anelastic characteristics. Thus, by definition, anelastic deformation has a linear dependence on stress and is reversible, that is fully recoverable with time; its variation with time can be described with the aid of expressions like eqns (1) and (2). This latter characteristic is unquestionably true for recovery data and is illustrated here by the data of test No. 16. The complete equation used to fit these data is:

$$\begin{aligned}\varepsilon_A = 25{\cdot}0614 &+ 0{\cdot}19695(1 - e^{-t/1000}) - 0{\cdot}51227(1 - e^{-t/200}) \\ &- 0{\cdot}38491(1 - e^{-t/40}) - 0{\cdot}31068(1 - e^{-t/8}) - 0{\cdot}33173(1 - e^{-t/1{\cdot}6}) \\ &- 0{\cdot}35388(1 - e^{-t/0{\cdot}32}) - 0{\cdot}20098(1 - e^{-t/0{\cdot}064}) \\ &- 0{\cdot}20940(1 - e^{-t/0{\cdot}0125}) + 0{\cdot}11054(1 - e^{-t/0{\cdot}0025}) \qquad (4)\end{aligned}$$

Here the zero point of creep recovery is a dial gauge reading of 25·0614 and subsequent calculations as a function of time give the new gauge reading, the difference being in mils (10^{-3} in). The comparison of the actual and calculated recovery curves over the time range 0·1–1000 hr is shown in Fig. 4, where the fit is seen to be excellent. Values of the multiple

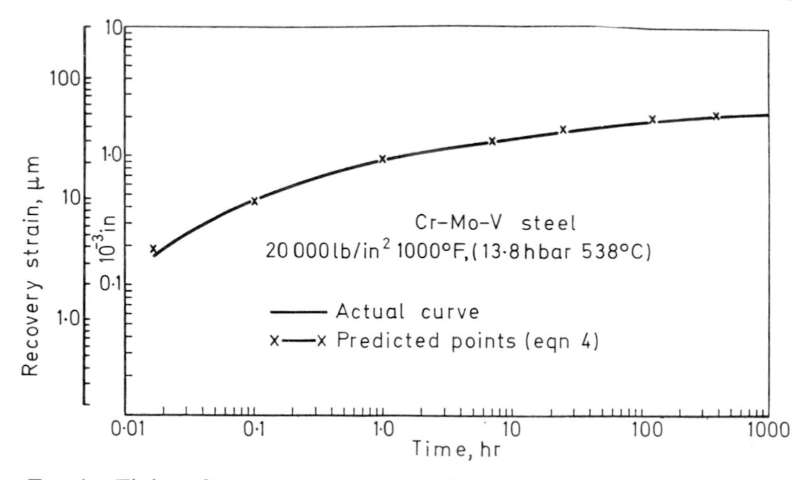

FIG. 4. Fitting of creep recovery data by the 'relaxation spectrum' equation.

regression correlation coefficient and the standard error (in mils) are shown in Table 1 for each test. In general, the agreement for each set of data, using the optimised equation, is good. Note, however, that the same equation did not apply to the 12 per cent Cr steel data. These latter data were processed by Lubahn's earlier technique.

Carrying the analogy further, note that creep recovery is linear with stress (Johnson *et al.* [2]; Johnson *et al.* [3]; Henderson and Snedden [5]; Lubahn and Felgar [4]; and Lubahn [6]). The data from this work have been treated similarly and are shown in Figs. 5 and 6; in all cases the linearity between ε_A and σ seems reasonable. Note that for the data used the duration of creep was at least 1000 hr and recovery periods were of

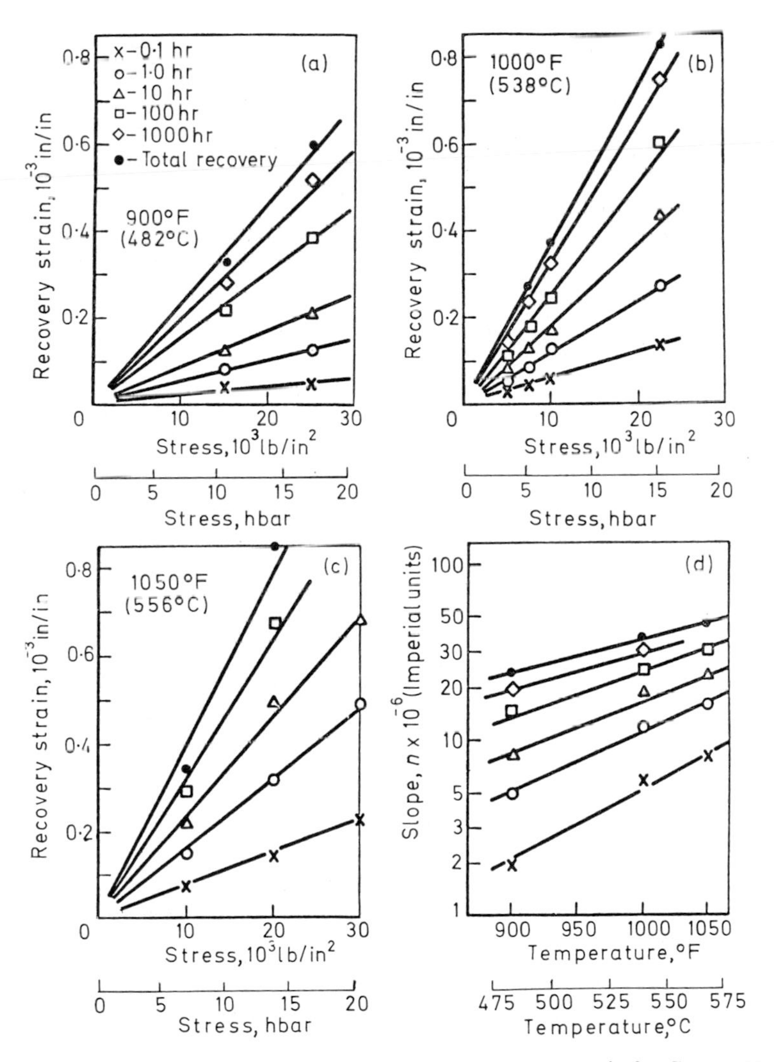

FIG. 5. Stress, time and temperature dependence of recovery strain for Cr–Mo–V steel.

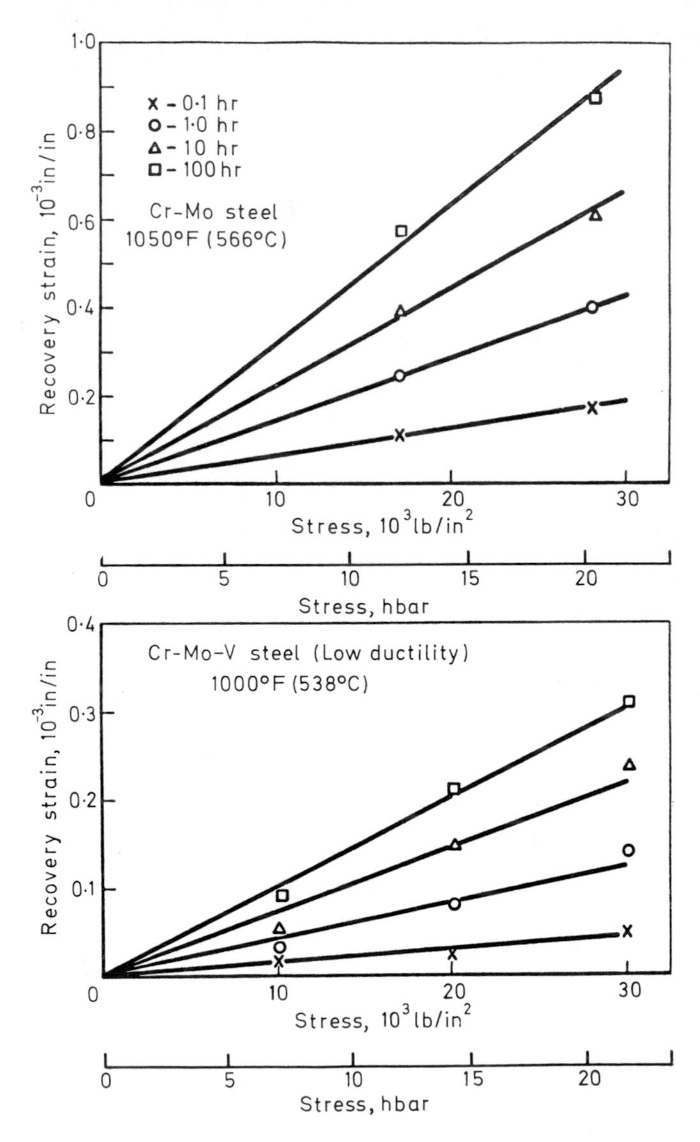

FIG. 6. Stress, time and temperature dependence of recovery strain for low alloy steels.

the same order. Linearity can only be expected for times following load removal roughly equal to the shortest duration of prior load. From Fig. 5a–c for Cr–Mo–V steel the stress, time, temperature (σ, t, T) dependence of ε_A can be derived. We can write

$$\varepsilon_A = n(T, t)\sigma \tag{5}$$

Plotting the values of slope n versus T as shown in Fig. 5d leads to

$$\log n = A(t) + B(t)T \tag{6}$$

and consequently

$$\log \varepsilon_A = A(t) + B(t)T + \log \sigma \tag{7}$$

This implies a linear relationship between $\log \varepsilon_A$ and $\log \sigma$ with slope of unity and intercept $[A(t) + B(t)T]$. Such a relationship does hold as shown in Fig. 7a–c.

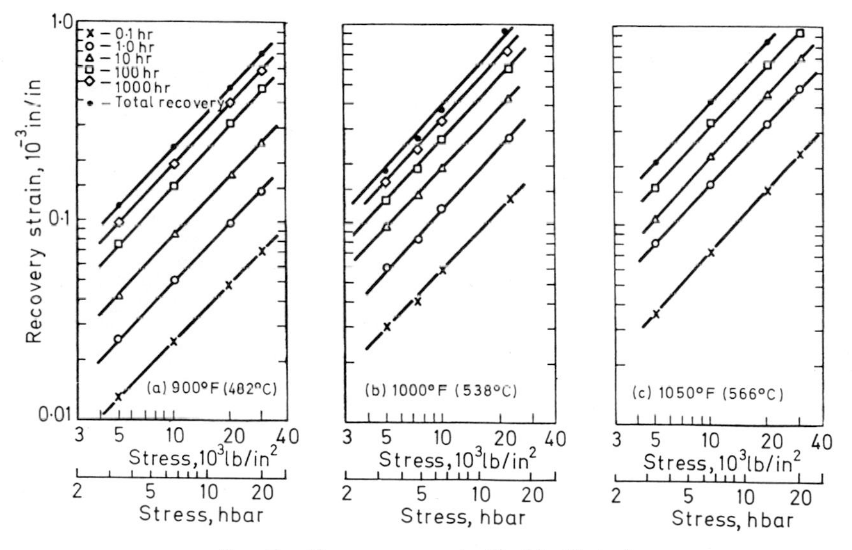

FIG. 7. Creep recovery in Cr–Mo–V steel.

Modulus effects and separation of strain components

Using the testing technique described earlier, a hot loading curve is established, but the elastic modulus obtained is always low because of the time dependent anelastic strain occurring along with the elastic strain. The curve obtained depends on the magnitude of load and the loading rate. Additionally, plastic strain may also occur on loading and a most difficult task arises when a separation of the total loading strain into component parts is attempted. Lubahn [6] suggests a method which depends on an accurate observation of the loading procedure and on reliable curve fitting. The analytical technique used here lends itself to a very rough estimate of the separation as described earlier. Table 1 shows three moduli compared for each test. The elastic modulus E_L is obtained directly from observation of the strain reading just prior to load removal and the strain

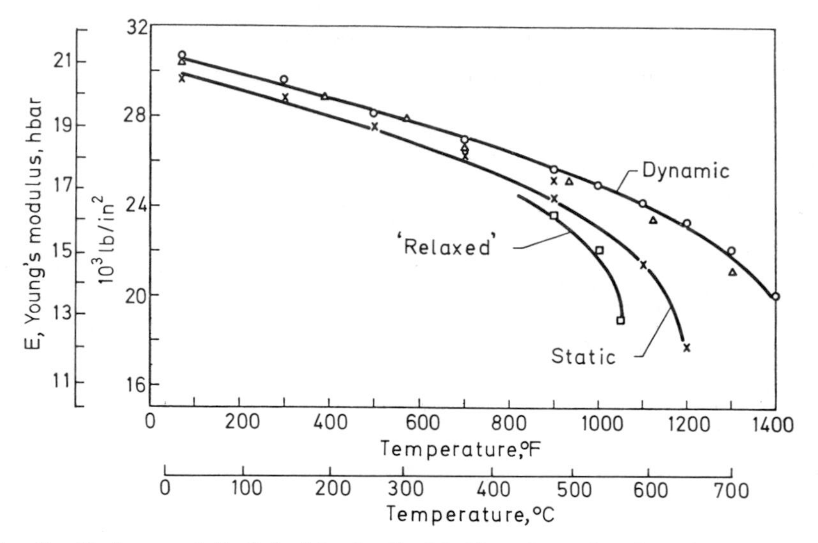

FIG. 8. Various moduli of elasticity for Cr–Mo–V steel as a function of temperature.

reading just prior to the start of recovery as given by the calculated constant A_0 of eqn (2). This can be compared to the experimentally determined dynamic moduli reported in the literature and, in general, the two are quite close. A 'relaxed' modulus E_R is also reported, and this is simply obtained by adding the estimated anelastic loading strain to the elastic strain before calculating the modulus by Hooke's law. The three moduli are compared in Fig. 8 for the Cr–Mo–V steel. For accurate modulus measurements, the dynamic measurements are undoubtedly to be preferred.

Because of the difficulties encountered in the strain separation, it is interesting to look at the total recoverable strain which is easily determined. The total recoverable strain is simply the sum of the elastic and anelastic components, *i.e.* all the strain recovered subsequent to load removal in the creep test. Calculated in this manner the total recoverable strain appears to be linear with time on a log-log plot. This is illustrated for the data of tests Nos. 8, 10 and 11 in Fig. 9. The curves are of the form:

$$\log \varepsilon_R = \log A(\sigma) + B \log t \tag{8}$$

or for the data shown, approximately:

$$\varepsilon_R = 3{\cdot}5 \times 10^{-4}\sigma t^{0{\cdot}05} (\varepsilon_R \text{ in mils, and } \sigma \text{ in } 1000 \text{ lb/in}^2) \tag{9}$$

This is an extremely good fit and could be quite useful in analysis. There is an anomaly, however, of no strain at zero time implying no elastic

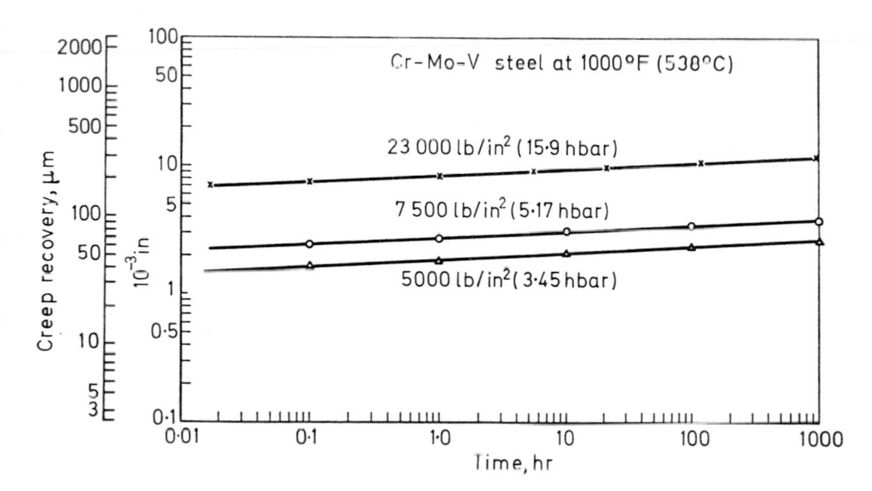

FIG. 9. Time and stress dependence of total recoverable strain for Cr–Mo–V steel.

deformation. If the elastic strain is taken as that read at 0·01 hr, then the modulus would be 35×10^6 lb/in^2 (2·4 Mbar). Values as high as this have been reported for this steel using dynamic loading tests at very high driving frequencies.

The 'plastic creep limit' concept

Lubahn's work led to the concept of a 'plastic creep limit' that is a stress below which, for a given time and temperature, the plastic creep strain is negligibly small in comparison with the anelastic creep strain. This is analogous to the 'elastic limit' in tensile testing. There are several ways of estimating this stress. In Fig. 10, plastic and anelastic strains are plotted versus stress as a function of load duration for the ductile Cr–Mo–V steel (tests Nos. 12–14). The latter are linear with stress while the former,

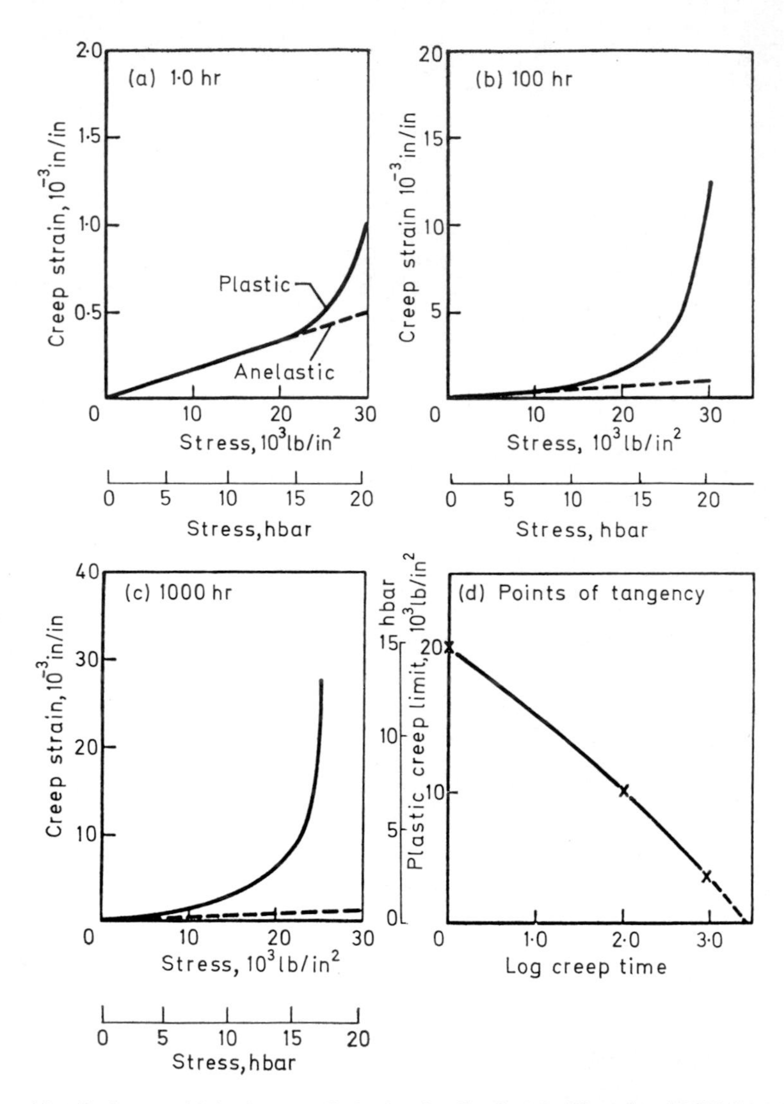

FIG. 10. Estimate of 'plastic creep limit' for ductile Cr–Mo–V steel at 566°C (1050°F).

derived from the usual creep curve, follow a more complex law, but they become tangent to the anelastic line at some low stress which depends on the creep time. This point of tangency occurs at lower stresses with longer times and is an estimate of the plastic creep limit. Figure 10d summarises these data showing that the plastic creep limit vanishes after about 5000 hr of creep; that is, after 5000 hr, no matter how low the stress, plastic creep will be significant. For very short times the limit is about 20 000 lb/in^2 (13·8 hbar). Figure 11 deals with the less ductile version of the

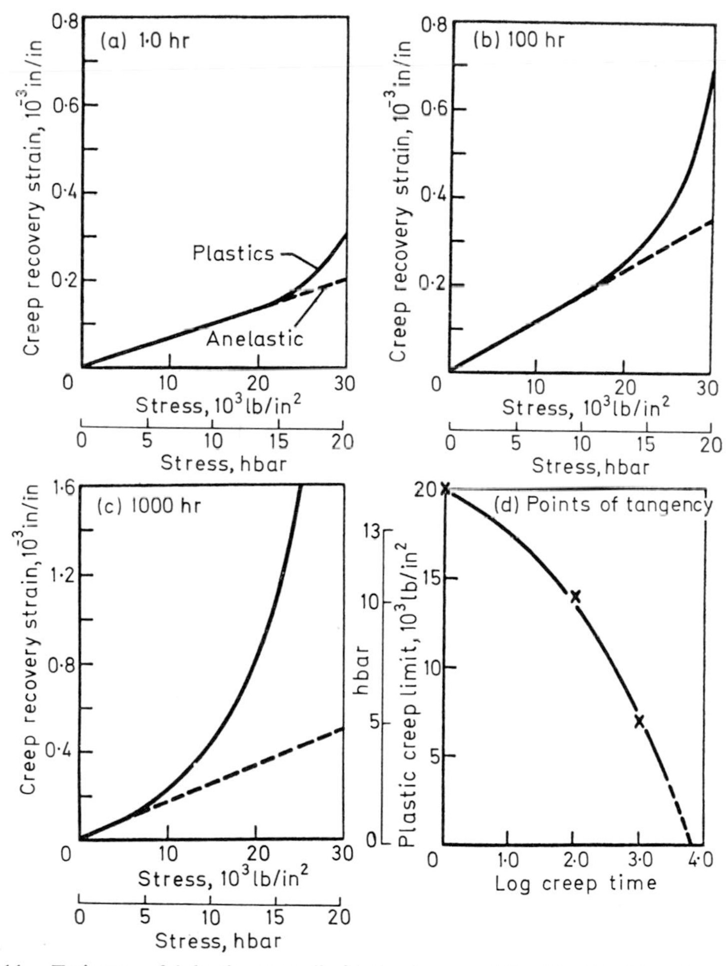

FIG. 11. Estimate of 'plastic creep limit' for low ductility Cr–Mo–V steel at 538°C (1000°F).

same steel where the comparable values are about 10 000 hr for the vanishing point and again about 20 000 lb/in^2 (13·8 hbar) for the upper bound of the plastic creep limit. Again, following Lubahn, the plastic creep limit can be estimated using a step-up stress test. Here tests were begun at 4000 lb/in^2 (2·76 hbar) and then sequentially loaded to 30 000 lb/in^2 (20·7 hbar) at about 200 hr intervals before obtaining recovery readings. The details will be evident from Fig. 12 where, for each stress, the values of total strain reached just prior to adding more stress are plotted. Using

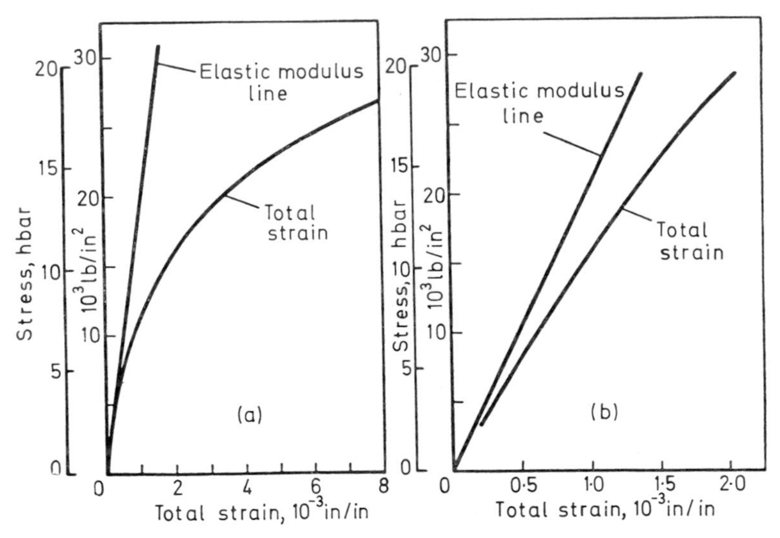

FIG. 12. Stress–strain curve from step-up tests. (a) Ductile Cr–Mo–V steel at 566°C (1050°F). (b) Low ductility Cr–Mo–V steel at 538°C (1000°F).

the assumption that the anelastic strain introduced is the sum of all preceding anelastic strains due to stress increments, and that these are indeed the same as the recovery strains, the plastic creep limit can be estimated. In Fig. 13a and c are plotted the calculated anelastic strain versus the measured creep strain just prior to each load increment. The point of deviation from the linear relationship represents the plastic creep limit. This junction can be related to the stress by plotting stress versus plastic strain which is obtained by subtracting the calculated anelastic strain from the creep strain at each point of Fig. 13a and c. The resulting data are shown in Fig. 13b and d. As shown, the limiting values are about 5000 lb/in^2 (3·45 hbar) at 1000 hr for the ductile steel and 7500 lb/in^2 (5·17 hbar) at 1000 hr for the low ductility steel, which is fair agreement with the previous result.

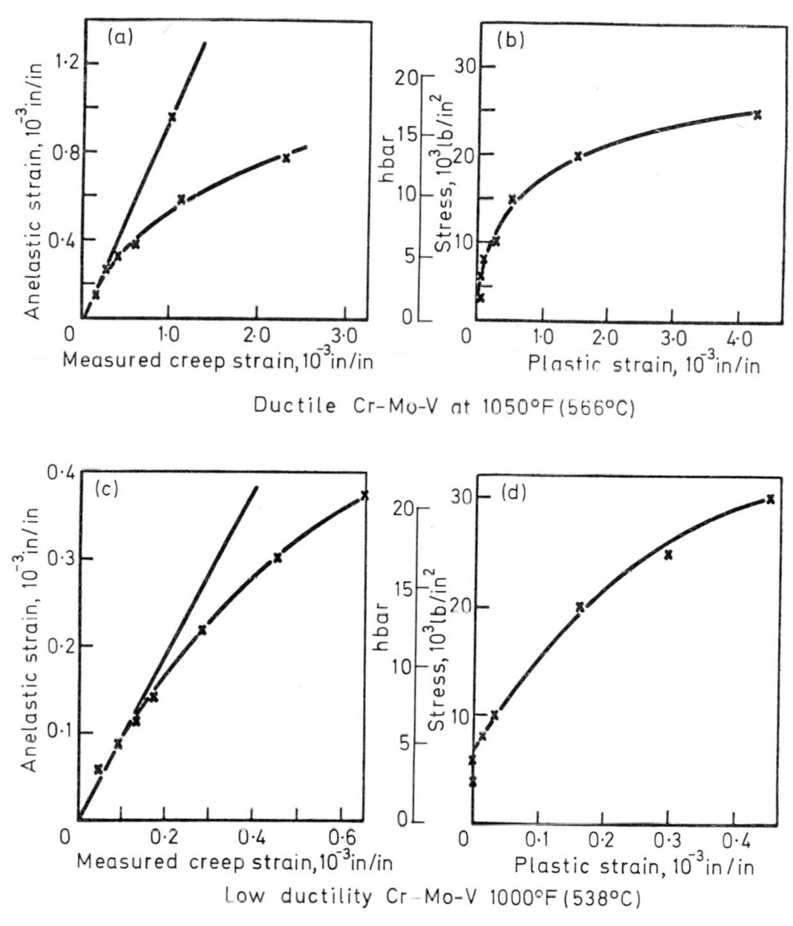

FIG. 13. Estimate of 'plastic creep limit' for Cr–Mo–V steel.

APPLICATIONS

Plastic creep analysis

It is quite obvious from the phenomenology of anelastic behaviour, that it is important at low stress and very low creep time, *i.e.* it is associated with transient or primary creep. However, the majority of conventional creep testing and plastic creep analyses emphasise the secondary or steady state creep regime. If the design problem is one of very limited permissible deformation, it seems likely that the usual plastic analysis will not be relevant. In this instance, the plastic strain is non-linear with stress and a common equation in use is

$$\varepsilon = K\sigma^n \tag{10}$$

where K and n are constants. We have already seen that anelastic strain has a linear stress dependence and can be described by

$$\varepsilon = K\sigma \tag{11}$$

where K is a constant.

Then, if the problem is basically one of anelastic strain rather than plastic strain, the analysis should be greatly simplified. It is of interest to examine the interface between the component strains and their dependence on time and stress.

Using the techniques and data described previously, it is possible to show an approximate separation of strains as indicated in Fig. 14 for a

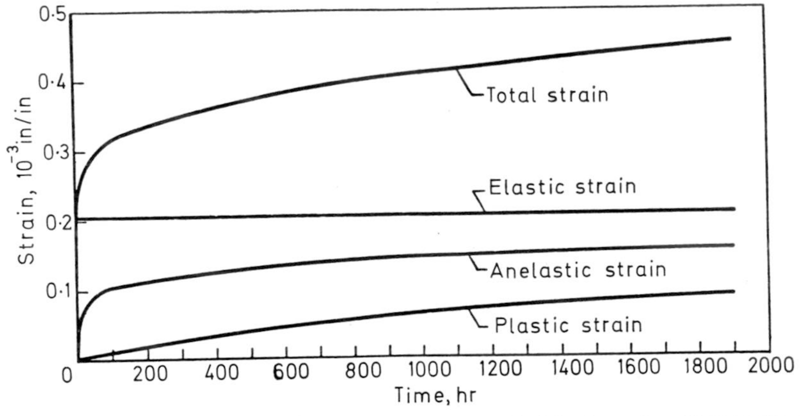

FIG. 14. The separation of strains for a creep test on Cr–Mo–V steel at 538°C (1000°F) and 5000 lb/in^2 (3·45 hbar).

Cr–Mo–V steel at 1000°F (538°C) and 5000 lb/in^2 (3·45 hbar). There is an obvious and important difference in the shape of the curves. In particular, the plastic strain curve has now an entirely different magnitude and character from the inelastic strain curve, which is of the same form as the total strain curve, and has the magnitude of total strain minus elastic strain. The inelastic strain is, of course, the strain measured in the conventional creep test and normally applied in analysis. Figure 15 shows the results of a similar, more comprehensive analysis of the tests shown in Fig. 5b. The plot of stress versus total non-recoverable strain is made on

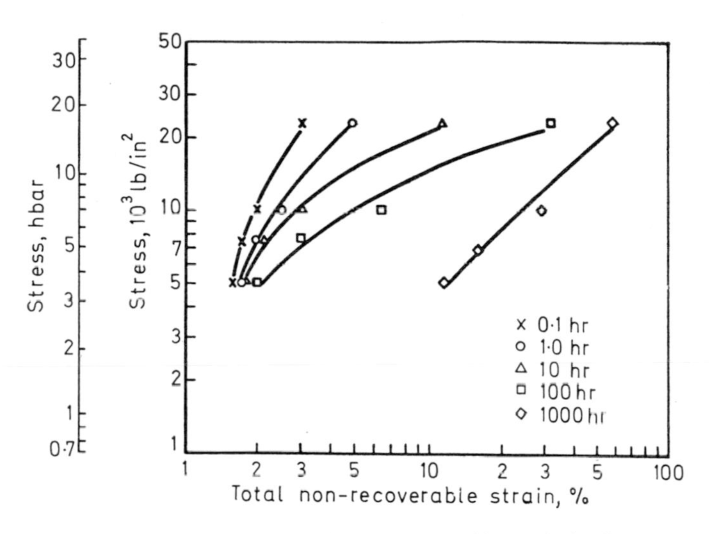

FIG. 15. Stress and time dependence of non-recoverable strain in Cr–Mo–V steel at 538°C (1000°F).

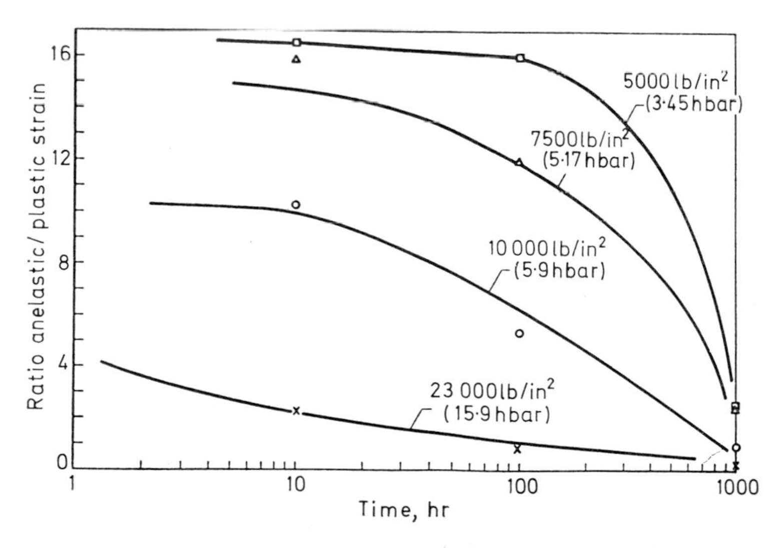

FIG. 16. Time and stress dependence of anelastic/plastic strain ratio for Cr–Mo–V steel at 538°C (1000°F).

logarithmic coordinates in order to expand the data in the low stress/short time regime. At 5000 lb/in^2 (3·45 hbar) and up to 100 hr some 98 per cent of the strain is recoverable while at 1000 hr 90 per cent is still recoverable. Even at 20 000 lb/in^2 (13·8 hbar) and 100 hr 50 per cent of the strain is recoverable. Another way to describe the important anelastic regime is to plot the ratio between anelastic and plastic strains. This is shown in Figs. 16 and 17 as a function of time, stress and, to a limited extent, temperature.

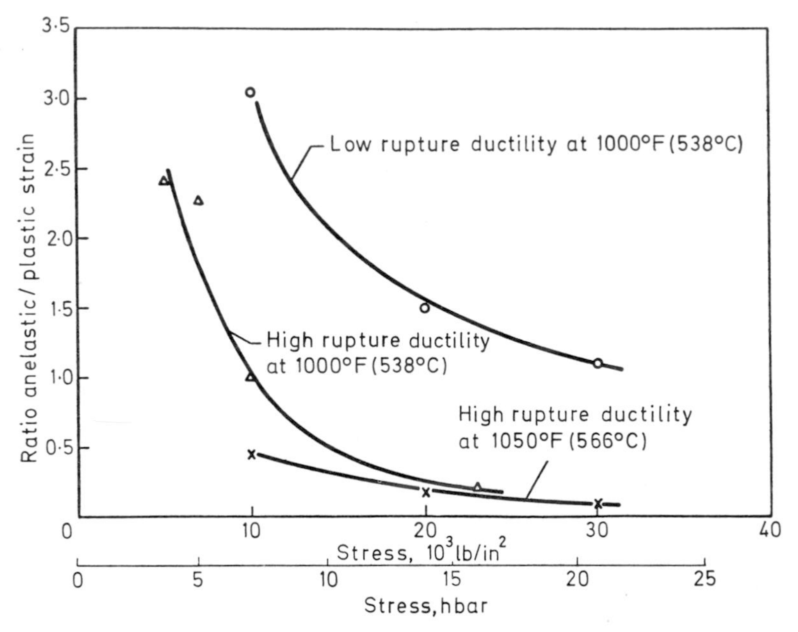

FIG. 17. Effect of ductility on recoverable strain for Cr–Mo–V steel after 1000 hr creep exposure.

Short times and low stress favour the dominance of anelastic strain, but Fig. 17 shows how highly creep resistance microstructure, *i.e.* the low ductility alloy, also tends to have a dominant anelastic component. This behaviour may have important damage effects as discussed later. In general, these data show that anelastic creep strain may be far more important than plastic creep strain at practical stress levels, for example, operating stresses of central power station turbines. Finally, it seems likely that adequate theories for creep will have to account for anelastic behaviour. This is particularly true with regard to the 'mechanical equation of state' and plastic creep behaviour under varying stress.

Creep under varying stress

When a creep test is partially unloaded a number of phenomena come into play as we have seen. Provided that the initial loading and subsequent creep have introduced elastic, plastic and anelastic strains, the following events take place on load removal [6]:

1 Elastic shortening occurs immediately.
2 Anelastic shortening (creep recovery) occurs gradually thereafter.
3 Because of the residual load, plastic strain continues gradually thereafter.

The engineering significance of anelasticity lies in these dimensional changes. Prestraining effects are of interest in this regard. The author has made some tests [8] on variable ductility Cr–Mo–V steel in which the effects of creep prestrain were studied in detail. Figure 18 (taken from that paper) shows that two effects are possible:

1 Subsequent creep at the reduced loads proceeds with virtually no transient stage.
2 Creep resistance may be greatly enhanced.

These effects are essentially due to the anelastic phenomena described and have been used in engineering practice. Note, however, the sensitivity to prestrain conditions and the possible subsequent effect of reduced rupture resistance [8]. The author has also conducted some tests on Cr–Mo–V steel [9], using sequential load changes which again illustrate anelastic effects. In Fig. 19a and b from that paper both the constant load creep data and the step test creep data are compared and again we note the virtual elimination of primary creep and increased creep resistance with lowered stress. Figure 19c shows the combined plastic–anelastic creep after each load reduction and illustrates the anelastic nature of creep due to prestraining history. It is our lack of quantitative understanding of this stress/strain/time history effect on mechanical behaviour of alloys that makes creep analysis so difficult. It seems obvious, also, that these circumstances must be a complicating factor in the mechanism and behaviour of alloys under high temperature, low cycle fatigue conditions.

Stress relaxation

In the general background section of this chapter, I have described very qualitatively the structural mechanisms giving rise to so-called after-effects. Stress relaxation is manifestly an after-effect. Phenomenologically, it is the process of converting initial elastic strain into non-recoverable strain of like amount by means of inelastic deformation which proceeds under decreasing stress. Freudenthal [10] in an elegant treatment of stress relaxation, points out that mechanistically this is not the same as creep

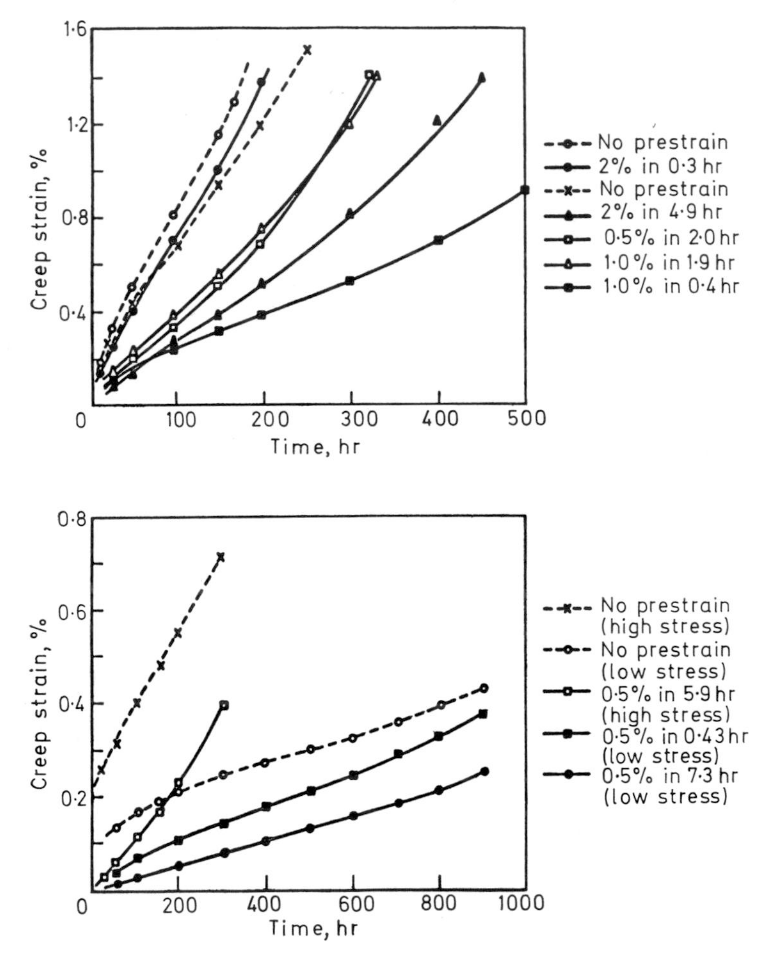

FIG. 18. Prestrain effects in Cr–Mo–V steel of variable ductility at 566°C (1050°F).

under decreasing stress because, among other things, of the behaviour of the recoverable anelastic component. Thus, the phenomenon of stress decreases with time in a specimen held at constant length after initial loading is due to *both* plastic and anelastic strain. This, in part, is why it

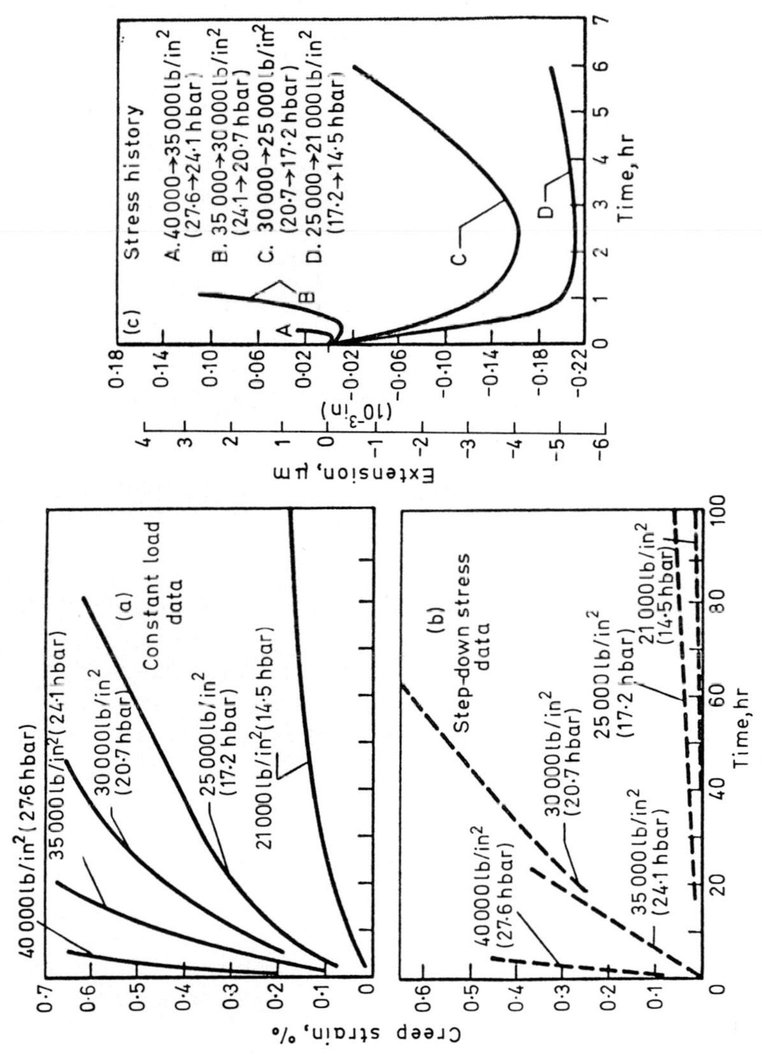

FIG. 19. Effect of stress decrement on creep and anelastic behaviour of Cr–Mo–V steel at 566°C (1050°F).

is difficult to derive relaxation behaviour from creep data. As a consequence, rigorous analytical treatment of the relaxation problem requires detailed knowledge of recovery characteristics.

In practice, the prestrain effects applied to relaxation could be useful in bolting closure applications. Since in relaxation, as in creep, the anelastic effects tend to counteract the plastic effects on partial unloading, it is possible to get 'uphill' relaxation, *i.e.* increasing stress with time. The interval of stress increase is usually followed by stress decrease but this depends on whether or not the residual stress after the load decrement is large enough to cause significant plastic creep rates. This effect has been

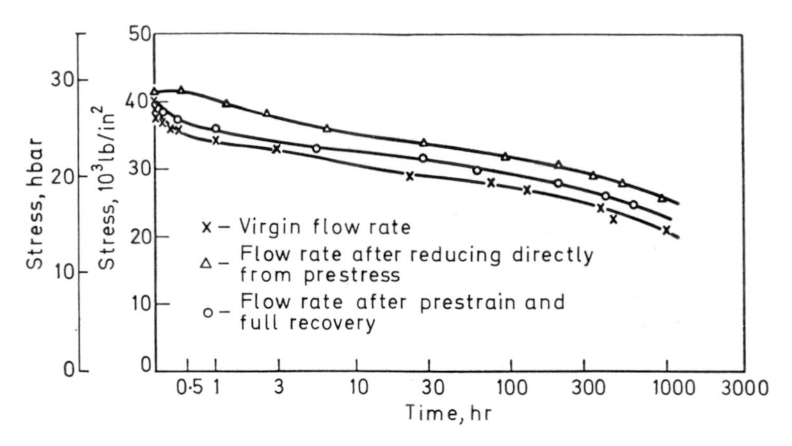

FIG. 20. Stress relaxation of 12% Cr steel at 538°C (1000°F).

well documented [6] and is further substantiated here in Fig. 20. These results were obtained on a 12 per cent Cr steel represented in Table 1 by the three tests Nos. 15–17 with a test procedure described by Goldhoff and Barker [11]. Test No. 1 is a simple relaxation from an original stress of 40 000 lb/in^2 (27·6 hbar); test No. 2 was slightly prestrained at 58 000 lb/in^2 (40·0 hbar) after which the stress was dropped to 40 000 lb/in^2 (27·6 hbar) and relaxation at constant strain allowed to develop; test No. 3 was similarly prestrained but fully recovered before reloading to 40 000 lb/in^2 (27·6 hbar) and then relaxing. Recovery readings were taken after 1000 hr of test in each case. It is obvious that the prestrained, unrecovered specimen displays a slight stress increase followed by a continuous stress decrease. It then maintains a parallelism with the virgin test after a few hours. The last test displays an intermediate behaviour which reflects the strain hardening initially built in by the prestress treatment. It is interesting that each test maintains its increment of residual stress conferred by the pretreatment even after the 1000 hr test. In addition, the total recovery strains after relaxation are in approximate proportion to the terminal residual stresses. Lubahn and Felgar [4, 6] have treated the engineering aspects of anelasticity and stress relaxation in great detail.

Damage

In the reference literature quoted in this chapter, mechanistic models for anelastic behaviour have been given. The essential argument for metallic

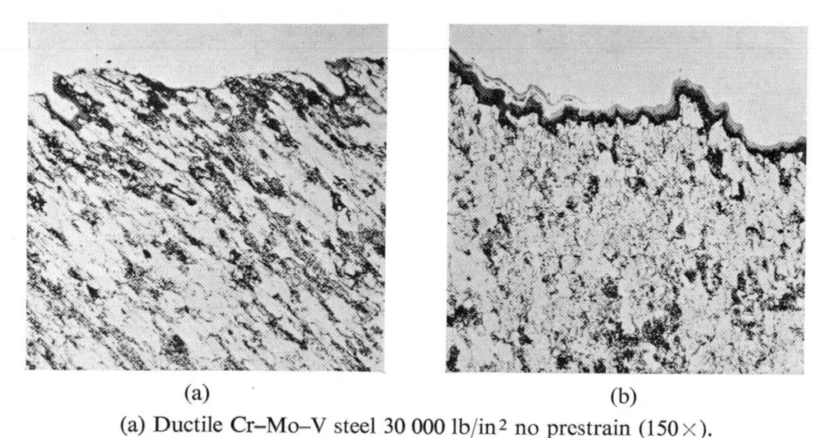

(a) (b)

(a) Ductile Cr–Mo–V steel 30 000 lb/in² no prestrain (150×).
(b) Ductile Cr–Mo–V steel 30 000 lb/in² 2·0 per cent prestrain (150×).

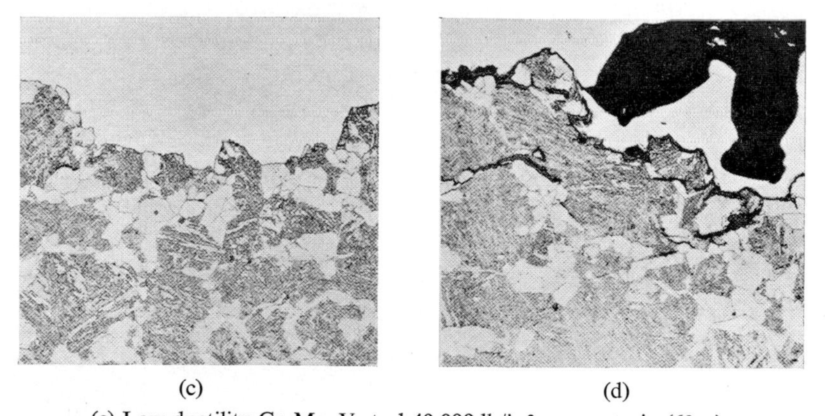

(c) (d)

(c) Low ductility Cr–Mo–V steel 40 000 lb/in² no prestrain (60×).
(d) Low ductility Cr–Mo–V steel 40 000 lb/in² 0·5 per cent prestrain (60×).

FIG. 21. Fracture characteristics of steels before and after prestrain (see Fig. 18).

alloys is the existence of viscous elements in an otherwise crystalline matrix. Zener [1] has pointed out two interesting features of such viscous zones; namely, a very large magnitude of anelastic effect may be produced by a very small amount of viscous material, and very high stress concentrations can occur in the elastic matrix due to the relaxation of shear stresses in the viscous regions. The viscous centres have long been attributed to slip bands and grain boundaries in metals. Although it is generally felt that the effects of anelastic straining in metallic alloys are negligible, Zener [1] seems to have realised a great potential for damage in their existence. Thus, he refers to the accumulation of stress concentration at the edges of slip bands and across grain boundaries as damage-producing mechanisms, citing delayed fracture phenomena and intercrystalline cracking as possible results of these mechanisms. It is, of course, well known that the conditions which lead to low fracture ductility in metals and alloys, *i.e.* high temperatures and low strain rates, are also the conditions which favour viscous flow at grain boundaries. In the work [8] which led to the results shown in Fig. 18 a decided change in the mode of fracture accompanied the prestrained specimens of both the high and low ductility steels. Figure 21 (taken from that work) illustrates these effects. In the case of the ductile steel the fracture after prestrain has a decidedly more intergranular character than the companion fracture with no prestrain. This is accompanied by lowered fracture ductility. For the low ductility steel both fractures are intergranular, but the prestrained fracture has many secondary cracks below the fracture face attesting to the larger role of intergranular processes. It is not outside the realm of possibility that anelastic straining dominated the prestrained creep tests and was the limiting feature of the final failure. Similar evidence has been presented in connection with the step-down stress tests [9].

It is to be realised that structural damage in metals, particularly at elevated temperatures, is a very complex matter and certainly not well understood. I do not suggest, therefore, that the effects discussed here are total or even dominant factors in damage mechanisms; rather I suggest that they may not be nearly as negligible as is usually supposed.

CONCLUSIONS

A brief review of the phenomenology and engineering significance of creep recovery has been given. I have not tried to treat the subject exhaustively, but rather have attempted to show how, with reasonable assumptions, the recovery characteristics of steels can be obtained and correlated in a useful fashion. I have further suggested that for problems involving small permissible deformations and particularly those involving significant stress changes during creep, these effects are highly important. While it is true that little heed is paid to this subject at present, it is also true that significant progress in understanding creep, relaxation and high temperature, low cycle fatigue will undoubtedly need to consider such effects in careful detail.

REFERENCES

1. ZENER, C. M. (1948). *Elasticity and Anelasticity of Metals.* University of Chicago.
2. JOHNSON, A. E., HENDERSON, J. and MATHUR, V. D. (1958). 'Pure torsion creep tests on Mg alloy at 20°C and on 0·2 per cent C steel at 450°C at low rates of strain.' *Metallurgia*, **347,** 109–117.
3. JOHNSON, A. E., HENDERSON, J. and KHAN, B. (1963). 'Pure torsion creep tests on Al alloy at 200°C at low rates of strain.' *Metallurgia*, **402,** 173–177.
4. LUBAHN, J. D. and FELGAR, R. P. (1961). *Plasticity and Creep of Metals.* London: John Wiley and Sons.
5. HENDERSON, J. and SNEDDEN, J. D. (1965). 'Creep recovery of aluminium alloy DTD 2142.' *Appl. Mater. Res.*, **4**(3), 148–168.
6. LUBAHN, J. D. (1953). 'The role of anelasticity in creep, tension and relaxation behaviour.' *Trans. Amer. Soc. Mech. Engrs*, **45,** 787–838.
7. HARVILL, L. R. (1966). 'Vito volterra and viscoelasticity.' *J. Franklin Inst.*, **282**(2), 85–91.
8. GOLDHOFF, R. M. (1962). 'The effect of creep prestrain on creep-rupture properties of variable notch-sensitivity Cr–Mo–V steel.' *Mater. Res. & Stand.*, **1**(1), 26–32.
9. GOLDHOFF, R. M. (1965). 'Uniaxial creep-rupture behaviour of low-alloy steel under variable loading condition.' *ASME Paper* No. 64-WA/MET-4. New York: American Society of Mechanical Engineers.
10. FREUDENTHAL, A. M. (1958). 'Stress relaxation in structural materials.' *Tech. Report* No. WADC 58–168, *ASTIA Document* No. 202497, 10/58. Cambridge, Mass.: Little Inc.
11. GOLDHOFF, R. M. and BARKER, L. B. (1962). 'Stress relaxation of a Cr–Mo–V steel heat treated to various notch sensitivities.' *Trans. ASTM*, **62,** 1176–1191.

Chapter 7

MATHEMATICAL RELATIONSHIPS FOR PRIMARY, SECONDARY AND TERTIARY CREEP AND THEIR USE IN EXTRAPOLATION OF TENSILE CREEP DATA

E. C. LARKE AND R. J. PARKER

SUMMARY

In the first part of the chapter, the use of empirically derived mathematical relationships, already established for primary and secondary creep, is extended to include the early stages of tertiary creep. In the second part of the chapter, the use of equations to represent experimental creep data is illustrated, and the application of the method to extrapolation of creep data for design is discussed.

NOTATION

A, m, n	Constants in creep law.
b, P	Constants in simplified primary creep law: eqn (2).
C	Constant of integration.
e	Tensile creep strain.
$\dot{e}$	Creep rate.
G	Constant in rational form of primary creep law: eqn (11).
J	Constant in rational form of tertiary creep law: eqn (16).
S	Secondary creep rate: eqn (4).
T	Constant in simplified tertiary creep law: eqn (6).
t	Time.
σ	Tensile stress.

Subscripts

p	Primary stage.
ps	Blending of primary and secondary stages.
s	Secondary stage.
st	Blending of secondary and tertiary stages.
t	Tertiary stage.

INTRODUCTION

In engineering design for prolonged service, it has always been necessary to estimate long term material properties without being able to carry out extended tests, a state of affairs which will undoubtedly continue to exist. The situation applies particularly to elevated temperature creep behaviour, and the present chapter describes a method for estimating long term tensile creep characteristics from the results of relatively short duration tests. It extends the use of an empirically derived mathematical relationship [1] which has been employed by many investigators to represent creep occurring under constant load at a fixed temperature. The practical value of the relationship has been demonstrated in published information dealing with primary and secondary creep [2, 3, 4], and in the present chapter, its use has been extended to the early part of tertiary creep.

The relationship (1) between stress σ, tensile creep strain e at constant temperature and creep rate $\dot{e}$ is:

$$\dot{e} = A\sigma^n e^{(m-1)/m}$$

which, after integrating with respect to time, becomes

$$e = (A/m)^m \sigma^{nm} t^m + C \tag{1}$$

It should be noted that e does not include any elastic or plastic strain which might occur during build-up of the applied load. In consequence, when eqn (1) is applied to primary creep strain, the value of the constant of integration C will be zero. On the other hand, as demonstrated below, it will have a finite value for secondary and tertiary creep. A, m and n are constants, the values of which will be different for the different stages of creep, and the present chapter gives details of an essentially non-subjective method of deriving these for each stage. Once the constants have been evaluated from a study of basic experimental data, it becomes possible to construct tensile creep curves for other stresses inside and outside the original range of data, and to use relationship (1) in engineering design analysis. The method involves linking the three stages of creep in a unique manner, and this enables an estimate to be made of when primary creep is likely to merge into the secondary stage, when tertiary creep is likely to predominate, and at what strains these changes are likely to occur.

The first step in the analysis is to consider each experimental curve individually and to smooth the measured data in accordance with simplified versions of eqn (1). Primary, secondary and tertiary creep are dealt with separately, but are linked by adjusting the various factors in order to make e and $\dot{e}$ continuous at transition times, this involving the retention of the experimental secondary phases in virtually their original forms. Finally, the influence of the different stresses is considered, these effects being smoothed by grouping all the experimental creep curves together.

ANALYSIS OF TENSILE CREEP DATA

The method of analysis† is demonstrated by making use of curves published by Skelton and Crossland [5] for the creep of a 0·19 per cent C steel at 450°C. A wide range of test stresses was covered by this work, but in this part of the present chapter, attention is deliberately restricted to three of these stresses: 10, 11 and 13 ton/in^2 (15·4, 17 and 20·1 hbar). In developing the analysis, so-called basic and rational equations are first derived for each of the selected tests, after which the material constants are evaluated. In the second part of the chapter, general equations, applicable to any test stress, are developed and used to calculate creep curves for other stresses considered by Skelton and Crossland, namely 6, 8, 9·09, 12 and 14 ton/in^2 (9·27, 12·35, 14·04, 18·55 and 21·6 hbar). This enables an assessment to be made of the validity of the proposed analysis.

Basic equations

The experimental curves referred to above for stresses of 10, 11 and 13 ton/in^2 (15·4, 17 and 20·1 hbar) are reproduced in Fig. 1 and it should be noted that although only three stresses were employed, certain repeat tests were carried out by Skelton and Crossland. Thus five curves were available for analysis, and each of these has been dealt with individually, the three stages of creep covered by each curve being considered separately as described in the sections immediately following.

† The present analysis is concerned with creep strain, and the experimental data reproduced in this chapter were obtained by deducting initial plastic strain data, reported by Skelton and Crossland, from their published curves.

PRIMARY STAGE

Considering first the primary stage of creep for any one test stress σ, eqn (1) can be written in a simplified or so-called basic form

$$e = Pt^b \tag{2}$$

where P takes the place of the two factors in eqn (1) which multiply t^m. Also, to facilitate subsequent explanation, the symbol b replaces m as the index of t. Regarding the magnitude of C in eqn (1), this, as mentioned above, will be zero for primary creep.

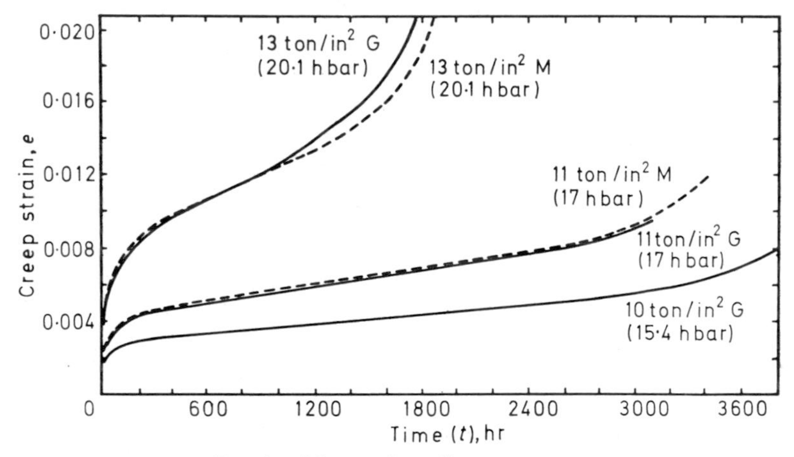

FIG. 1. Measured tensile creep curves.

Equation (2) can be written as

$$\log e = b \log t + \log P$$

and if $e{:}t$ values read from the primary portions of the five tensile creep curves in Fig. 1 are plotted on logarithmic scales, the majority of coordinates will approximate to straight lines which blend into the secondary phase; *see* Fig. 2. Although, according to eqn (1), each line should have the same slope m, in practice, differences will generally be found. However, when each straight line has been drawn, the value of theslope b for each test, and the corresponding magnitude of log P, are readily determined, after which the basic primary equations can be written down. For the five tests in Fig. 1, they are:

$$e = 1{\cdot}277 \times 10^{-3} t^{0{\cdot}1477} \text{ for } 10 \text{ ton/in}^2 \quad \text{G}\dagger$$
$$e = 1{\cdot}984 \times 10^{-3} t^{0{\cdot}1447} \text{ for } 11 \text{ ton/in}^2 \quad \text{G}$$
$$e = 1{\cdot}609 \times 10^{-3} t^{0{\cdot}1845} \text{ for } 11 \text{ ton/in}^2 \quad \text{M} \tag{3}$$
$$e = 2{\cdot}408 \times 10^{-3} t^{0{\cdot}2344} \text{ for } 13 \text{ ton/in}^2 \quad \text{G}$$
$$\text{and } e = 2{\cdot}472 \times 10^{-3} t^{0{\cdot}2313} \text{ for } 13 \text{ ton/in}^2 \quad \text{M}$$

SECONDARY STAGE

In the secondary stage, the relationship between creep strain and time is linear and in its basic form eqn (1) becomes

$$e = St + C_s \tag{4}$$

In this equation, S is the secondary creep rate which remains constant with time, and values of this rate, as determined from the experimental

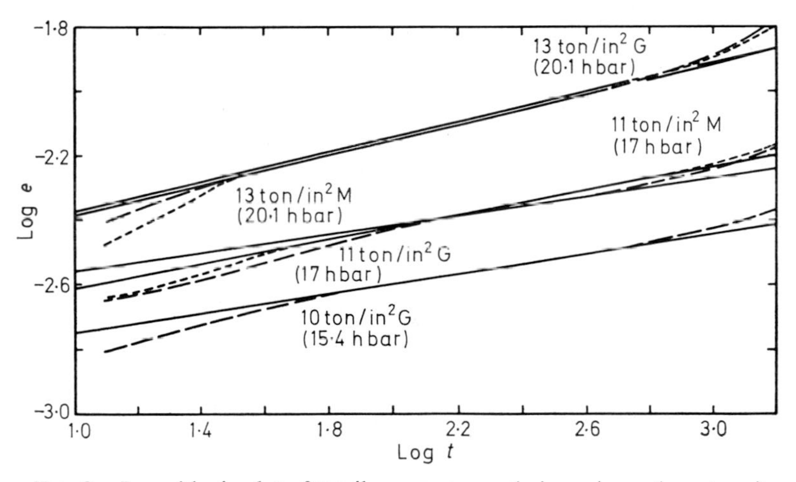

FIG. 2. Logarithmic plot of tensile creep curves (primary/secondary stages).

† It should be noted that certain abbreviations have been employed throughout the present chapter. Thus:

10 ton/in² G denotes 10 ton/in² (15·4 hbar) G4/2/CD
11 ton/in² G denotes 11 ton/in² (17 hbar) G3/2/CD
11 ton/in² M denotes 11 ton/in² (17 hbar) M14/1/CD
13 ton/in² G denotes 13 ton/in² (20·1 hbar) G3/1/CD
and 13 ton/in² M denotes 13 ton/in² (20·1 hbar) M14/3/CD
the latter identifications being used by Skelton and Crossland.

curves in Fig. 1, are recorded in Table 1. After determining S, the constant C_s can readily be calculated in the usual manner, *i.e.* using eqn (4). Values of C_s for the curves in Fig. 1 are also recorded in Table 1, and the various basic secondary equations are:

$$\begin{aligned} e &= 9{\cdot}524 \times 10^{-7}t + 0{\cdot}002721 \text{ for } 10 \text{ ton/in}^2 \quad \text{G} \\ e &= 1{\cdot}544 \times 10^{-6}t + 0{\cdot}004106 \text{ for } 11 \text{ ton/in}^2 \quad \text{G} \\ e &= 1{\cdot}483 \times 10^{-6}t + 0{\cdot}004351 \text{ for } 11 \text{ ton/in}^2 \quad \text{M} \\ e &= 5{\cdot}102 \times 10^{-6}t + 0{\cdot}007788 \text{ for } 13 \text{ ton/in}^2 \quad \text{G} \\ \text{and } e &= 4{\cdot}800 \times 10^{-6}t + 0{\cdot}008000 \text{ for } 13 \text{ ton/in}^2 \quad \text{M} \end{aligned} \tag{5}$$

TERTIARY STAGE

The basic form of eqn (1) for the tertiary stage of creep is

$$e = Tt^b + C_t \tag{6}$$

where, for convenience, b again replaces m. Differentiating this with respect to time gives the creep rate as

$$\dot{e} = bTt^{(b-1)}$$

or, taking logs of both sides,

$$\log \dot{e} = (b - 1) \log t + \log (bT)$$

Using e and t values read from the experimental curves in Fig. 1 which relate to the tertiary stage, rates $\dot{e}$ are determined for a series of consecutive time intervals. Log $\dot{e}$ is then plotted against log t (where t is the mean of each time interval) and a straight line, determined using linear regression analysis, is drawn through the coordinates, the slope of this line defining the magnitude of $(b - 1)$. With the value of $(b - 1)$ thus established for each test, the corresponding values of log (bT), and hence T, are then readily found.

Regarding the constant C_t in eqn (6), its magnitude depends on the test stress, and the method of evaluating it is as follows. The tertiary creep rate at the time t_{st}, which is the time when secondary creep blends into the tertiary stage, is put equal to the secondary creep rate S, that is

$$bTt_{st}^{(b-1)} = S \tag{7}$$

and from this

$$t_{st} = (S/bT)^{1/(b-1)} \tag{8}$$

Table 1

Data relating to the tensile creep tests

σ (ton/in^2)	10 G	11 G	11 M	13 G	13 M
b (primary)	0·1477	0·1447	0·1845	0·2344	0·2313
S (in/in hr)	$9{\cdot}524 \times 10^{-7}$	$1{\cdot}544 \times 10^{-6}$	$1{\cdot}483 \times 10^{-6}$	$5{\cdot}102 \times 10^{-6}$	$4{\cdot}800 \times 10^{-6}$
C_s (in/in)	$2{\cdot}721 \times 10^{-3}$	$4{\cdot}106 \times 10^{-3}$	$4{\cdot}351 \times 10^{-3}$	$7{\cdot}788 \times 10^{-3}$	$8{\cdot}000 \times 10^{-3}$
b (tertiary)	4·543	3·001	4·840	6·700	6·390
t_{st} hr (basic)	2497	2032	2226	1294	1339
C_t (in/in)	$4{\cdot}575 \times 10^{-3}$	$6{\cdot}198 \times 10^{-3}$	$6{\cdot}972 \times 10^{-3}$	$13{\cdot}403 \times 10^{-3}$	$13{\cdot}421 \times 10^{-3}$

Defining the creep strain at the time t_{st} as e_{st}, then from eqns (6), (4) and (7), it follows

$$C_t = St_{st}\left(\frac{b-1}{b}\right) + C_s \tag{9}$$

Values for t_{st} and C_t, calculated using eqns (8) and (9), are also recorded in Table 1.

Substituting T and C_t values together with the appropriate tertiary b values into eqn (6) enables the following basic tertiary equations to be derived:

$$\begin{aligned}
e &= 1{\cdot}926 \times 10^{-19} t^{4{\cdot}543} + 0{\cdot}004575 \text{ for } 10 \text{ ton/in}^2 \quad \text{G} \\
e &= 1{\cdot}233 \times 10^{-13} t^{3{\cdot}001} + 0{\cdot}006198 \text{ for } 11 \text{ ton/in}^2 \quad \text{G} \\
e &= 4{\cdot}266 \times 10^{-20} t^{4{\cdot}840} + 0{\cdot}006972 \text{ for } 11 \text{ ton/in}^2 \quad \text{M} \\
e &= 1{\cdot}388 \times 10^{-24} t^{6{\cdot}700} + 0{\cdot}013403 \text{ for } 13 \text{ ton/in}^2 \quad \text{G} \\
\text{and } e &= 1{\cdot}057 \times 10^{-23} t^{6{\cdot}390} + 0{\cdot}013421 \text{ for } 13 \text{ ton/in}^2 \quad \text{M}
\end{aligned} \tag{10}$$

Rational equations

In the previous section, basic equations for primary, secondary and tertiary creep were determined for each of the individual experimental curves in Fig. 1. Two sets of these basic equations, (3) and (10) above, indicate that as the stress varied, the values of b varied for primary and for tertiary creep in a somewhat erratic manner, this being a regular observation in tensile creep testing. However, to employ eqn (1) usefully, the time index for any one material and test temperature must be taken as constant for all test stresses; that is m in eqn (1) must have single values for primary, secondary and tertiary creep.

The proposition is therefore made that, for the primary and tertiary stages of creep, the average of the individual b values recorded in Table 1 defines the required m values. Thus, for primary creep,

$$m = \Sigma(b)/5 = 0{\cdot}1885$$

and for tertiary creep

$$m = \Sigma(b)/5 = 5{\cdot}095$$

It should be noted that no problem arises with the secondary stage because the time index is always unity: *see* eqn (4).

Following this, the next step towards deriving the other constants in eqn (1) is to transform the basic equations into so-called rational equations, in which these constant time index values m are incorporated. The necessary method is described below, and it should be noted that since for secondary creep m equals unity, the basic secondary equations are retained unaltered. Hence, for the data of Fig. 1, the basic eqns (5) become the rational equations for secondary creep strain.

PRIMARY STAGE

Let the form of the rational equation for the primary stage of creep be

$$e = Gt^m \tag{11}$$

The strain at the time t_{ps} when primary creep blends into the secondary stage will be

$$e_{ps} = G{t_{ps}}^m$$

Utilising eqn (4),

$$e_{ps} = G{t_{ps}}^m = St_{ps} + C_s \tag{12}$$

Also, differentiating eqn (11) with respect to time gives the creep rate at time t_{ps} as

$$\dot{e}_{ps} = mG{t_{ps}}^{(m-1)} = S$$

from which

$$G = \left(\frac{S}{m}\right) {t_{ps}}^{(1-m)} \tag{13}$$

Also, from eqn (12)

$$G = (St_{ps} + C_s)/{t_{ps}}^m \tag{14}$$

and, after equating (13) and (14),

$$t_{ps} = \frac{mC_s}{S(1 - m)}$$

Substituting this into eqn (13) gives

$$G = \left(\frac{S}{m}\right)^m \left(\frac{C_s}{1 - m}\right)^{(1-m)}$$

Using this equation, values of G were calculated for each test in Fig. 1 by substituting the S and C_s values recorded in Table 1, and using the m

value of 0·1885 given above. Hence, using eqn (11), the rational equations for primary creep strain are:

$$
\begin{aligned}
e &= 9{\cdot}851 \times 10^{-4} t^{0{\cdot}1885} \text{ for } 10 \text{ ton/in}^2 \quad \text{G} \\
e &= 1{\cdot}507 \times 10^{-3} t^{0{\cdot}1885} \text{ for } 11 \text{ ton/in}^2 \quad \text{G} \\
e &= 1{\cdot}568 \times 10^{-3} t^{0{\cdot}1885} \text{ for } 11 \text{ ton/in}^2 \quad \text{M} \\
e &= 3{\cdot}137 \times 10^{-3} t^{0{\cdot}1885} \text{ for } 13 \text{ ton/in}^2 \quad \text{G} \\
\text{and } e &= 3{\cdot}205 \times 10^{-3} t^{0{\cdot}1885} \text{ for } 13 \text{ ton/in}^2 \quad \text{M}
\end{aligned}
\tag{15}
$$

SECONDARY STAGE

As discussed above, the basic equations for secondary creep strain are taken as being the rational equations, hence the latter equations are the same as those numbered (5) above.

TERTIARY STAGE

Let the form of the rational equation for the tertiary stage of creep be

$$e = Jt^m + C_t \tag{16}$$

so that the strain e_{st}, at the time t_{st}† when secondary creep blends into tertiary creep, is

$$e_{st} = Jt_{st}^{\ m} + C_t \tag{17}$$

At this time, the creep rate will be S, so from eqn (17)

$$mJt_{st}^{\ (m-1)} = S \tag{18}$$

and

$$J = \left(\frac{S}{m}\right) \Big/ t_{st}^{\ m-1}$$

The constant C_t in eqn (17) can be expressed as

$$C_t = e_{st} - Jt_{st}^{\ m}$$

or, making use of eqn (4),

$$C_t = (St_{st} + C_s) - Jt_{st}^{\ m}$$

† Following the previous section, the magnitude of t_{st} is taken as being equal to its basic value: see Table 1.

From eqn (18),

$$Jt_{st}{}^{m} = \left(\frac{S}{m}\right) t_{st}$$

from which it follows that

$$C_t = St_{st}\left(\frac{m-1}{m}\right) + C_s$$

Following the previous section, the magnitude of C_s is taken as being equal to its basic value.

Values of S, b (tertiary), t_{st} and C_s for each test in Fig. 1 are recorded in Table 1, and since m equals 5·095, J and C_t can be determined. Substituting these values into eqn (16) gives the rational equations for tertiary creep strain as:

$$\begin{aligned}
e &= 2{\cdot}287 \times 10^{-21}t^{5\cdot095} + 0{\cdot}004632 \text{ for } 10 \text{ ton/in}^2 \quad \text{G}\\
e &= 8{\cdot}620 \times 10^{-21}t^{5\cdot095} + 0{\cdot}006627 \text{ for } 11 \text{ ton/in}^2 \quad \text{G}\\
e &= 5{\cdot}697 \times 10^{-21}t^{5\cdot095} + 0{\cdot}007004 \text{ for } 11 \text{ ton/in}^2 \quad \text{M}\\
e &= 1{\cdot}806 \times 10^{-19}t^{5\cdot095} + 0{\cdot}013095 \text{ for } 13 \text{ ton/in}^2 \quad \text{G}\\
\text{and } e &= 1{\cdot}479 \times 10^{-19}t^{5\cdot095} + 0{\cdot}013165 \text{ for } 13 \text{ ton/in}^2 \quad \text{M}
\end{aligned} \tag{19}$$

Derivation of the constants n and A

PRIMARY STAGE

For primary creep, eqn (1) can be written as

$$\log e = nm \log \sigma + m \log (A/m) + m \log t$$

and it is clear that nm, the slope of the line connecting the coordinates of $\log e$ and $\log \sigma$, will be the same irrespective of the time chosen to compute e. To effect this computation, the rational primary eqns (15) must be used, and since any time can be chosen to calculate the strain e, it is convenient to choose 1 hr, the strains then being the coefficients which multiply t in eqns (15).

The required plot for the five tests under consideration is shown in Fig. 3, and from this, nm and $m \log (A/m)$ can easily be found. With m already known as 0·1885, the magnitude of n and A are readily determined. Hence, for the primary stage of creep, the constants are

$$\begin{aligned}
m &= 0{\cdot}1885\\
n &= 23{\cdot}597\\
\text{and} \quad A &= 5{\cdot}700 \times 10^{-41}
\end{aligned}$$

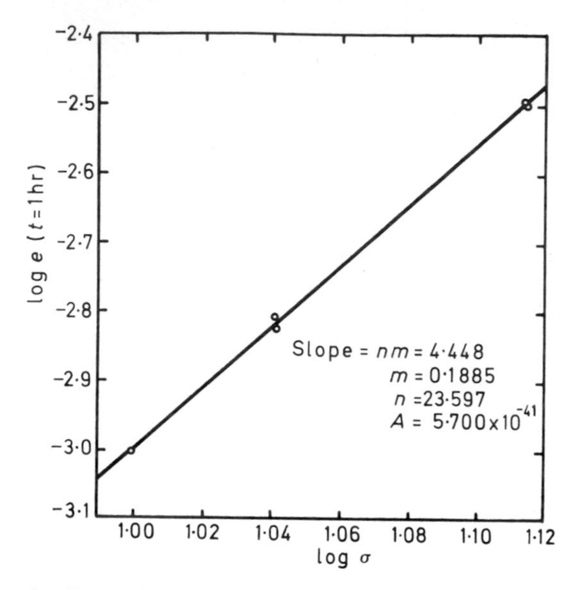

FIG. 3. Determination of constants n and A for primary creep.

SECONDARY STAGE

Throughout the secondary stage of creep m is equal unity and in consequence eqn (1), after differentiating with respect to time, can be written as

$$\log \dot{e} = n \log \sigma + \log A = \log S$$

Using the values of S in Table 1 together with the associated test stresses σ, $\log S$ has been plotted against $\log \sigma$ in Fig. 4, and the slope of the line drawn through the coordinates gives the value of n. Knowing this, A is easily computed and therefore, for the secondary stage, the values of the constants are

$$m = 1{\cdot}000$$
$$n = 6{\cdot}528 \text{ and}$$
$$A = 2{\cdot}583 \times 10^{-13}$$

TERTIARY STAGE

When dealing with the tertiary stage, where m is greater than unity, eqn (1) is first differentiated with respect to time, and then written as

$$\log \dot{e} = nm \log \sigma + \log [m(A/m)^m] + (m - 1) \log t \qquad (20)$$

From this, it will be clear that the magnitude of nm is the slope of the line connecting the coordinates of log $\dot{e}$ and log σ, and that this slope will be the same whatever time is chosen to compute $\dot{e}$ values.

If 1 hr is chosen, the values of $\dot{e}$ are simply the coefficients which multiply t in the following rational creep rate equations, which were obtained by differentiating eqns (19) with respect to time:

$$\dot{e} = 1{\cdot}165 \times 10^{-20} t^{4{\cdot}095} \text{ for 10 ton/in}^2 \quad \text{G}$$
$$\dot{e} = 4{\cdot}392 \times 10^{-20} t^{4{\cdot}095} \text{ for 11 ton/in}^2 \quad \text{G}$$
$$\dot{e} = 2{\cdot}903 \times 10^{-20} t^{4{\cdot}095} \text{ for 11 ton/in}^2 \quad \text{M}$$
$$\dot{e} = 9{\cdot}202 \times 10^{-19} t^{4{\cdot}095} \text{ for 13 ton/in}^2 \quad \text{G}$$
$$\text{and } \dot{e} = 7{\cdot}536 \times 10^{-19} t^{4{\cdot}095} \text{ for 13 ton/in}^2 \quad \text{M}$$

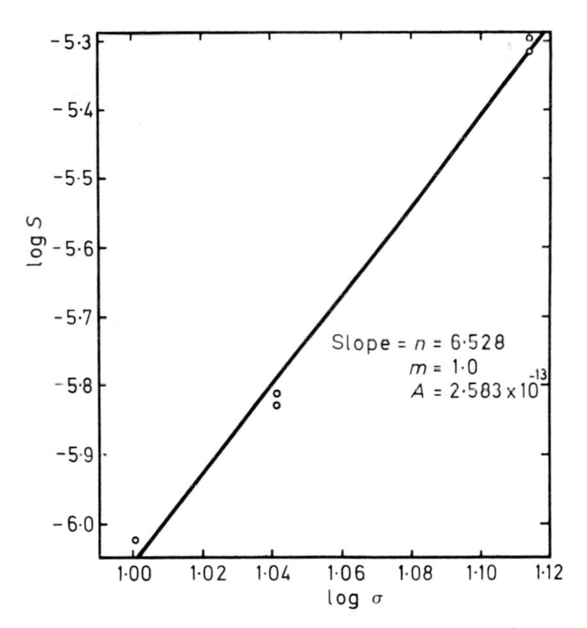

FIG. 4. Determination of constants n and A for secondary creep.

The logs of these coefficients are plotted in Fig. 5 against the logs of the corresponding σ values, and the slope of the drawn line gives the magnitude of nm. With m having the value of 5·095, n can then be found. Since, in plotting eqn (20), t is taken as unity, the third term on the right hand side of that equation becomes zero and, in consequence, with the value of m

known, the magnitude of A is readily determined. Hence, the values of the constants for the tertiary stage are

$$m = 5{\cdot}095$$
$$n = 3{\cdot}350 \text{ and}$$
$$A = 1{\cdot}915 \times 10^{-7}$$

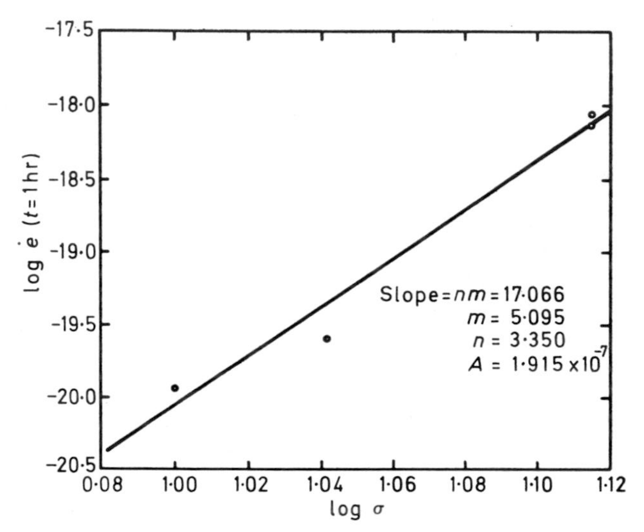

FIG. 5. Determination of constants n and A for tertiary creep.

THE EXTRAPOLATION OF TENSILE CREEP DATA

General equations

With the constants m, n and A now established for all three stages of creep, it is possible to extrapolate the original basic data, and to calculate tensile creep strain/time curves for any stresses of interest. The first stage in this process is to write down so-called general equations which relate to the various stages of creep. Thus, using the values of the constants previously determined and eqn (1), the general equation for primary creep strain is

$$e = 3{\cdot}553 \times 10^{-8}\sigma^{4{\cdot}448}t^{0{\cdot}1885} \tag{21}$$

Similarly, for secondary creep strain

$$e = 2{\cdot}583 \times 10^{-13}\sigma^{6{\cdot}528}t + C_s \tag{22}$$

the corresponding general equation for tertiary creep strain being

$$e = 1{\cdot}477 \times 10^{-38}\sigma^{17{\cdot}066}t^{5{\cdot}095} + C_t \qquad (23)$$

Having established these equations, it is necessary to determine the periods of time for which each is applicable, and this is readily achieved as follows. Equation (21) is assumed to apply until the primary creep rate becomes equal in magnitude to the secondary creep rate, which can be calculated from eqn (22). Secondary creep behaviour is then assumed to continue until tertiary creep rate, calculated from eqn (23), also equals the secondary creep rate from eqn (22).

Having determined the time at which primary creep merges into secondary creep, denoted, by t_{ps}, it is a simple matter to calculate the corresponding creep strain from eqn (21). This readily enables the value of C_s in eqn (22) to be evaluated. With this done, and with a knowledge of the time t_{st} when tertiary creep begins to predominate, the creep strain e_{st} associated with this change of behaviour is calculated using eqn (22). Constant C_t in eqn (23) can then be determined. Expressed mathematically, this means that

$$t_{ps} = 2{\cdot}747 \times 10^5/\sigma^{2{\cdot}563}$$

$$e_{ps} = 3{\cdot}763 \times 10^{-7}\sigma^{3{\cdot}965}$$

$$C_s = 3{\cdot}055 \times 10^{-7}\sigma^{3{\cdot}965}$$

$$t_{st} = 9{\cdot}809 \times 10^5/\sigma^{2{\cdot}573}$$

$$e_{st} = 2{\cdot}534 \times 10^{-7}\sigma^{3{\cdot}954} + C_s$$

$$\text{and } C_t = 2{\cdot}049 \times 10^{-7}\sigma^{3{\cdot}954} + C_s$$

Application to the data of Skelton and Crossland

It will be recalled that the values for the constants m, n and A were determined by reference to creep strain/time curves established from data published by Skelton and Crossland [5] for test stresses of 10, 11 and 13 ton/in^2 (15·4, 17 and 20·1 hbar). Using the methods described above, general equations, embodying these constants, have been developed and used to calculate creep strain/time curves for other test stresses considered by Skelton and Crossland, namely 6, 8, 9·09, 12 and 14 ton/in^2 (9·27, 12·35, 14·04, 18·55 and 21·6 hbar). In this way both interpolation and extrapolation on either side of the basic experimental data were carried out. To complete this comparison work, creep strain/time curves were also calculated using the general equations for the basic stresses of 10, 11 and 13 ton/in^2, and all the data, both experimental and calculated, have been plotted in Fig. 6.

It is evident that agreement between the experimental and calculated curves is good, and in this connection certain observations are of interest relating to the lines denoting t_{ps} and t_{st} in Fig. 6. The positions of these lines were determined using the equations for t_{ps}, e_{ps}, t_{st} and e_{st} given above, calculations being made for all the test stresses considered. In this way it was found that the majority of creep at 6 ton/in^2 (9·27 hbar) was primary in nature, whereas most of the test at 8 ton/in^2 (12·35 hbar) was of the secondary type, with tertiary creep just beginning to manifest itself

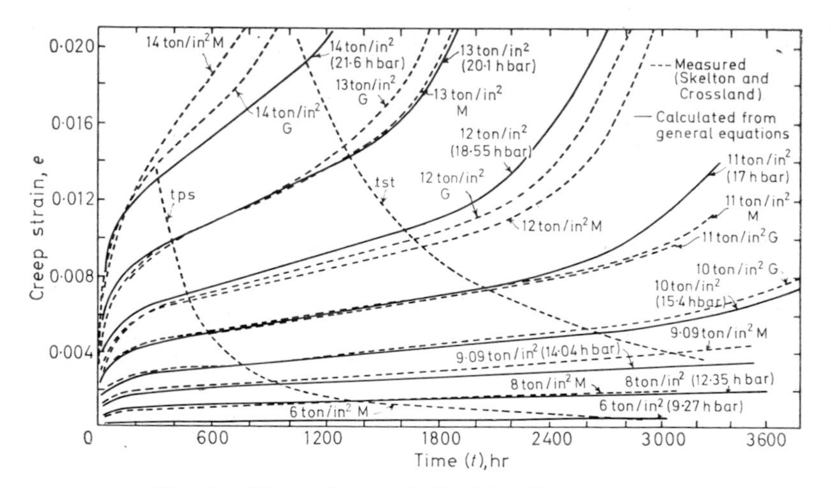

FIG. 6. Measured and calculated tensile creep curves.

towards the end of the 9·09 ton/in^2 (14·04 hbar) test. At the higher stresses, significant contributions from primary, secondary and tertiary creep were found to occur. The proposed method of analysis, therefore, in addition to enabling creep strain data to be calculated, is also capable of indicating the type of creep which is likely to be dominant at a given time.

CONCLUSIONS

Using the method of analysis described in the first part of the present chapter, mathematical equations for primary, secondary and tertiary tensile creep can be established with greater certainty than has been possible hitherto. A comparison with published creep data for 0·19 per cent C steel has shown an excellent standard of agreement between calculated and experimental data, and it seems reasonable to conclude that the precision with which designers can extrapolate from existing information should be increased by use of the new method.

An important practical design feature of the method is that it enables the time when primary creep merges into the secondary stage to be estimated with reasonable precision. A similar estimate can also be made of when secondary creep is likely to merge into the tertiary stage. In addition, it is a simple matter to calculate the creep strains at which these changes can be expected to take place.

REFERENCES

1. BAILEY, R. W. (1951). 'Creep relationships and their applications to pipes, tubes and cylindrical parts under internal pressure.' *Proc. Instn. Mech. Engrs*, **164**(4), 425–431.
2. LUBAHN, J. D. and FELGAR, R. P. (1961). *Plasticity and Creep of Metals.* London: J. Wiley.
3. INGLIS, N. P. and LARKE, E. C. (1958). 'Strength at elevated temperatures of aluminium and certain aluminium alloys.' *Proc. Instn. Mech. Engrs.*, **172,** 991–999.
4. OHJI, K. and MARIN, J. (1963). 'Creep of metals under non-steady conditions of stress.' *Thermal Loading and Creep in Structures and Components. Proc. Instn. Mech. Engrs*, 1963–64, **178**(Pt 3L), 126–134.
5. SKELTON, W. J. and CROSSLAND, B. (1967). 'Results of high sensitivity tensile creep tests on a 0·19 per cent carbon steel used for thick-walled cylinder creep tests.' *Conf. High Pressure Engng. Proc. Instn Mech. Engrs*, 1967–68, **182**(Pt 3C), 151–158.

Chapter 8

THE COMPARISON OF TORSION AND TENSION CREEP DATA FOR A 0·18 PER CENT CARBON STEEL

B. CROSSLAND, R. G. PATTON AND W. J. SKELTON

SUMMARY

The possible advantages of carrying out torsion creep tests on thin tubular specimens compared with constant load tension creep tests are discussed, and it is concluded that they are significant. A theory based on strain hardening to allow the derivation of constant load tension creep data from constant shear stress tests on thin tubes is presented.

The torsion machine designed and developed by the authors is described, and torsion data on a 0·18 *per cent C steel at* 400°*C are given. Comparable constant load tension creep data from creep machines of NEL design are also reported.*

The correlation between the constant load tension creep curves derived from torsion creep data, allowing for the change of stress, and the experimental constant load tension creep data is excellent. The discrepancy between the constant load tension creep curves, deduced from torsion creep data but ignoring the change of stress, and the curves of experimental data is significant at all stress levels, but particularly at the higher values of stress.

NOTATION

A	Original cross-sectional area of specimen.
A'	Current cross-sectional area.

B, m, n	Constants in eqn (11).
l	Original gauge length of specimen.
l'	Current gauge length.
s	True stress.
s_0	Initial true stress.
t	Time.
γ	Equivalent shear strain.
γ_c	Equivalent creep shear strain.
ε	Tensile (engineer's) strain.
ε^*	Effective strain.
ε_c	Creep strain.
η	True or logarithmic strain.
σ	Tensile (nominal) stress.
σ^*	Effective stress.
τ	Equivalent shear stress.

Subscripts

t	At time t.
1, 2, 3	Principal stress directions.

INTRODUCTION

Even in conventional material testing the torsion test has many advantages over the tension test [1] and these advantages apply equally well to creep testing. In tension testing, axiality of loading is difficult to achieve even under ideal conditions especially if the specimens are screwed at their ends. With screwed ends the problem is made even worse under creep conditions because of oxidation. If new shackles were made for each specimen there would still be the problem of ensuring that the axis of the load was coincident with the axis of the specimen. In creep tests the axiality of the load is important particularly during the primary creep region [2].

In tension tests the area reduces as the specimen elongates and unless the load is reduced the stress must increase. However, it is difficult to provide for an automatic reduction of load as the specimen strains, and nearly all the creep data recorded have been at constant load even though at larger strains this has a significant effect, as will be shown in this chapter. Another problem is that at moderately small strains a tension specimen becomes unstable and a neck is formed and it then becomes virtually impossible to continue collecting data which are meaningful.

With torsion testing it is very easy to design couplings which ensure that only a pure couple is applied to the specimen [1]. Also, as there are no significant dimensional changes, the stress condition can be maintained constant if a thin walled tube is used. Another advantage is that greater sensitivity of strain measurement can be obtained in torsion compared with tension. Dr Johnson [3] claimed that strain rates of the order of 10^{-9} per hour could be achieved in torsion compared with 10^{-7} in tension.

A major disadvantage of torsion is that a solid specimen cannot be used as the strain rate would vary with the radius. Consequently, it cannot be assumed that the material behaves in a homogenous way, and, as a result, the Nadai construction [1] for deducing the shear stress–strain curve from a test on a solid specimen is not applicable. The only alternative is to use a tube which has a wall thickness that is small compared with the radius, so that it can be assumed that the stress and the strain rate are uniform. However, these tubes are expensive to manufacture and they suffer buckling at larger strains which affect the results rather like tertiary creep [4]. Up to the onset of buckling the results appear to be of more fundamental significance than constant load tension tests.

In problems involving plasticity the fundamental material characteristic involved is the shear stress–strain properties, and consequently it seems preferable to obtain these data in as direct a manner as possible. This would indicate a preference for torsion tests on thin walled tubes. In particular, in the analysis of the creep of closed end thick walled cylinders, which is of importance in the chemical and power producing industries, the state of stress can be reduced to a simple shear stress with a superimposed hydrostatic tensile stress. Consequently, if the hydrostatic stress has no significant effect on creep, which appears probable, shear stress creep data can be directly applied in the design of cylinders.

The object of this chapter is to report shear and constant load tension creep data obtained for a 0·18 per cent C steel, and to investigate the problems of correlating data from these two forms of test.

THEORETICAL ANALYSIS

In order to correlate tension and torsion creep data it is necessary to define an effective stress and effective strain and a stress–strain–time relationship.

The von Mises criterion has been accepted to define the effective stress and effective strain, namely:

$$\sigma^* = \frac{1}{\sqrt{2}}[\Sigma(\sigma_1 - \sigma_2)^2]^{\frac{1}{2}}$$
$$\varepsilon^* = \frac{\sqrt{2}}{3}[\Sigma(\varepsilon_1 - \varepsilon_2)^2]^{\frac{1}{2}} \qquad (1)$$

For simple tension and torsion:

$$\sigma^* = \sigma = \sqrt{3}\tau$$
$$\varepsilon^* = \varepsilon = \frac{\gamma}{\sqrt{3}} \qquad (2)$$

These equations are only applicable if the elastic strains are small compared with the plastic or creep strains, though they can be rewritten in a more complex form to allow for the elastic component. At large strains the tensile expression is only approximate if the nominal tensile stress is employed, owing to the significant rise in the true stress resulting from a reduction of area.

The current gauge length of a specimen with an original length l is given by:

$$l' = l(1 + \varepsilon) \qquad (3)$$

where ε is the engineer's strain. The true or logarithmic strain is:

$$\eta = \int_l^{l'} \frac{dl}{l} = \ln\frac{l'}{l} = \ln(1 + \varepsilon) \qquad (4)$$

The equivalent shear strain from eqn (2) will be

$$\gamma = \sqrt{3}\eta = \sqrt{3}\ln(1 + \varepsilon)$$

or

$$\varepsilon = \exp(\gamma/\sqrt{3}) - 1 \qquad (5)$$

When the elastic strains are negligible it can be assumed that the volume is constant, which leads to:

$$Al = A'l'$$

where A is the original cross-sectional area. Thus,

$$\frac{A}{A'} = \frac{l'}{l} = (1 + \varepsilon) \tag{6}$$

The true stress s will be

$$s = \frac{\text{Load}}{A'}$$

whereas the nominal or engineer's stress will be

$$\sigma = \frac{\text{Load}}{A}$$

Therefore

$$s = \frac{A}{A'}\sigma = (1 + \varepsilon)\sigma \tag{7}$$

By von Mises, the equivalent shear stress is

$$\tau = \frac{s}{\sqrt{3}} \tag{8}$$

To apply the true stress–strain relationships to creep conditions, consider a tensile creep test where the initial true stress after loading is s_0. Combining eqns (7) and (8), the equivalent shear stress at any time t is thus given by:

$$\tau_t = \frac{s_0}{\sqrt{3}}(1 + \varepsilon_{c,t}) \tag{9}$$

where $\varepsilon_{c,t}$ is the engineer's creep strain at time t. Using eqn (5) with eqn (9) gives:

$$\tau_t = \frac{s_0}{\sqrt{3}} \exp(\gamma_{c,t}/\sqrt{3}) \tag{10}$$

Therefore, if the equivalent creep shear strain is known for time t, the shear stress at that time can be calculated. To allow for the continuously increasing equivalent shear stress, a strain hardening method has been adopted, as in general this gives a more accurate prediction of creep under increasing stress than the time hardening law [5, 6]. It is assumed that a set of constant shear stress creep curves can be represented by an expression of the form:

$$\gamma = B\tau^n t^m \tag{11}$$

The stress in the tension creep test is regarded as remaining constant for small intervals of time. At the end of each increment the equivalent shear stress is recalculated. By the strain hardening theory considering time t, when the total equivalent shear strain is $\gamma_{c,t}$, it is first necessary to find the time at which a torsion creep test, at a constant shear stress τ_t, would have reached a shear strain of $\gamma_{c,t}$. From eqn (11) this time is given by:

$$t = \left(\frac{\gamma_{c,t}}{B\tau_t^{\,n}}\right)^{1/m} \tag{12}$$

Hence, the equivalent shear strain for time, $t + \Delta t$, at a shear stress τ_t is:

$$\gamma_{c,t+\Delta t} = B\tau_t^{\,n}\left[\left(\frac{\gamma_{c,t}}{B\tau_t^{\,n}}\right)^{1/m} + \Delta t\right]^m \tag{13}$$

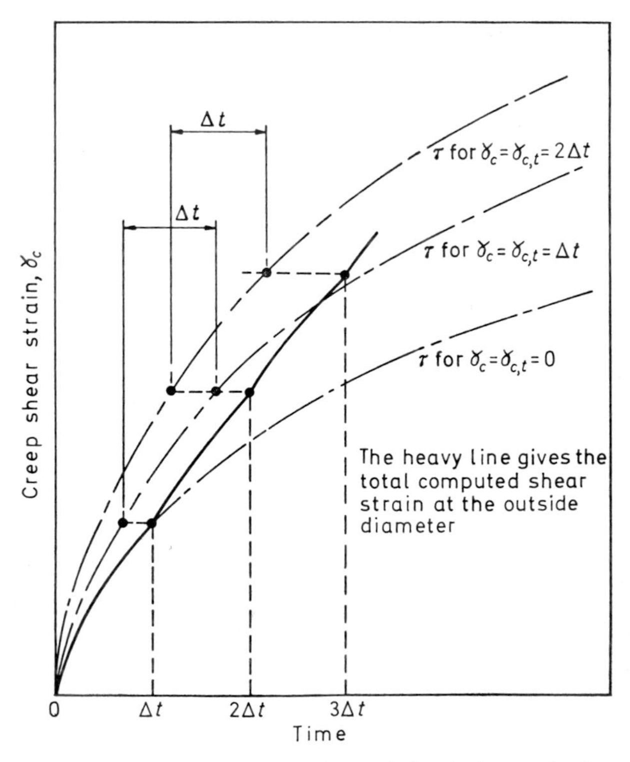

FIG. 1. Exaggerated diagram to illustrate the strain hardening method used to predict constant load tensile creep.

The total engineer's strain at time $(t + \Delta t)$ for the tension creep test is therefore, from eqn (5):

$$\varepsilon_{t+\Delta t} = \exp\,(\gamma_{c(t+\Delta t)}/\sqrt{3}) - 1 \tag{14}$$

and the shear stress $\tau_{t+\Delta t}$ for the next interval is found from eqn (10).

In this manner, starting with the true tensile stress at zero time (s_0), when the creep strain is zero, it is possible, using torsion creep data, to calculate the engineer's strain occurring in a tensile creep test. Reference to Fig. 1 will illustrate the first three steps of the above procedure. As the calculations involved for many time intervals are both numerous and repetitive the technique is obviously suited to a digital computer.

CREEP MACHINES

Figures 2 and 3 show a general arrangement drawing and photograph of the torsion creep machine designed and developed for this research. The design owes much to the torsion creep machine used by the late Dr Johnson [3], but it differs from this machine in many important respects.

A tangential load is applied to a large diameter plate pulley by wires which pass over small pulleys and are connected to an equalising beam to which the load in the form of weights is applied. The pulley is keyed to a Nimonic 80A torque tube which in turn is keyed to the specimen. The weight of the pulley, torque tube and half a specimen is supported by a spherically seated thrust bearing, while the top torque bar, fittings and half a specimen are supported by a counterbalance arrangement. The reaction torque at the top end is taken by a self-aligning torsion grip [1]. All the keying used Kennedy keys to ensure a rigid torsional connection and within the furnace these keys, like the torque tube and torque bar, are made of Nimonic 80A.

The torsion creep machine is designed for a maximum shear stress of 22 tonf/in^2 (34 hbar) in the tubular specimen shown in Fig. 4, though its capacity could be considerably increased without difficulty. The dimensions of the thin walled tubular specimen were chosen so that on the one hand the variation of shear stress across the wall was not too large, and on the other hand that buckling would not occur at too small a shear strain. Under elastic conditions the variation of shear stress from that at mean radius is $\pm 6{\cdot}25$ per cent, but under creep conditions with the material tested, the variation is less than $\pm 1{\cdot}3$ per cent. It was found that oxidation virtually ceased after 24 hr when an oxide layer of 0·0005 in (13 μm) had been produced. As this oxide layer is brittle and flakes off, its load bearing capacity was considered negligible, so that the tube wall thickness was taken as being reduced by 0·001 in (25 μm) which raises the stress by 2 per

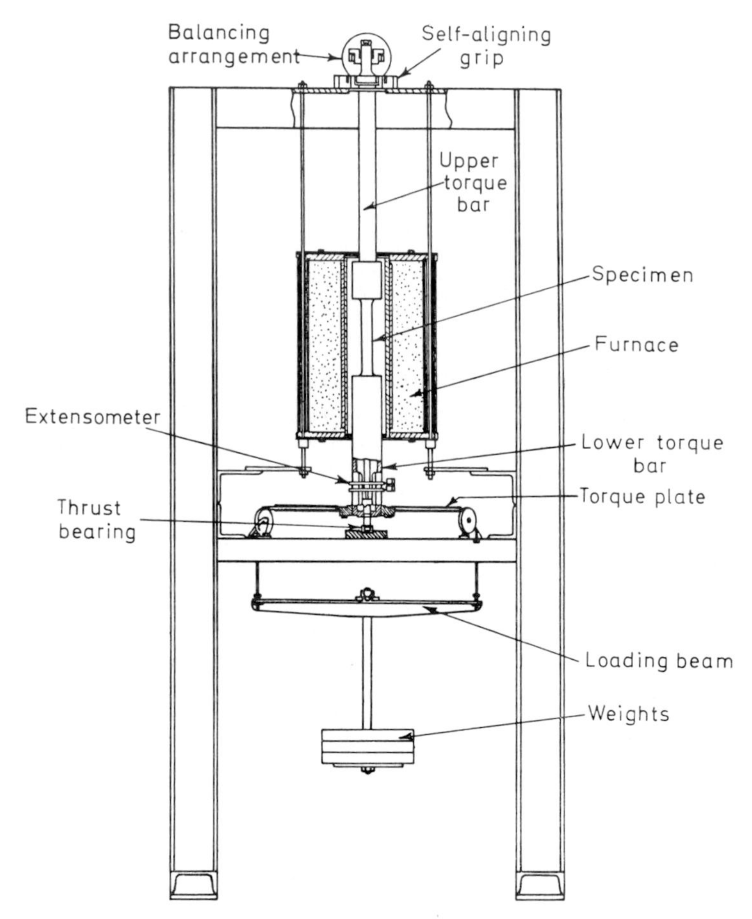

FIG. 2. Torsion creep machine.

cent. Buckling of the torsion specimens has to date given rise to no problems and several tubes have been taken up to a 10 per cent shear strain without evidence of buckling. During manufacture great care was taken to ensure concentricity of the bore with the outer diameter to a high degree of accuracy. The bore was machined and honed, and the specimen was then

FIG. 3. Torsion creep machine.

fitted on a tight fitting mandrel and the outside surface was machined and polished.

With the exception of Everett [7], previous investigators have measured the angular twist between the ends of the specimen and not over a specified gauge length. This introduces errors in the strain readings which cannot be accurately estimated; consequently the torsiometer shown in Fig. 5 was designed, which measures the creep strain over a fixed 5 in (127 mm) gauge length in the parallel walled portion of the specimen. The torsiometer consists of a concentric rod and tube which are fixed at a predetermined

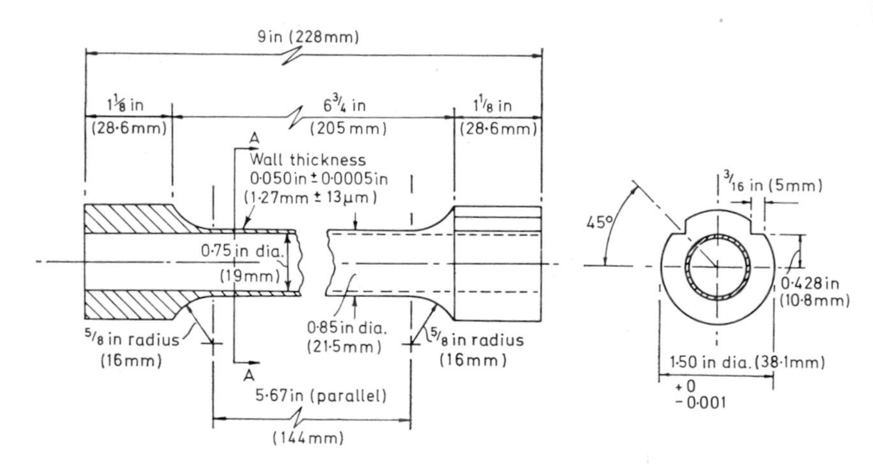

FIG. 4. Torsion creep specimen.

distance apart by three points which can be forced radially outwards to contact the bore of the tube, in a manner similar to an internal micrometer. A small bearing at the bottom of the torsiometer ensures that no side movement can occur. The rod and tube, which are of stainless steel, are coupled to light aluminium mirror rings, at the bottom of the torsiometer, to which mirrors can be attached. Using a telescope and horizontal centimetre scale at a distance of 108 in (2·72 m), a scale movement of 1 cm represents a shear strain of 0·0145 per cent. The mirrors have to be occasionally reset.

The furnace is 16¼ in (413 mm) long and it is wound in two zones to enable the temperature distribution to be adjusted. Temperature control is achieved by a platinum resistance thermometer in conjunction with a CNS proportional controller and saturable reactor. The temperature distribution is measured by four platinum/10 per cent platinum rhodium thermocouples tied at intervals along the gauge length. The temperature fluctuation throughout the tests was within ±0·5°C and the temperature variation along the length of the specimen was within ±1°C.

The tension creep machines used in this investigation are of the NPL design as modified by the Electrical Research Association, and have been described by Skelton and Crossland [8].

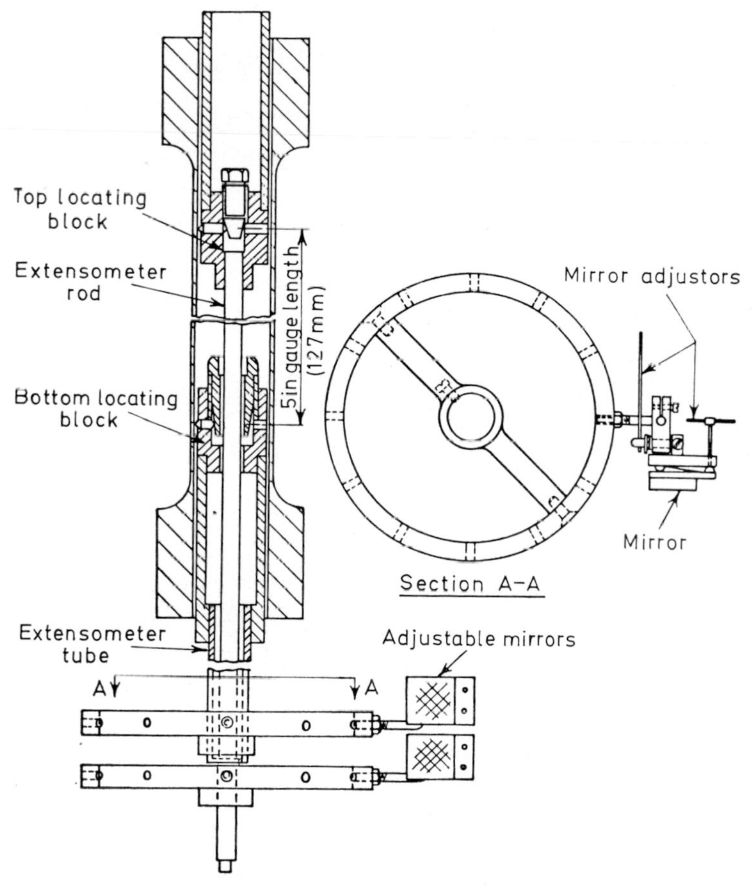

FIG. 5. Torsion creep extensometer.

MATERIAL

The material, a wrought carbon steel, was supplied by English Steel Corporation Ltd in the form of a bar 15 in (380 mm) diameter and 74 in

(1·88 m) long. This had been trepanned from the centre of a large forging for a chemical reaction vessel. Before delivery, ESC Ltd performed ultrasonic tests on the material, and also sulphur printed the end faces to check for segregation. The material was originally rather coarse grained and the suppliers gave the entire core a refining heat treatment. Before testing, all specimens were given a stress relieving heat treatment in a vacuum. This

14·4 in (366mm) | 14·4 in (366mm) | 14·4 in (366mm) | 14·4 in (366mm) | 14·4 in (366mm)

15 in dia. (380mm)

AB BC CD DE EF

A B B C C D D E E F

74 in (1·88)

Initial cutting diagram

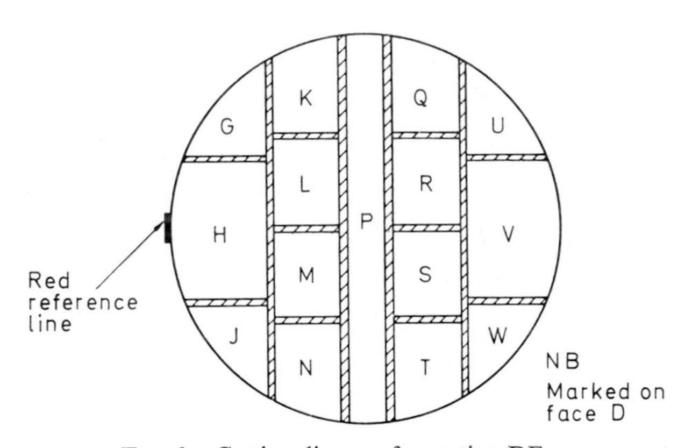

FIG. 6. Cutting diagram for section DE.

consisted of heating slowly to 610°C, soaking for 1 hr at this temperature and then cooling, still within the vacuum, over a period of about 18 hr. In this way any residual stresses induced by the machining operations were reduced to a uniform magnitude or removed altogether. It was anticipated that the thin walled tubular form of the torsion creep specimen might have distorted after stress relieving. A more complicated multistage heat treatment might therefore have been necessary to overcome this problem. This fear, however, proved unfounded as careful checks on the torsion specimens after stress relieving showed no dimensional change or distortion.

Figure 6 shows the manner in which the material was cut up. All the tests reported in this chapter were performed on material from section DE at a test temperature of 400°C. Also shown in Fig. 6 is the way this section was further cut up.

Chemical analysis was carried out on four samples taken from pieces J, M, R and U. Table 1 shows the results of these analyses. Microscopic examination of the grain structure in both the longitudinal and transverse directions showed that the grains were orientated randomly which suggested a minimum of anisotropy and difference in directional properties.

TABLE 1

Percentage composition by weight

Element	Specimen number			
	J1/3A	M1/1A	R1/3A	U1/3A
Carbon	0·17	0·17	0·17	0·18
Sulphur	0·029	0·025	0·019	0·020
Phosphorus	0·023	0·020	0·019	0·021
Silicon	0·27	0·26	0·25	0·26
Manganese	0·71	0·72	0·69	0·69

TEST PROGRAMME

The creep tests at 400°C on section DE of the steel have in general not exceeded 3000 hr. Eight torsion creep tests have been completed at shear stresses ranging from 6·5–10·5 tonf/in^2 (10–16 hbar) while another test at a shear stress of 11 tonf/in^2 (17 hbar) is still in progress. To date, six tension creep tests covering the nominal stress range 12–17 tonf/in^2 (19–26 hbar) have been completed. Tables 2 and 3 show the details of these tests. Also included is the initial plastic strain associated with each stress level. Because of the difficulty of measuring the instantaneous strain at the instant of loading, the initial plastic strain is taken as the non-elastic strain which occurs up to 1 min from the instant when the test load is reached. Thereafter all the time dependent strain is taken as creep strain. The results of the torsion creep tests are plotted in Fig. 7 and the tension creep tests in Fig. 8. The results are plotted as total strain (initial plastic + creep) versus time. Experimental points are not shown as in every case they lay on the smooth curves drawn.

TABLE 2

Torsion creep tests at a test temperature of 400°C

Shear stress (tonf/in²)	Shear stress (h bar)	Specimen number	Initial plastic shear strain (%)
6·52	10·1	G1/1/DE	0·594
7·51	11·6	W1/1/DE	1·074
8·00	12·4	M1/3/DE	1·382
8·51	13·2	R1/2/DE	1·910
8·96	13·8	J1/1/DE	2·275
9·51	14·7	U1/1/DE	2·395
10·00	15·4	M1/2/DE	3·403
10·49	16·2	R1/1/DE	3·722
11·04	17·1	P1/3/DE[a]	4·041

[a] Still running.

TABLE 3

Tension creep tests at a test temperature of 400°C

Shear stress (tonf/in²)	Shear stress (h bar)	Specimen number	Initial plastic strain (%)
12	18·5	M2/5/DE	0·588
13	20·0	M2/1/DE	0·838
14	21·6	M2/3/DE	1·051
15	23·2	M2/2/DE	1·320
16	24·7	M2/4/DE	1·656
17	26·2	M2/6/DE	1·935

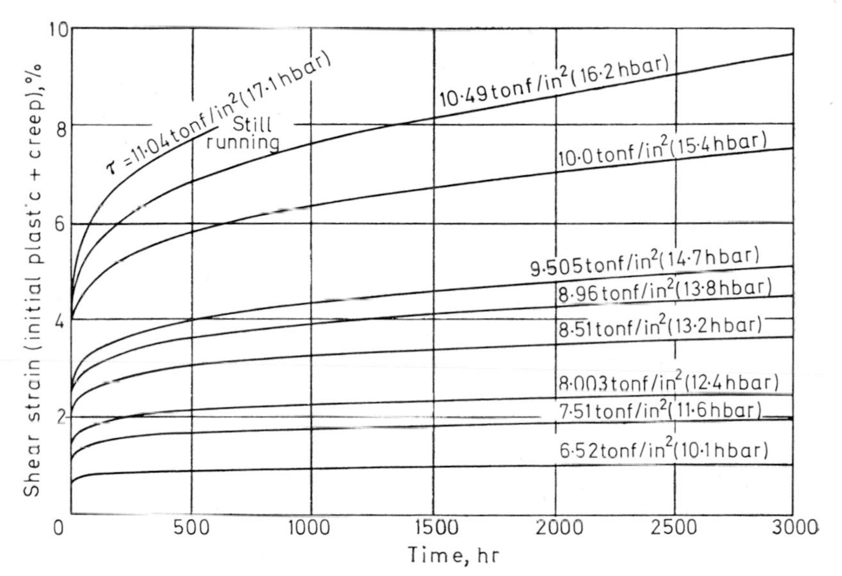

FIG. 7. Torsion creep curves for 0·18 per cent C steel at 400°C.

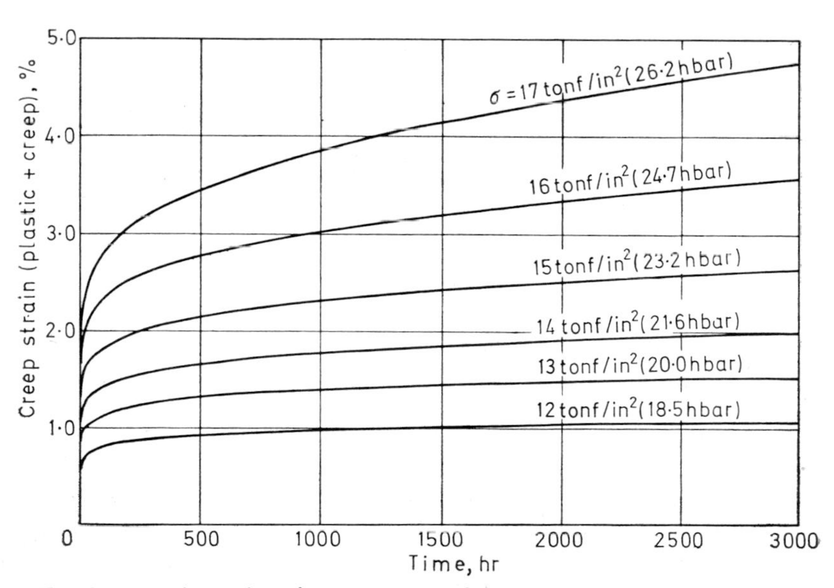

FIG. 8. Experimental tension creep curves for 0·18 per cent C steel at 400°C.

DISCUSSION AND CORRELATION OF RESULTS

There was some experimental scatter between the results for one stress level and another, especially in the torsion creep tests, and as the torsion data were to form the basis of correlation work, it was deemed necessary, before analysing them, to eliminate this scatter. This was done by cross-plotting from the torsion creep curves shear stress–strain curves for

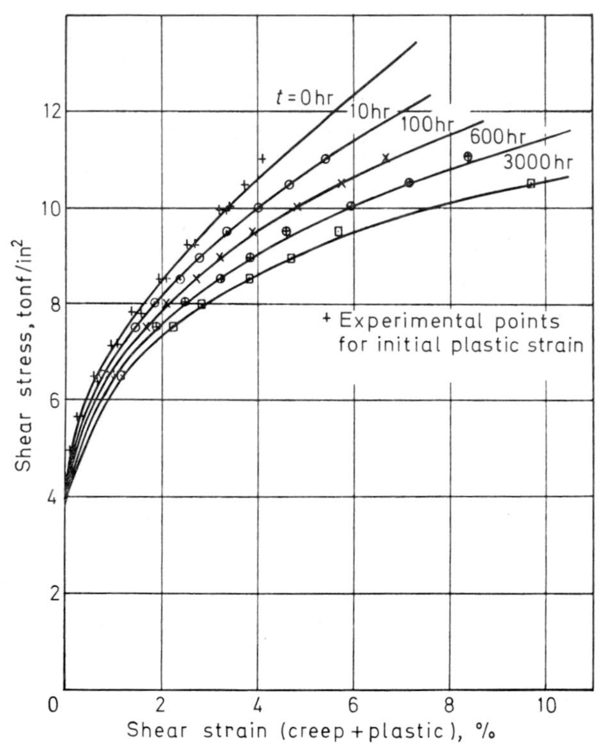

FIG. 9. Typical isochronous curves drawn from torsion creep results (0·18 per cent C steel at 400°C).

various values of time, or what are termed isochronous curves. Smooth curves were drawn through the points on the isochronous curves, and to assist in this it was found helpful to plot the initial curve ($t = 0$) using not only the initial plastic shear strain values but also values taken during the complete loading procedure for each test. Some of the isochronous stress–strain curves obtained in the above manner are shown in Fig. 9.

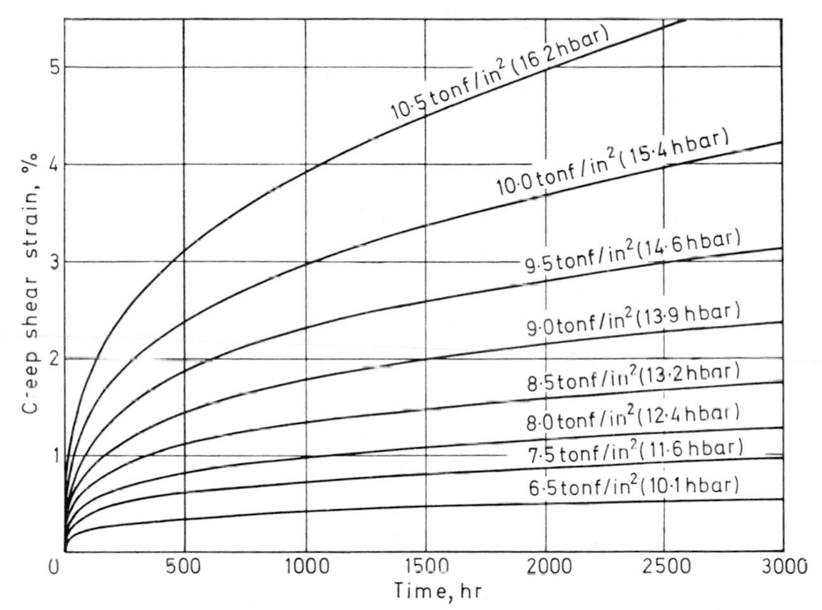

FIG. 10. Modified torsion creep curves (0·18 per cent C steel at 400°C).

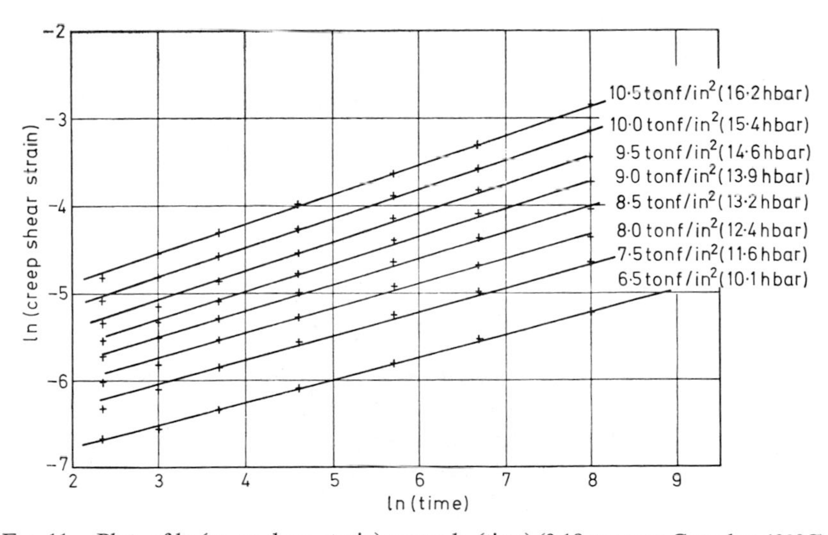

FIG. 11. Plots of ln (creep shear strain) versus ln (time) (0·18 per cent C steel at 400°C).

From these it was a simple matter to redraw the modified torsion creep curves of Fig. 10, which shows plots of creep shear strain versus time. The shear stresses in the modified curves were chosen with an equal increment of 0·5 tonf/in^2 (0·8 hbar) between each, which is an added advantage of drawing isochronous curves.

It can be seen from Figs. 7 and 8 that there was no evidence up to 3000 hr of tertiary creep, or more specifically an increase of strain rate, in any of the tests so far conducted.

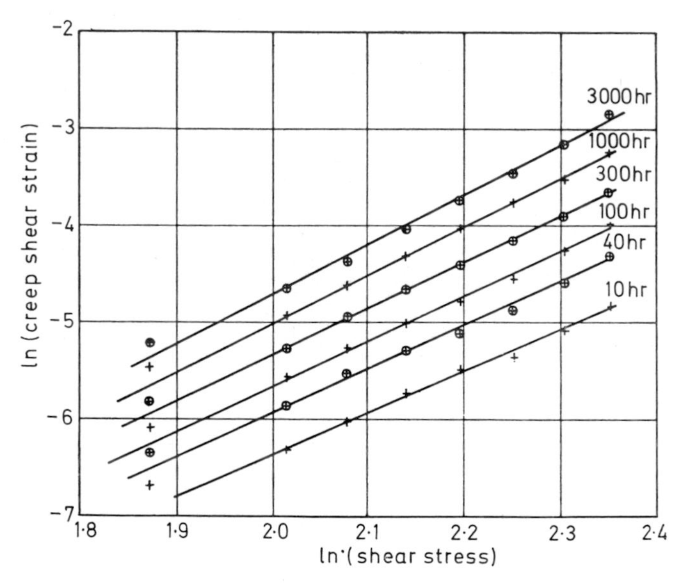

FIG. 12. Plots of ln (creep shear strain) versus ln (shear stress) (0·18 per cent C steel at 400°C).

On the assumption that the modified torsion creep curves could be represented by eqn (11) the following procedure was adopted to evaluate the constants B, n and m. Natural logs of both sides of eqn (11) were taken, giving:

$$\ln \gamma = \ln B + n \ln \tau + m \ln t \tag{15}$$

If a graph of $\ln \gamma$ is plotted against $\ln t$ with τ constant it should be a straight line of slope m. This has been done in Fig. 11 for the eight shear stress levels from 6·5–10·5 tonf/in^2 (10·0–16·0 hbar). As can be seen, it is possible over the full range of stress to draw a straight line through each set of points, and the lines are reasonably parallel. The average value for the slopes of all the lines drawn was:

$$m = 0{\cdot}313$$

The next step was to plot a series of graphs of $\ln \gamma$ against $\ln \tau$ for various constant values of t. The resultant curves should be straight lines of slope n. This has been done in Fig. 12 and, except for the strains corresponding to the lowest stress levels, straight lines can be drawn through

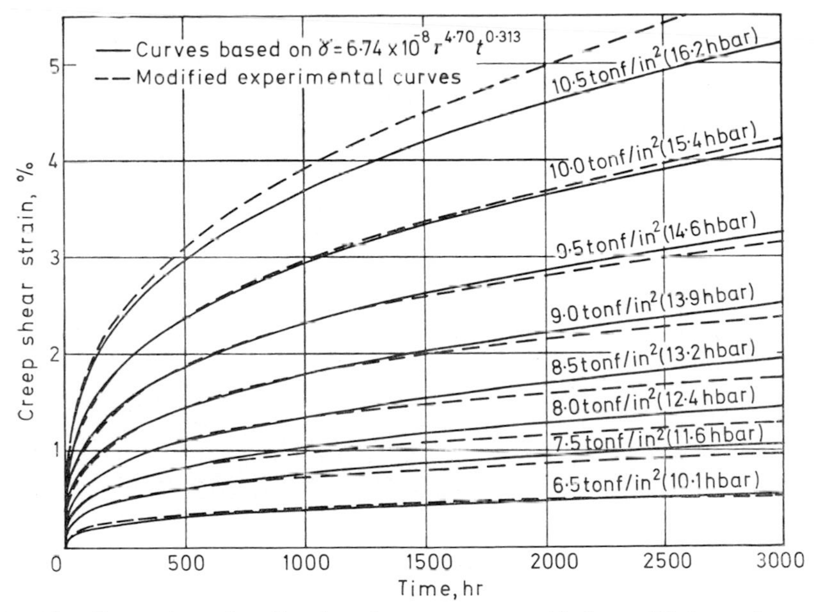

FIG. 13. Comparison of predicted torsion creep curves with the modified experimental curves (0·18 per cent C steel at 400°C).

the points for a given time. The average value for the slope of all the lines drawn was:

$$n = 4{\cdot}70$$

Substituting these values of m and n into eqn (11) and using numerous corresponding shear strain, stress and time values taken from Fig. 10, the average value of B finally found was:

$$B = 6{\cdot}74 \times 10^{-8}$$

Consequently, eqn (11) becomes:

$$\gamma = 6{\cdot}74 \times 10^{-8}\tau^{4{\cdot}70}t^{0{\cdot}313} \tag{16}$$

As a check on how well this equation represents the original data the curves predicted by it are compared with the modified experimental curves in Fig. 13. A reasonable fit has been achieved for the majority of stress levels and at worst the deviation for any curve is less than 10 per cent at 3000 hr.

A comparison of the initial true effective stress and true effective strain for both torsion and tension data, on the basis of the von Mises flow rule, is shown in Fig. 14. The correlation is remarkably good and the slight

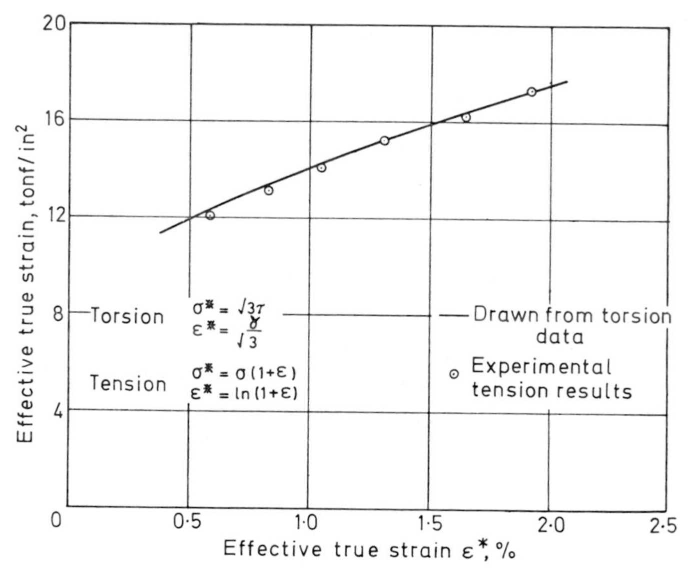

FIG. 14. Comparison of effective true stress and initial plastic true strain for torsion and tension data based on the von Mises flow rule (0·18 per cent C steel at 400°C).

deviation at the lower end of the curve could be attributed to the elastic strains becoming less negligible in comparison with plastic strains. In this region the assumption of constancy of volume becomes less valid.

Using the theory developed in this chapter the tension creep curves for nominal tensile stress from 12–17 tonf/in^2 (18·5–26·2 hbar) have been computed (Fig. 15). Tension curves have also been predicted ignoring the reduction in cross-sectional area of the specimen. It can be seen that when

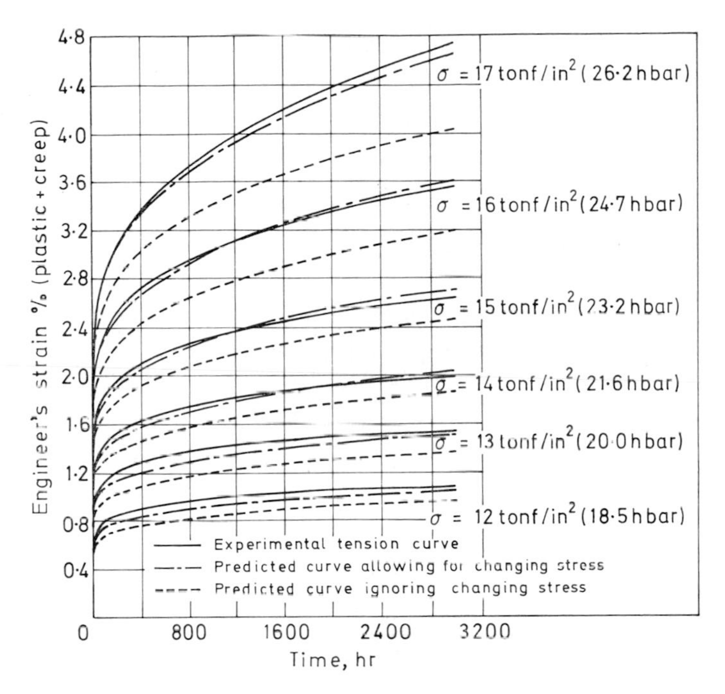

FIG. 15. Comparison of experimental tension creep curves with those predicted from torsion creep data on the basis of the von Mises flow rule. (0.18% carbon steel at 400°C.)

the increasing stress is taken into account a very much better correlation is achieved, especially for the higher values of nominal stress, though the correlation is still good even for the 12 and 13 tonf/in^2 (18·5 and 20 hbar) tests. The increased discrepancy for the low stress data can be attributed to the increasing significance of elastic strains in these tests. In all cases the difference between theories allowing for and ignoring the increase in stress is quite marked, and at higher stresses the difference is considerable.

CONCLUSIONS

It is concluded that the change of stress in a constant load tensile creep test is of significance, and it is desirable to allow for this in the analysis of constant load tension data. Alternatively, a constant stress tensile creep machine should be employed, or perhaps preferably a torsion creep machine using a thin tubular specimen as described in this chapter. It would appear that the torsion creep data obtained are of a more fundamental nature than constant load tension data.

ACKNOWLEDGEMENTS

The authors wish to acknowledge the Science Research Council who financed this work and also the generosity of English Steel Corporation Ltd which donated the material.

In the early stages of this work the late Dr Johnson was particularly helpful and encouraging, and we would take this opportunity of acknowledging his assistance.

This project was carried out in the Mechanical Engineering Department of the Queen's University of Belfast, and the assistance of the Departmental Workshop Staff and Laboratory Staff, in particular that of Mr R. H. Agnew and Mr N. Alexander, is gratefully acknowledged.

REFERENCES

1. Crossland, B. (1965). 'Torsion testing machines.' *Machines for Materials and Environmental Testing. Proc. Instn Mech. Engrs*, **180** (Pt 3A), 243–254.
2. Penny, R. K. and Leckie, F. A. (1968). 'The mechanics of tensile testing.' *Int. J. Mech. Sci.*, **10**, 265–273.
3. Johnson, A. E. (1950). 'A high sensitivity torsion creep unit.' *J. Sci. Instru.*, **27,** 74–75.
4. Finnie, I. (1958). 'Creep buckling of tubes in torsion.' *J. Aero. Sci.*, **25,** 66–67.
5. Ohji, K. and Marin, J. (1963). 'Creep of metals under non-steady conditions of stress.' *Thermal Loading and Creep in Structures and Components. Proc. Instn Mech. Engrs*, **178** (Pt 3L), 126–134.
6. Warren. J. W. L. (1966). 'A survey of the mechanics of uniaxial creep deformation of metals.' Aero. Res. Counc. CP No. 919. London: HMSO.
7. Everett, F. L. (1931). 'Strength of materials subjected to shear at high temperatures.' *Trans. Amer. Soc. Mech. Engrs*, **53,** APM-53-10.
8. Skelton, W. J. and Crossland, B. (1967). 'Results of high sensitivity tensile creep tests on a 0·19 per cent carbon steel used for thick-walled cylinder creep tests.' *Conf. on High Pressure Engng. Proc. Instn Mech. Engrs*, **182** (Pt 3C), 151–155.

Chapter 9

MATERIALS PROPERTIES IN COMPLEX STRESS THEORY

A. GRAHAM

SUMMARY

Complex stress theories should preferably take account of rapid changes of practical concern in the uniaxial performance of materials. Constitutive equations adopted by Dr Johnson are of wide application and extend, with interpretation, to two major types of change which determine life to failure under uniaxial loading. One of the types appears to be directly associated with the development of anisotropy.

NOTATION

T Temperature.
t Time.
ε Strain.
ε_0 Time independent plastic strain on loading.
σ Stress.

Subscripts
m, n Exponents in eqn (1).

RAPID CHANGES IN UNIAXIAL PROPERTIES

To a designer concerned with the safety of critical structures, like hot airframes or turbine rotors, the predictions of theories of complex stress can be accepted only with reserve because of their oversimplified treatment of the properties of materials. The detailed studies of metallurgical engineering and materials science present metals as so complex and individual that a formal mathematical expression for the behaviour of even a simple metal under the simplest loading can be regarded only as an approximation tenable in a limited range over which a single structural mechanism predominates. Metals used for the construction of heavily loaded structures are far from simple and, over the wide ranges of the mechanical variables which may prevail in service, a variety of metallurgical mechanisms are certainly active. The theories embody assumptions which allow the relationship between complex stress, complex strain, time and temperature at any point in a material to be expressed in terms of the theoretically simpler and experimentally more accessible relationship between the corresponding quantities for a uniaxial specimen, respectively σ, ε, t and T. There is little agreement however on the correct mathematical expression to adopt for the uniaxial relationship. Nevertheless, the practical successes of complex stress theory are considerable. The apparent inconsistency is resolved if the many differences of metallurgical detail average out on an engineering scale of magnitude. On these grounds, together with the grounds that predictions need only be accurate to within certain factors of safety, stress analysis would claim a freedom to disregard the multitudinous factors to which attention is directed in other disciplines. It is not, of course, directly concerned with metals in all conceivable circumstances, but with metals that have been successfully developed as marketable products to give reliable and consistent performance. The freedom claimed tends to be supported by the families of curves which represent σ, ε, t, T relationships as measured for each metal in various standard types of test. Any particular type, *e.g.* 'creep' at different constant σ and T, or 'tensile' at different constant $\mathrm{d}\varepsilon/\mathrm{d}t$ and T, will provide a family of curves. Between families of curves measured for different well developed metals in any one type of test, a distinct family likeness may usually be recognised. The feature is witnessed by the various phenomenological formulae which have been advocated for general use. However, that freedom cannot be taken for granted. A noteworthy reason is the well recognised occurrence of rapid and not easily predictable changes of trend of experimental curves for developed alloys, changes which have been attributed to a variety of metallurgical causes. Of considerable importance are the onset of accelerating creep and breaks, as studied especially by N. J. Grant, in the trend of increase of life with reduced level of loading. They may occur within the working range of a component or may themselves be responsible for determining the working range.

Reliable stress analysis of a component of a long life machine or structure, involving cold starts, hot running and wide ranges of loading and temperature, demands an extensive range of data on the properties of materials, a range too great to be covered by direct experiment without aid from prediction. The problems of predicting uniaxial data have been the subject of much discussion, and there is general agreement that predictions attempted without close attention to these changes of trend can be liable to serious error [1, 2]. The device adopted in stress analysis, in order to avoid the use of an untrustworthy σ, ε, t, T relationship, of basing numerical calculations directly upon measured curves from uniaxial experiments, does not avoid these difficulties. A mathematical formalism which takes essential microscale factors into account appears therefore to be indispensable. Which of the multitudinous factors are essential when assessed from the standpoint of practical engineering is a question for stress analysis which microscale arguments appear unable to answer.

MATHEMATICAL REPRESENTATION OF UNIAXIAL PROPERTIES

In the literature few exact discussions are to be found of the mutual interactions between different micromechanisms in their statistical relationship to behaviour on the engineering scale. A noteworthy exception mentioned by Dr Johnson is the reasoning of G. Sachs, who showed that the von Mises criterion of flow in a crystal aggregate is equivalent to the maximum shear stress criterion in a single crystal. On the other hand, phenomenological studies, like those of Dr Johnson, actually performed on engineering materials on an engineering scale, avoid most of the detailed problems of interactions.

Conclusions from a number of studies of this kind are in substantial agreement. Collectively, they point to the significance of a basic type of structural process.

Dr Johnson confirmed an observation reported in 1936 by Soderberg [3] that creep curves obtained at different constant stresses are often related by simple scaling, so that stress and time enter an expression for strain as a product of independent functions of each variable separately. He was one of the many to have supported an early view that the independent functions of stress, strain and time are power functions. He also adopted the view that creep strain comprises a sum of components, each represented by a power term, and was one of the few to have used a power term to represent the accelerating stage of creep. He found further that the dependence of creep strain on stress in the decelerating and, in a number of instances, the accelerating stages of creep might comprise a sum in each stage of at least two components. These findings led him to adopt expressions for creep of the form

$$\varepsilon = \Sigma_{m,n} C_{mn} \sigma^m t^n \tag{1}$$

in which the constants m and n were assigned decimal values dependent on the temperature. Such dependence becomes doubtful, however, if his reasoning is extended to recognise the well known, even though qualified, success of time–temperature parameters. Their prediction that a change of temperature merely alters the scale of time requires that m and n shall be independent of temperature. In these circumstances, temperature will be involved only in the C_{mn}.

It is not without significance that Dorn [4, 5] and his associates, on the basis of extensive structural and phenomenological studies of metals, were led to favour formulae of similar general type, namely

$$\varepsilon = \Sigma f(\sigma) g(t) h(T) \tag{2}$$

in which f, g and h are different functions. Physical arguments resting upon structural considerations of the kind advocated by Dorn usually lead to exponential forms of function. Exponential forms of f and g have not always been found adequate: the alternative of power functions has frequently been suggested and supported. It is not easy to decide conclusively between the two. An overriding factor involved is the occurrence within a metal—which no one questions—of several mechanisms of deformation each liable to have an individual influence on the shape of an experimental curve. Evidently, it borders on the futile to dispute the fine details of formulae without reference to clearly defined proposals for distinguishing between the contributions made by different mechanisms. The mathematical expressions here discussed formally propose the simplest possibility for this purpose, namely, that the effects that are of engineering interest are simply additive. In view of the number of variables, mechanisms and associated features demanded of any representative formula, sound conclusions can only rest upon experiments wide enough in range to provide the requisite number of definitive conditions. Detailed assessments of the scatter in a material's performance are also essential.

Profiting from the work of Dr Johnson and Dorn, the writer and his colleagues were led to investigate the suitability of formulae of the above type for describing the performance of heat resistant steels. Of particular concern were the above-mentioned changes of trend, particularly those which occur under the simple conditions in which Dr Johnson was interested, namely constant stress constant temperature creep. The expressions showed promise but, in the forms advocated by Dr Johnson and Dorn, proved inadequate to relate uniaxial data, obtained for convenient durations of testing, to the durations of long time service. The failure drew attention to the deficiencies of familiar time–temperature parameters and to the significance of the researches of Andrade [6], Grant [2] and their associates. An assembly of wide ranging data for heat resistant alloy steels together with some careful experiments confirmed that exponents equivalent to Dr Johnson's m and n were indeed independent of temperature. These quantities were presented as being regular fractions, the n values being integer powers of $\frac{1}{3}$ or 3, and the ratios n:m being integer powers of $\frac{1}{2}$. The metallurgical 'mechanisms', to which, as discussed below, Grant's conclusions may be taken to refer, corresponded to different values of the ratio n:m. The emergence of these regularities offered firm support for the validity of the power form of function.

Grant's breaks in the trend of increase of life with decrease of stress were presented by these results not as actual breaks, but as continuous transitions with curvature dependent upon the m and n values concerned. The appearance of breaks could be attributed to a combination of the circumstances of sharp curvature with scatter in the points. The other change of trend—the progressive development of accelerating creep—also appeared to be adequately represented by the formula and to be predictable by extrapolation with use of the formula. The progressive nature of the transitions, which cause points to be off lines of standard slopes, provided an adequate explanation of why the regular features just described are seldom apparent to casual inspection. Brief summaries of the research to a recent date are given in refs. 7 and 8.

WORKED AND VIRGIN MATERIAL

The evidence there summarised for the systematic nature of various aspects of the mechanical performance of complex metals sets the question of the nature of the metallurgical processes responsible. An answer in general terms is suggested by features of the work of Andrade [6], Cottrell and Aytekin [9], Grant [2] and Dr Johnson.

Andrade's equation for the decelerating stages of creep of many pure metals, effectively

$$\varepsilon = \varepsilon_0 + at^{\frac{1}{3}} + bt$$

supplied the first evidence that the exponents n in eqn (1) had constant standard magnitudes. Andrade's equation has received rather qualified acceptance on account of the ill defined term ε_0, representing the 'time independent' plastic strain on loading. Without this term, the equation, like its single term competitor

$$\varepsilon = ct^n$$

represents the fact of observation that the initial stages of creep in any experiment may be so rapid that they escape detailed observation owing to mechanical and instrumental limitations peculiar to the experiment. It will also be recalled that, when a metal is plastically deformed, deliberately or incidentally during manufacture or prior to testing, some arbitrary part of the deformation curve is traversed. Owing to the occurrence of strain hardening, that part is not traversed again during the measured progress of the subsequent testing. To secure the simplest results, it is clearly essential to ensure as far as practicable that test specimens are initially in a virgin condition, and to recognise also, through the term ε_0, an experimental inability to commence observations of performance under constant stress immediately from that condition. Andrade was particularly careful to anneal his specimens thoroughly, the judgement of thoroughness being no preconceived idea of the structure of the metal, but the pragmatic test that further annealing, either for a longer time or at a higher temperature, made no difference to results.

Andrade's viewpoint received firm confirmation from the study by Cottrell and Aytekin [9] of the creep of single crystals of zinc. Their method of growing the crystals by passing a band of temperature along the metal, of magnitude just above the melting temperature, would appear to represent the ultimate limit of annealing. They were surprised to find that the Andrade equation provided the best, and indeed a very close fit to their results. The surprise was due to the absence of any adequate microstructural explanation of the Andrade formula.

Related evidence of significance in the present context is that which stemmed from an initial observation made by K. F. A. Walles in connection with the creep of a complex nickel–chromium alloy with hardening additions of aluminium and titanium. He found that the Andrade equation, with its predetermined exponents, gave a rather better fit to the decelerating creep than the simple power law alternative with freely adjustable exponent. The observation led to the discovery of the above-mentioned regularities in m and n. The subsequent work firmly supported the conclusion which followed, namely that complex dispersion hardened alloys with optimum treatment for long life and reliability in creep are governed by the same law of decelerating creep as that advocated by Andrade and his supporters for pure metals. On this evidence it appears that optimum hardening may bring a metal to a similar virgin condition to that of a pure metal single crystal. The conclusion does not appear to be subject to questions of the fine degree of exactitude of the fitting, because dispersion hardening is a major factor, indeed is one of the few major factors, which influence the strength of a metal. Other major factors are annealing and strain hardening.

Grant and Bucklin [2], concerned with the second type of change of trend, based their conclusions on detailed studies of Monel and the high strength cobalt base alloy S590. They found that the apparent breaks in the relation of stress to time-to-rupture in creep corresponded to changes of structural mechanism resulting in changes from transgranular to intergranular failure, the onset of grain growth or, more generally, from low temperature to high temperature mechanisms. The correspondence was most clearly displayed on a log–log plot of the experimental points. The curves through the points for each constant temperature then consisted of linear sections of differing slope with rather sudden transitions between them. Each linear section corresponded to the power law relationship represented by a single term of eqn (1). The authors' conclusion that such transition was associated with a progressive change of mechanism is supported by a result from the research summarised in refs. 7 and 8, namely that each linear section (associated with a different standard value of n/m) differed in time–temperature relationship. In contrast, different values of n, corresponding to progression along a creep curve, appeared not to be associated with change of time–temperature relationship. The second observation is consistent with the findings of Dorn [5] and his co-authors, and also with the comments of Cottrell and Aytekin on the absence of a structural explanation of the Andrade creep equation.

In a later paper, Servi and Grant [10] extended the conclusions drawn for Monel and S590 to commercial aluminium and even to high purity aluminium. In these metals, the transitions were accompanied by metallurgical evidence for the beginning of grain boundary flow, for changes in the relative strengths of coarse and fine grained structures, and change from transgranular to intergranular failure. The creep curves were found to be consistent with the Andrade equation. On the basis of the two papers,

Grant claimed to have shown that similar creep and creep rupture behaviour prevailed, at least as far as the transition from low temperature to high temperature deformation is concerned, for a high purity metal, two less pure metals and two complex alloys. The results of the writer and his associates are consistent with this conclusion. Unfortunately, accurate data as comprehensive and extensive in range as those that have been provided, as a result of commercial interest for a rather small number of highly developed complex alloy steels, do not appear to be available for the creep of high purity metals; thus, similar stringent checking does not at present appear to be practicable. The fairly extensive data for high purity aluminium obtained by Servi and Grant, although too limited to be conclusive, are in support, as also are Carreker's data [11] (with the support of other data given in a private communication from K. F. A. Walles to the author) for pure platinum. When data are reviewed, a question difficult to answer is whether the prior treatment of the material and manner of conducting the tests are such that test results truly refer to material in a basic condition. On the evidence available, it appears not unreasonable to conclude, as a tenable economical hypothesis, that high purity metals follow, on an engineering scale of magnitude, laws no different in form from those of dipersion hardened alloys.

ORDERING OF THE STRUCTURE AND ANISOTROPY

The evidence here summarised that components of creep corresponding to different ratios $n:m$ derive from different 'mechanisms' of deformation is supported by the evidence summarised in refs. 7 and 8 that these ratios form a regular sequence (namely 2^{-p}). The conclusion follows that the mechanisms, whatever they may be in detail, if not directly related, must at least have some statistical feature in common. Well known aspects of deformation with the support of some results of Dr Johnson suggest the common feature.

An annealed uniaxial specimen, when progressively extended, tends to develop a preferred orientation. Two opposite poles of this behaviour are the structurally ordered single crystal, to be regarded as an ultimate end product of annealing, and the highly oriented overdrawn wire, presumably indicating an end product of deformation. The two ordered states are undoubtedly different. These gross features suggest that the traditional view of deformation as a balance between strain hardening and recovery should be amended into a balance between two competing tendencies towards opposite poles of order. Deformation encourages progression from

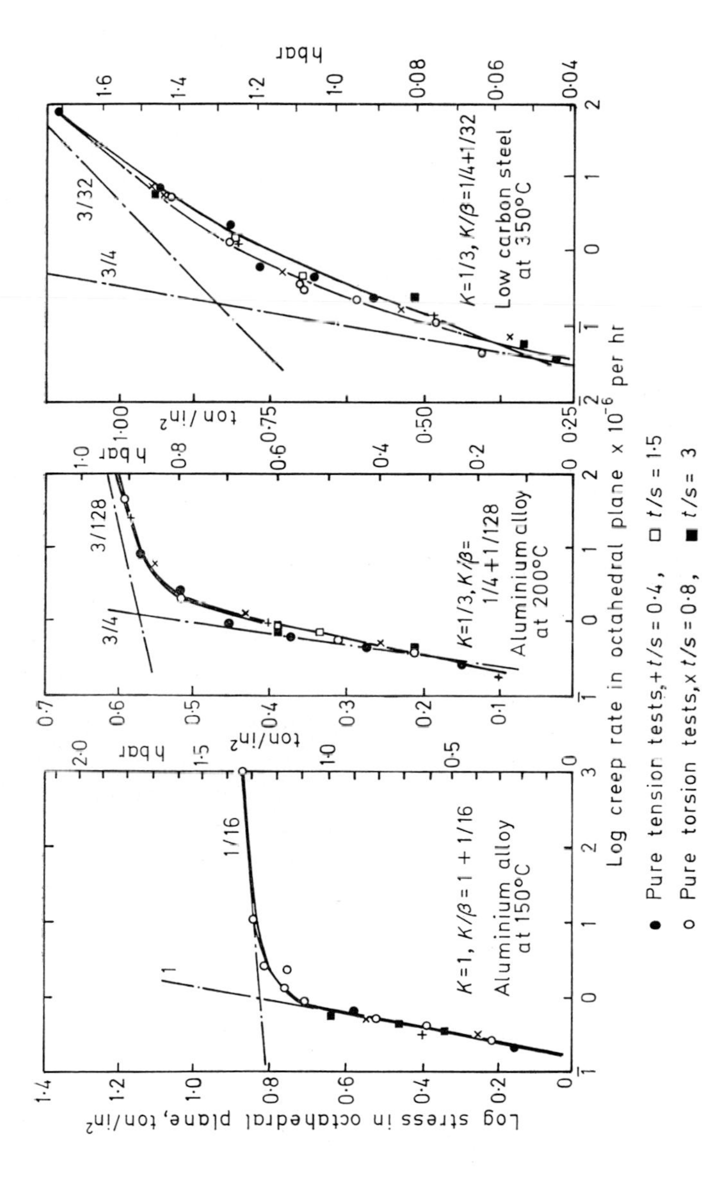

FIG. 1. Log stress–log creep rate curves for the octahedral plane. Data of Dr Johnson [12].

the first towards the second against the thermally encouraged reverse progress from the second towards the first. Many states of order are conceivable in an assembly of atoms.

Dr Johnson's results [12] relate to the development of anisotropy during the complex creep deformation of both an aluminium alloy and a low carbon steel. Figure 1 reproduces his experimental relation between the octahedral creep rate at 150 hr and the octahedral stress. It illustrates changes of trend of the type studied by Grant. The figure suggests the presence, in each instance, of two terms of eqn (1) so that, at the constant times considered, the equation to each curve should be

$$\dot{\varepsilon} = \frac{\mathrm{d}}{\mathrm{d}t}(C_1\sigma^{m_1}t^{n_1} + C_2\sigma^{m_2}t^{n_2})$$

or

$$\dot{\varepsilon} = C_1'\sigma^{m_1} + C_2'\sigma^{m_2}$$

in which the $n{:}m$ have values from the sequence 2^{-p} and C_1' and C_2' are constants. The appropriate values of n may be found from the uniaxial creep curves which he gives in his paper. The values which are relevant are $\frac{1}{3}$ and 1. The curves shown in broken line in Fig. 1 demonstrate the close fitting of this equation after selection of the standard values of $n{:}m$ there indicated. The resulting free choice of m_1 and m_2 is not so free that successful fitting is a foregone conclusion. The fitting of the next most suitable values, namely values for which the asymptotes have slopes of twice or half those indicated, is significantly worse: also it is seen that the fitting with the values chosen is as good as that achieved by Dr Johnson with his complete freedom to choose the exponents. From his observed relationships between the creep rates in the three principal directions, Dr Johnson concluded that the trend towards a smaller slope at the larger strains was associated with an increase of anisotropy of the deformation. This took the form of a redistribution of the creep rates in the secondary and minor directions, the minor rate tending gradually to zero while the major rate remains unaffected. The observation suggests that no error of principle has been introduced by an argument based, as in the present note, on uniaxial experiments without reference to creep in the secondary and minor directions.

CONCLUSIONS

From the viewpoint of stress analysis, the overall evidence of the results reviewed may be considered to support the disregard by stress analysis of the details of structural mechanisms within a metal. It is not the mechanisms as such which appear significant, but the statistical states, and changes of state, of a metal that the various mechanisms produce. The evidence suggests, however, that if reliable predictions are to be made of effects arising from changes of trend of the type which are of major concern in the metallurgical engineering of critical structures, it will need to take close account of departures from the condition of isotropy. Dr Johnson's use of eqn (1) was a step in that direction.

The common laws which appear to govern the mechanical performance of different compositions, structures and states of anisotropy include the onset of tertiary creep and the rapid changes in the trend of dependence on stress of life to rupture. The laws include, as a simple special case, the Andrade creep equation. Cottrell and Aytekin's support for that equation by the creep of, presumably anisotropic, single crystal zinc suggests that the basic condition of the metal in which it may be expected to perform, in the simplest manner, is not necessarily a condition of isotropy. Dr Johnson's results on initially isotropic metals suggest that changes in the trend of stress with life are associated with the development of anisotropy. The considerable care usually taken in complex stress experiments to adjust prior treatment to ensure that the initial material is isotropic may therefore not lead in the long run either to the simplest or the most practically significant results.

REFERENCES

1. GLEN, J. (1958). 'A new approach to the problem of creep.' *J. Iron & Steel Inst.*, **189,** 333–343.

2a. GRANT, N. J. and BUCKLIN, A. G. (1950). 'On the extrapolation of short-time stress-rupture data.' *Trans. Amer. Soc. Metals*, **42,** 720–761.

2b. GRANT, N. J. and BUCKLIN, A. G. (1952). 'Creep-rupture and recrystallisation of Monel from 700 to 1700F. *Trans. Amer. Soc. Metals*, **45,** 151–176.

3. SODERBERG, C. R. (1936). 'Interpretation of creep tests for machine design.' *Trans. Amer. Soc. Mech. Engrs*, **58**(8), 733–743.

4. DORN, J. E. (1956). 'Some fundamental experiments on high-temperature creep.' *Creep and Fracture of Metals at High Temperature,* pp. 89–138. London: HMSO.

5. SHERBY, O. D. and DORN, J. E. (1952). 'Creep correlations in alpha solid solutions of aluminium.' *J. Metals*, **4,** 959–964.

6. ANDRADE, E. N. DA C. (1952). Sixth Hatfield Memorial Lecture. 'The flow of metals.' *J. Iron & Steel Inst.*, **171,** 217–228.

7. GRAHAM, A. (1967). No. 7. Descriptive approach to mechanical properties of materials. (Under *Materials for Engineers*. Ed. N. L. Parr.) *Engineer Lond.*, **224,** 192–194.

8. WALLES, K. F. A. and TILLY, G. P. (1967). 'Creep and fatigue behaviour of materials.' *Engineer Lond.*, **224,** 551–554.
9. COTTRELL, A. H. and AYTEKIN, V. (1950). 'The flow of zinc under constant stress.' *J. Inst. Metals*, **77,** 389–422.
10. SERVI, I. S. and GRANT, N. J. (1951). 'Creep and stress rupture behaviour of aluminium as function of parity.' *J. Metals*, **3** (10), 909–916.
11. CARREKER, R. P. (1950). 'Plastic flow of platinum wires.' *J. Appl. Phys.*, **21,** 1289–1296.
12. JOHNSON, A. E. (1949). 'The plastic creep and relaxation properties of metals.' *Aircr. Engng.*, **21** (239), 2–8, 13.

Chapter 10

COMPLEX STRESS CREEP RELAXATION OF COMMERCIALLY PURE COPPER AT 250°C

J. HENDERSON AND J. D. SNEDDEN

SUMMARY

Combined tension and torsion relaxation tests on thin walled tubular specimens, and tensile relaxation tests on solid cylindrical specimens have established that, for the copper tested at 250°C, the results of pure tensile relaxation tests can be used to predict complex stress relaxation behaviour on the basis of octahedral shear stress equivalence. Thus, complex stress relaxation data can be obtained from conventional double lever tensile creep machines manually controlled for relaxation tests, and this method has the advantage of requiring virtually no mathematical computation.

Where only tensile creep data are available, it has been shown that, using the strain hardening form of the mechanical equation of state, they may provide a fairly accurate prediction of complex stress relaxation.

NOTATION

A, A', A_1, A_2 Constants.
C_a Axial creep rate.
C_{ij} Creep rate tensor.
C_o Octahedral shear creep rate.
C_1 Maximum principal creep rate.
E Young's modulus of elasticity.
J_2 Second order invariant of stress deviation tensor $= \frac{1}{6}\Sigma(\sigma_1 - \sigma_2)^2$.
$K, m, n, n_1, n_2, p_1, p_2$ Constants.

S Shear stress in tension torsion tests.
S_1 Major principal stress deviator.
S_{ij} Stress deviation tensor.
T Axial stress in tension torsion tests.
t Time.
δ_{ij} The Kronecker delta which equals unity when $i = j$ and zero when $i \neq j$.
ε_a Axial creep strain.
ε_{ij} Creep strain tensor.
ν Poisson's ratio.
σ Stress or tensile stress in tests on solid cylindrical specimens.
σ_{ij} Stress tensor.
σ_v Hydrostatic stress or spherical stress tensor.
$\sigma_1, \sigma_2, \sigma_3$ Principal stresses.
τ_o Octahedral shear stress.

INTRODUCTION

Many engineering components, such as bolted flanges, press fit elements and springs are subject to creep relaxation, the initial loading strain remaining constant with time, and subsequent creep strain manifesting itself by an interchange of elastic strain for creep strain, resulting in diminishing stress with time. Further, it is recognised that understanding of the behaviour of components under conditions of thermal stress, stress relieving treatment and peak stress redistribution in complex elements, must involve a knowledge of creep relaxation under complex stress conditions.

Some years ago, a study was initiated by Dr Johnson [1] at the National Engineering Laboratory of complex stress creep relaxation behaviour of engineering materials at elevated temperatures. Apart from the work of Namestnikov [2], and a few tests by Griffiths and Marin [3], the programme at NEL represents virtually the only experimental investigations which have been undertaken in this field.

In the present study on copper two questions were primarily considered:

1 Could simple tensile relaxation tests provide an acceptable prediction of relaxation under complex stress conditions?
2 Could simple tensile creep data form the basis of a similar prediction of relaxation under complex stress conditions?

Combined tension–torsion relaxation tests were made, representative of complex stress conditions in general, with the conventional assumption of volumetric constancy or ineffectiveness of hydrostatic stress under creep conditions. Material isotropy, under creep conditions, for this type of copper, has been demonstrated in a previous paper [4].

Where numerical values of constants are quoted these relate to stress in tons per square inch and time in hours unless otherwise stated.

DETAILS OF MATERIAL AND TESTING TECHNIQUES

A commercially pure copper, fully annealed and similar to that used in previous investigations, was used in the tests. Blanks of 1·5 in (38·1 mm) diameter rolled bar were heat treated at 400°C (752°F) for 1 hr and furnace cooled. Thin walled tubular specimens, 0·56 in (14·23 mm) and 0·50 in (12·7 mm) external and internal diameter respectively, were tested in the tension–torsion relaxation machine described by Dr Johnson and Frost [5] (Fig. 1). The solid tensile specimens, 0·357 in (9·07 mm) diameter and 2·5 in (63·5 mm) gauge length, were tested in a conventional double lever tensile creep unit. The extensometers in the tensile units, and the torsionmeter in the complex stress unit had strain sensitivities of the order of 10^{-6}.

Five complex stress relaxation tests, six tensile creep tests and five tensile relaxation tests were performed. The tests were conducted at 250°C $\pm$ 0·5°C with a temperature gradient of <2°C. Complex stress relaxation tests of approximately 500 min duration, and tensile relaxation tests of 400 min to 50 hr duration were manually controlled, except for one repeat tensile relaxation test in which load conditions were automatically controlled. The tensile creep tests, used in the derivation of the primary creep equation, were of approximately 30 hr duration. Details of the tests are given in Table 1.

Relaxation tests of relatively short duration were used since observations show that the higher relaxation rates occur at the beginning of the tests. Furthermore, in the rapid initial stages, greater sensitivity of unloading control can be achieved more readily by manual unloading than by conventional electronically controlled unloading mechanisms which often, in the early stages, require manual control.

The stress range, in the experiments, was confined to that producing primary creep; at the higher stresses slight inelastic loading strain was

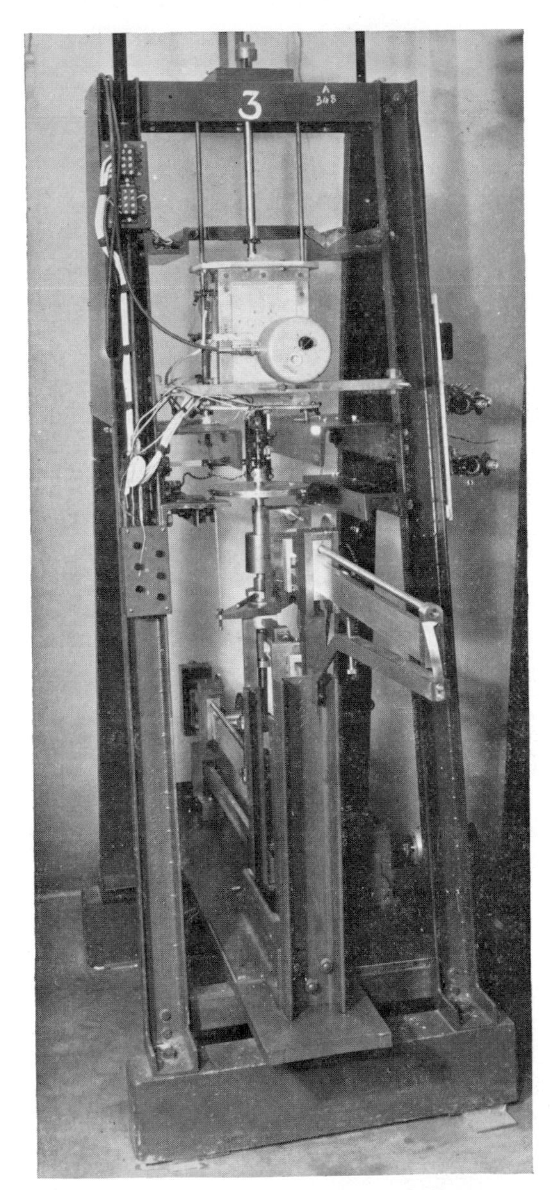

FIG. 1. Tension–torsion creep relaxation unit.

Table 1

Results of complex stress relaxation tests

Type of test	Tensile stress, T (ton/in^2)	(hbar)	Shear stress, S (ton/in^2)	(hbar)	Octahedral shear stress, τ_o (ton/in^2)	(hbar)	Test duration (min)	Final octahedral shear stress, τ_o (ton/in^2)	(hbar)	Final creep strain ($\times 10^{-4}$)	Ratio T/S
Complex	0·6	(0·93)	1·5	(2·32)	1·26	(1·95)	500	0·845	(1·31)		0·4
stress	0·6	(0·93)	1·5	(2·32)	1·26	(1·95)	500	0·854	(1·32)		0·4
relaxation	1·225	(1·89)	1·53	(2·36)	1·375	(2·12)	500	0·885	(1·37)		0·8
tests	2·0	(3·09)	1·33	(2·05)	1·44	(2·22)	500	0·907	(1·40)		1·5
tension–	3·0	(4·63)	1·00	(1·54)	1·65	(2·55)	500	1·125	(1·74)		3·0
torsion											
Tensile	2·67	(4·12)			1·26	(1·95)	1800	0·640	(0·99)		
relaxation	2·92	(4·51)			1·375	(2·12)	1800	0·710	(1·10)		
tests	3·05	(4·71)			1·44	(2·22)	400	0·985	(1·52)		
(solid	3·50[a]	(5·40)			1·65	(2·55)	3000	0·730	(1·13)		
specimens)	3·50	(5·40)			1·65	(2·55)	3000				
Tensile	1·71	(2·64)			0·81	(1·25)	1800			4·8	
creep	2·29	(3·54)			1·08	(1·67)	1800			10·4	
tests	2·90	(4·48)			1·365	(2·11)	1800			13·7	
(solid	3·05	(4·71)			1·44	(2·22)	1800			24·0	
specimens)	3·50	(5·40)			1·65	(2·55)	1800			30·0	
	4·00	(6·18)			1·885	(2·91)	25200			Ruptured	

Octahedral shear stress $\tau_o = (\sqrt{2}/3)(T^2 + 3S^2)^{\frac{1}{2}}$ for tension torsion tests. $\tau_o = (\sqrt{2}/3)\sigma$ for pure tensile tests on solid specimens.

[a] Automatic control: repeat test.

encountered. While a relatively narrow range of stresses is involved, it should be considered relative to that producing creep fracture. This is indicated by a creep test at $\sigma = 4$ ton/in^2 (6·18 hbar) which fractured in 420 hr.

The observation that, for primary creep conditions on this material, the ratio of axial to shear strains maintains a constant ratio with the axial to shear stresses, was used in the complex stress relaxation tests. Thus, maintenance of constant shear strain by unloading shear stress would also maintain approximately constant axial strain, provided that reduction in the axial load was proportionate. In the tension–torsion relaxation tests, after applying appropriate loads in suitable increments, the shear strain was kept constant by torsionmeter strain observation. (The shear strain is more accurate for control purposes since it is relatively unaffected by minor temperature fluctuation.) Manual control of the shear stress was achieved by slowly winding back the jockey weight on the shear load lever. Proportional reduction of the tensile load was obtained automatically. A record of shear load against time was made and axial and shear relaxation curves thus derived.

THEORY

In earlier investigations Dr Johnson, Henderson and Khan [6] showed that, for several engineering metals and temperatures including the present copper, primary creep under constant complex stress systems could be represented by the creep rate equation

$$C_{ij} = F(J_2)S_{ij}\varphi(t)$$

where

C_{ij} = Creep rate tensor

J_2 = Second invariant of stress deviation tensor $= \frac{1}{6}\Sigma(\sigma_1 - \sigma_2)^2$

S_{ij} = Stress deviator tensor $= (\sigma_{ij} - \delta_{ij}\sigma_v)$

t = Time, and

$\sigma_1, \sigma_2, \sigma_3$ = Principal stresses.

$F(J_2)$ was usually found to be of the form $A'(J_2)^n$ or $[A_1(J_2)^{n_1} + A_2(J_2)^{n_2}]$ and $\varphi(t) = mt^{(m-1)}$, where A', A_1, A_2, n, n_1, n_2 and m are constants for a given temperature.

Thus, complex stress and tensile relaxation tests were compared on the basis of equivalent J_2 values or, more conveniently, octahedral shear stress τ_o

$$\tau_o = \tfrac{1}{3}[\Sigma(\sigma_1 - \sigma_2)^2]^{\frac{1}{2}} = (\tfrac{2}{3}J_2)^{\frac{1}{2}}$$

A tension–torsion and a solid tensile relaxation test, starting at the same initial octahedral shear stress, should develop the same octahedral-shear-stress/time curves. Tests were conducted to check the validity of this theory.

The mechanical theories of creep, on the other hand, purport to provide a means of converting constant stress creep equations or data to a form representing changing stress conditions, such as are encountered in relaxation.

The equation

$$C_{ij} = F(J_2)S_{ij}\varphi(t)$$

known as the 'age hardening' equation, may be rewritten in the form:

$$C_{ij} = \frac{f_1[F(J_2)S_{ij}]}{f_2(\varepsilon_{ij})}$$

The latter form is known as the 'strain hardening' equation. The first equation assumes, for isothermal conditions, that creep rate is dependent solely on stress and time, and the second assumes that creep rate depends only on current stress and strain. Neither equation makes allowance for the path by which either the stress or strain was reached, or the effect of creep recovery. For constant stress, both give identical results.

The general creep rate equation

$$C_{ij} = F(J_2)S_{ij}\varphi(t)$$

may be related to the octahedral plane, where the general stress system is represented by shear stress, to give the equation

$$C_o = 2(\sqrt{\tfrac{2}{3}})F(J_2)(\sqrt{J_2})\varphi(t)$$

The normal stress in this plane, being equal to the hydrostatic stress, is ineffective under primary creep conditions. A degree of error in the predictions of complex stress creep relaxation by the mechanical equations of state might be attributed to the assumption of $\nu = 0{\cdot}5$ for total strain; this was discussed in a previous paper [1]. By similar reasoning, it could be shown that the error for the present copper, for which Poisson's ratio at 250°C was 0·44, would be of the order of 4 per cent. Since normal creep strain for the octahedral plane is proportional to the sum of the principal creep strains, and volumetric constancy is assumed, the normal strain is therefore zero. The analysis can therefore be confined to the shear–stress/shear–strain/time relationships in this plane, provided proportional unloading occurs.

The complex stress–time relationship given by the age hardening theory, as derived by Dr Johnson, Henderson and Mathur [1], leads to:

$$t^m = \frac{2}{E}\left(\sqrt{\frac{2}{3}}\right)(1+v)\int_{\sqrt{J_{2,t}}}^{\sqrt{J_{2,0}}} [A_1 J_{2,t}{}^{(2p_1+1)/2} + A_2 J_{2,t}{}^{(2p_2+1)/2}]^{-1}\, \mathrm{d}(\sqrt{J_{2,t}})$$

and for the strain hardening theory:

$$t = \frac{K^{(1/m)}}{m}\int_{\sqrt{J_{2,t}}}^{\sqrt{J_{2,0}}} [A_1 J_{2,t}{}^{(2p_1+1)/2} + A_2 J_{2,t}{}^{(2p_2+1)/2}]^{-1/m} \times [(\sqrt{J_{2,0}}) - (\sqrt{J_{2,t}})]^{([1/m]-1)}\, \mathrm{d}(\sqrt{J_{2,t}})$$

where

$$K = \frac{2}{E}\left(\sqrt{\frac{2}{3}}\right)(1+v)$$

The constants in the complex stress creep equation were derived from simple tensile creep test curves (Appendix 1). The octahedral-shear-stress/time curves, derived by both theories using the tensile creep data, were compared with the experimental complex stress relaxation curves.

RESULTS OF EXPERIMENTS

The experimental tensile relaxation/time curves obtained on solid test pieces are shown in Fig. 2. The tensile creep curves and derivation of the primary creep equation are shown respectively in Figs. 3 and 4. Finally, Fig. 5 a–d shows the complex stress creep relation/time curves together with the predictions afforded by:

1 The tensile relaxation tests.
2 The tensile creep tests via the age and strain hardening equations.

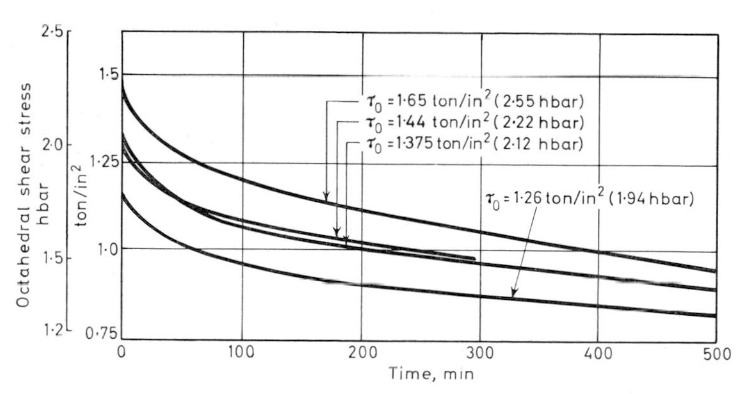

FIG. 2. Tensile relaxation tests.

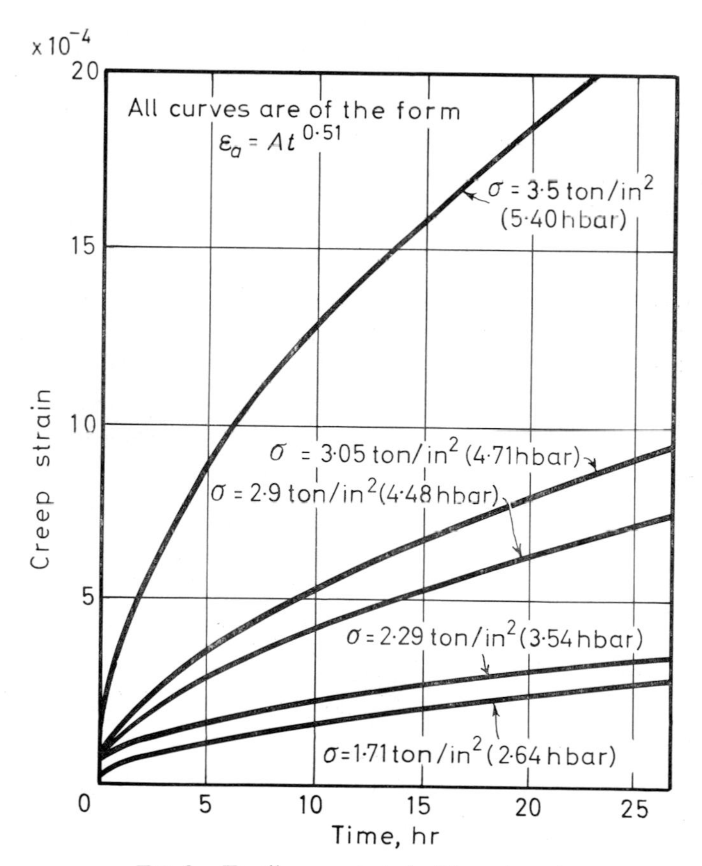

FIG. 3. Tensile creep tests (solid specimens).

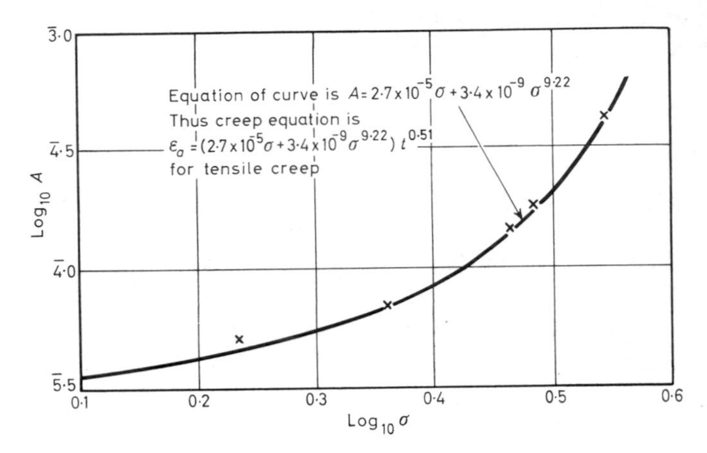

FIG. 4. Tensile creep stress dependence.

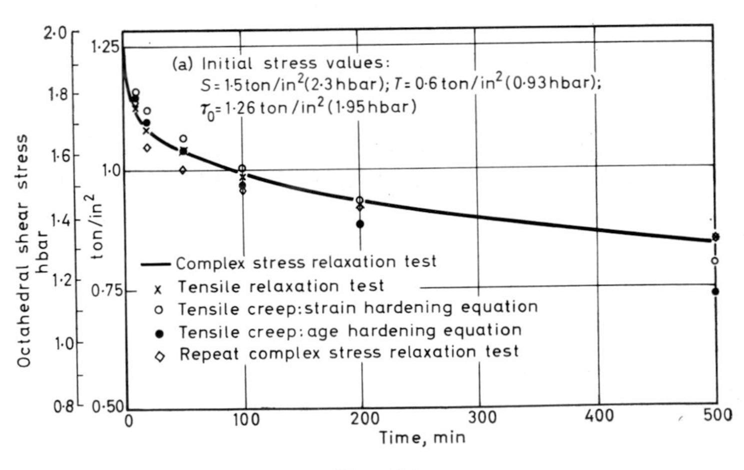

FIG. 5(a)

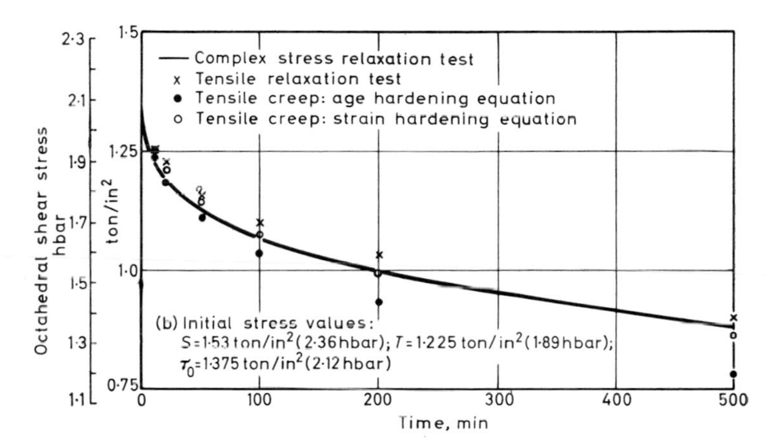

Complex stress relaxation test
x Tensile relaxation test
• Tensile creep: age hardening equation
o Tensile creep: strain hardening equation
(b) Initial stress values:
S=1·53 ton/in² (2·36 hbar); T=1·225 ton/in² (1·89 hbar);
τ₀=1·375 ton/in² (2·12 hbar)
Octahedral shear stress hbar
ton/in²
Time, min
2·3
2·1
1·9
1·7
1·5
1·3
1·1
1·5
1·25
1·0
0·75
0
100
200
300
400
500

FIG. 5(b)

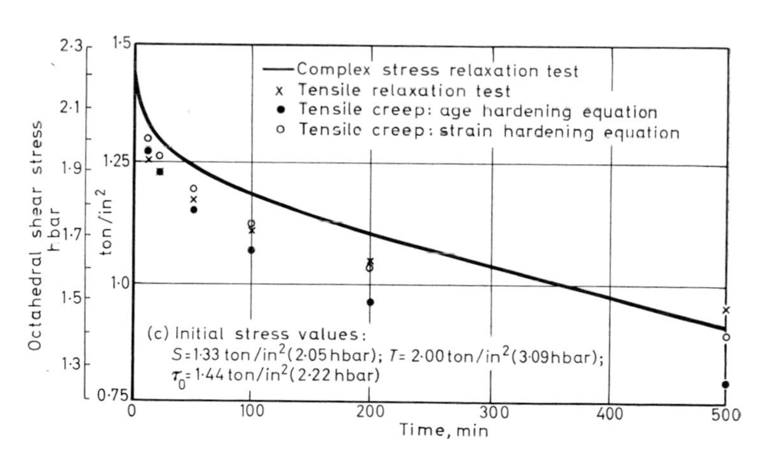

Complex stress relaxation test
x Tensile relaxation test
• Tensile creep: age hardening equation
o Tensile creep: strain hardening equation
(c) Initial stress values:
S=1·33 ton/in² (2·05 hbar); T= 2·00 ton/in² (3·09 hbar);
τ₀= 1·44 ton/in² (2·22 hbar)
Octahedral shear stress hbar
ton/in²
Time, min
2·3
2·1
1·9
1·7
1·5
1·3
1·5
1·25
1·0
0·75
0
100
200
300
400
500

FIG. 5(c)

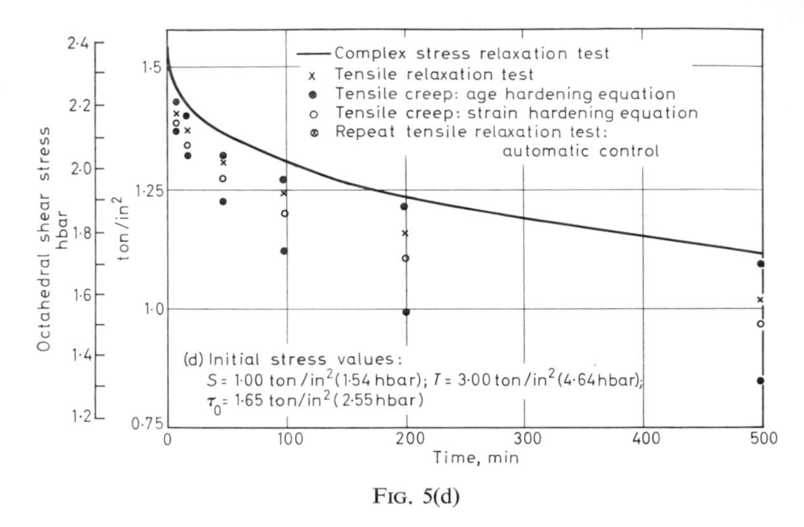

FIG. 5(d)

FIG. 5. Complex stress relaxation tests.

DISCUSSION

The four complex stress creep relaxation tests covered a range of stress ratios T/S from 0·4 to 3·0. The results of a repeat test at $T = 0{\cdot}6$ ton/in^2 (0·93 hbar), $S = 1{\cdot}5$ ton/in^2 (2·32 hbar) differed from those of the original test by an average of 1 per cent over the 500 min of testing, Fig. 5a. The predictions of complex stress test results by the three methods considered are illustrated in Fig. 5 a–d. The average percentage errors, maximum deviations and range of errors, over the period of each test, and their predicted values are given in Table 2. The errors are based on consideration of experimental and predicted stresses at 10, 20, 50, 100, 200 and 500 min as shown in Fig. 5 a–d.

Examination of the predictions of the tension–torsion relaxation test results (*see* Table 2) obtained from the tensile relaxation tests on solid cylindrical specimens, on the basis of octahedral shear stress equivalence, shows an overall average error of -2 per cent, a maximum deviation from the experimental curve of 5 per cent and a range of 6 per cent; this represents an extremely satisfactory predictability. Individual specimen differences could well lead to variations of this order and evidence to this effect is, in fact, available in the tensile relaxation test at $\sigma = 3{\cdot}5$ ton/in^2 (5·40 hbar), $\tau_o = 1{\cdot}65$ ton/in^2 (2·55 hbar), which was duplicated in another

TABLE 2

Summary of errors in prediction of complex stress relaxation tests

Method of prediction	$\tau_o = 1{\cdot}26$ ton/in² (1·94 hbar)	$\tau_o = 1{\cdot}375$ ton/in² (2·12 hbar)	$\tau_o = 1{\cdot}44$ ton/in² (2·22 hbar)	$\tau_o = 1{\cdot}65$ ton/in² (2·55 hbar)	Overall error average
Average errors over 500 min (per cent)					
Tensile relaxation test	0	−2	−4	−6	−2
Age hardening equation	−3	−4	−9	−14	−8
Strain hardening equation	+1	0	−3	−9	−3
Maximum deviation from experimental curve (per cent)					
Tensile relaxation test	1	3	6	9	5
Age hardening equation	12	12	13	25	15
Strain hardening equation	5	2	5	14	6
Range of errors (per cent)					
Tensile relaxation test	2	3	10	9	6
Age hardening equation	15	12	13	25	16
Strain hardening equation	9	4	5	14	8

unit using automatic load reduction control. Both tests had durations of about 50 hr (*see* Fig. 6). Despite the differences in test piece, testing machine and load reduction technique, the relaxation curves were identical to within 2 per cent over the 500 min and to within 5 per cent over 50 hr. The repeat test predicted the complex stress relaxation test curve even more closely than that used in the analysis. Thus, the prediction of complex stress relaxation, from manually controlled tensile relaxation tests based on octahedral shear stress equivalence, was shown to be valid. Furthermore, the error involved was negative for two of the tests, while the other two

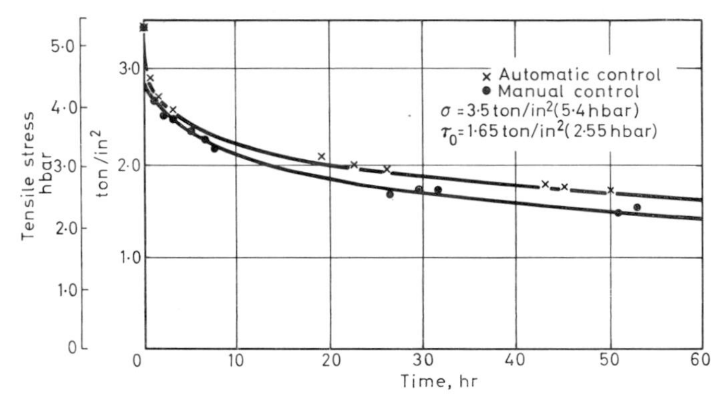

FIG. 6. Tensile relaxation tests.

gave a zero and +2 per cent error. A slightly greater degree of relaxation was, therefore, forecast, which would tend to represent a conservative estimate for conditions where relaxation was undesirable.

The obvious advantages of this method of prediction of complex stress relaxation, apart from accuracy, are (a) little or no mathematical computation is involved, (b) the tests can be carried out on conventional tensile creep units, and (c) no unusual skills, on the part of the operator, other than those associated with performing creep tests are required.

The predictions of the complex stress tests, from tensile creep tests and the mathematical equations of state, are now considered. Results of tensile creep tests (*see* Fig. 3), plotted on a log creep/log time basis, displayed a considerable linear range and could thus be represented by the equation

$$\varepsilon_a = At^{0.51}$$

where t is in hours and A is a constant dependent on stress. Values of A

are plotted against stress on a log–log plot (*see* Fig. 4) and a continuous relationship derived such that

$$A = 2{\cdot}7 \times 10^{-5}\sigma + 3{\cdot}4 \times 10^{-9}\sigma^{9{\cdot}22}$$

Accepting the von Mises relationship, which has been well established in previous work and on this type of copper, the multiaxial creep equation could be obtained (*see* Appendix 1). The equation for shear creep rate on the octahedral plane was thus given by:

$$C_o = [3{\cdot}37 \times 10^{-5} + 3{\cdot}90 \times 10^{-7}(J_2)^{4{\cdot}11}](\sqrt{J_2})t^{-0{\cdot}49}$$

The relaxation/time relationship given by the age hardening theory was:

$$t^m = \frac{2}{E}\left(\sqrt{\frac{2}{3}}\right)(1 + \nu)\int_{\sqrt{J_{2,t}}}^{\sqrt{J_{2,0}}} [3{\cdot}37 \times 10^{-5}(\sqrt{J_{2,t}}) + 3{\cdot}90 \times 10^{-7}(J_{2,t})^{4{\cdot}61}]^{-1}\, \mathrm{d}(\sqrt{J_{2,t}})$$

For the strain hardening theory:

$$t = \frac{K^{1/m}}{m}\int_{\sqrt{J_{2,t}}}^{\sqrt{J_{2,0}}} [3{\cdot}37 \times 10^{-5}(\sqrt{J_{2,t}}) + 3{\cdot}90 \times 10^{-7}(J_{2,t})^{4{\cdot}61}]^{-1/m} \times [(\sqrt{J_{2,0}}) - (\sqrt{J_{2,t}})]^{([1/m]-1)}\, \mathrm{d}(\sqrt{J_{2,t}})$$

where

$$K = \frac{2}{E}\left(\sqrt{\frac{2}{3}}\right)(1 + \nu),$$

$m = 0{\cdot}51$,

$E = 6070$ ton/in^2 (9374 hbar), and

$\nu = 0{\cdot}44$.

From an initial test value of $\sqrt{J_{2,0}}$, *i.e.* $(\sqrt{\frac{3}{2}})\tau_o$, graphical integration gave times for progressively reduced values of $\sqrt{J_{2,t}}$. A curve of τ_o against t for each complex stress test condition was derived and drawn; specific values of τ_o are compared in Fig. 5 with those values obtained in the tension–torsion relaxation tests. The average percentage errors, over 500 min, are given in Table 2. Undoubtedly, the better of the two mechanical theory predictions is that of the strain hardening equation with an overall average error of -3 per cent, a maximum deviation of 6 per cent and a range of errors of 8 per cent. In comparison, the age hardening theory shows an overall error of -8 per cent and correspondingly higher values of maximum deviation and range.

Three of the tensile relaxation tests were extended to about 30 hr so that evidence of longer time predictions of the mechanical equations of state could be noted.

Indication of the predictions provided by the strain hardening equation is given for the three tensile relaxation tests (*see* Fig. 7); an overall error of −4 per cent, a maximum deviation of 8 per cent and a range of 8 per cent is shown.

For an aluminium alloy at 200°C and magnesium alloy at 50°C [1], the age hardening theory provided the closest prediction of the course of complex stress relaxation/time curves. In these tests, while an average

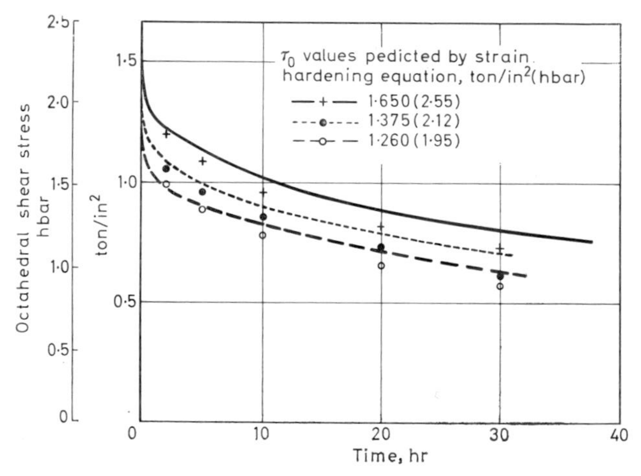

FIG. 7. Tensile relaxation tests and predictions by strain hardening equation.

error of 1 or 2 per cent was obtained for the age hardened equation, here the most accurate of the equations investigated, nevertheless, the range of errors was 36 per cent for aluminium and 58 per cent for magnesium. Complex stress relaxation in copper and aluminium by tension–torsion relaxation tests at 150°C were investigated by Namestnikov [2] and deviation from constancy in the tension–torsion stress ratios noted where both the axial and shear strains were held constant; the stress ratios are maintained constant in the present work. However, Namestnikov considered that the correspondence of theory with experiment was satisfactory. The results of the present work, however, agree with the findings of Davies [7] and of Roberts [8] on copper at 150°C where tensile relaxation tests were well described by the strain hardening equations.

CONCLUSIONS

Manually controlled tensile relaxation tests can provide a satisfactory prediction of the complex stress relaxation behaviour of commercially pure copper at 250°C. This, combined with the advantage of involving no complicated mathematical computation and the simplicity of testing technique, makes this a most attractive method of providing complex stress relaxation predictions.

Tensile primary creep data and the strain hardening equation may be used to predict complex stress relaxation to a reasonable degree of accuracy.

ACKNOWLEDGEMENTS

This chapter is published by permission of the Director of the National Engineering Laboratory, Department of Trade and Industry. It is Crown copyright and is reproduced by permission of the Controller of HM Stationery Office.

REFERENCES

1. JOHNSON, A. E., HENDERSON, J. and MATHUR, V. D. (1959). 'Complex-stress creep relaxation at elevated temperatures.' *Aircr. Engng*, **31**(361), 75–79; **32**(362), 113–118.
2. NAMESTNIKOV, V. S. (1963). 'Combined stress creep under changing loads.' *Joint Int. Conf. on Creep. Proc. Instn Mech. Engrs*, **178**(3A), 2-109–2-116.
3. GRIFFITHS, J. E. and MARIN, J. (1956). 'Creep relaxation for combined stresses.' *J. Mech. Phys. Solids*, **4**(4), 283–293.
4. JOHNSON, A. E., HENDERSON, J. and MATHUR, V. D. (1956). 'Combined stress creep fracture of a commercial copper at 250°C.' *Engineer*, **202**(5248), 261–265; **202**(5249), 299–301.
5. JOHNSON, A. E. and FROST, N. E. (1954). 'Combined tension and torsion machine for relaxation tests.' *Engineer*, **198**(5160), 834–835.
6. JOHNSON, A. E., HENDERSON, J. and KHAN, B. (1962). *Complex-stress Creep, Relaxation, and Fracture of Metallic Alloys*. Edinburgh: HMSO.
7. DAVIES, E. A. (1943). 'Creep and relaxation of oxygen-free copper.' *J. Appl. Mech.*, 65A-101.
8. ROBERTS, I. (1951). 'Prediction of relaxation of metals from creep data.' *Proc. Amer. Soc. Test. Mater.*, **51**, 811–831.

Appendix 1

MULTIAXIAL FORM OF THE PRIMARY CREEP EQUATION

The multiaxial form of the primary creep equation obtained from tensile creep tests is derived as follows.

Tensile creep strain ε_a is given (*see* Figs. 3 and 4) by the equation

$$\begin{aligned}\varepsilon_a &= (2{\cdot}70 \times 10^{-5}\sigma + 3{\cdot}40 \times 10^{-9}\sigma^{9\cdot22})t^{0\cdot51}\\ &= \tfrac{2}{3}[4{\cdot}05 \times 10^{-5} + 4{\cdot}65 \times 10^{-7}(J_2)^{4\cdot11}]\sigma t^{0\cdot51}\end{aligned}$$

where, for pure tension,

$$J_2 = \tfrac{1}{6}\Sigma(\sigma_1 - \sigma_2)^2 = \frac{\sigma_1{}^2}{3}$$

and

$$C_a = \tfrac{2}{3}[2{\cdot}07 \times 10^{-5} + 2{\cdot}38 \times 10^{-7}(J_2)^{4\cdot11}]\sigma t^{-0\cdot49}$$

or

$$C_1 = [2{\cdot}07 \times 10^{-5} + 2{\cdot}38 \times 10^{-7}(J_2)^{4\cdot11}](S_1)t^{-0\cdot49}$$

where, for pure tension,

$$S_1 = \left(\sigma_1 - \frac{\Sigma\sigma_1}{3}\right) = \frac{2\sigma_1}{3}$$

The general equation is thus given by

$$C_{ij} = [2{\cdot}07 \times 10^{-5} + 2{\cdot}38 \times 10^{-7}(J_2)^{4\cdot11}](S_{ij})t^{-0\cdot49}$$

and is the form of equation found to satisfy several materials and temperatures [6].

Therefore, since

$$\begin{aligned}C_o &= \tfrac{2}{3}[\Sigma(C_1 - C_2)^2]^{\frac{1}{2}}\\ C_o &= [3{\cdot}37 \times 10^{-5} + 3{\cdot}90 \times 10^{-7}(J_2)^{4\cdot11}](\sqrt{J_2})t^{-0\cdot49}\end{aligned}$$

when C_o is the octahedral shear creep rate which is the form used in the analysis.

Chapter 11

DETERMINATION OF FACTORS CAUSING EMBRITTLEMENT IN TIME-TO-RUPTURE TESTS

W. SIEGFRIED

SUMMARY

Physical and metallurgical factors tending to reduce rupture life are considered, particularly the influence of notches, grain-boundary precipitation and the formation of zones low in alloy content. The findings are applied, in Appendices, to practical cases.

NOTATION

A, B	Coefficients in eqn (43).
A_1, n_1	Parameters in eqn (11).
a	Mean distance between fissure nuclei.
D	Diffusion coefficient.
F	Area.
k	Boltzmann's constant.
L	Mean grain diameter.
l	Length.
m	Slope of time-to-rupture curve, defined by eqn (9).
$N(t)$	Number of fissure nuclei at time t.
$n(r_0)$	Number of particles per centimetre with radius r_0.
n_0	Total number of precipitated particles per centimetre.
p	$a^3kT/\Omega D\delta$.
p_0	Cottrell distribution of fissure nuclei $1/n_0(kT/\Omega D\delta)$.
p_0'	Gaussian distribution of particles, *see* eqn (52).

r	Radius of a circular fissure.
r_m	Mean value of r_0.
r_0	Radius of a particle in the grain boundaries.
T	Absolute temperature.
t	Time.
t_B	Time to rupture.
v_k	Critical volume of a fissure nucleus.
v_0, σ_1	Parameters in eqn (10).
v_{sec}	Speed of secondary creep.
z	$\sigma_0 \exp \varepsilon$.
α	χ/t_B.
α_1, γ_1	Parameters in eqn (57).
γ	Surface stress.
δ	Thickness of grain boundary.
ε	Elongation.
$\dot{\varepsilon}$	Creep rate.
ε_B	Linear elongation at fracture.
ε_{kg}	Sliding of the grain boundaries.
ρ	Deviation of Gaussian distribution.
σ	Stress.
σ_0	Nominal stress.
ψ_B	Reduction of area at fracture (per cent).
Ω	Volume of an atom (or hole).
erf(x)	Error function $= 2/\sqrt{\pi} \int_{\infty}^{x} e^{-\xi^2} \, d\xi$.
erfc(x)	$1 - \text{erf}(x)$.

$$J(\sigma_0/\sigma_1, \varepsilon_B) = v_0 t_B = \int_0^{\varepsilon_B} [d\varepsilon/\sinh(\sigma_0/\sigma_1, \exp \varepsilon)].$$

INTRODUCTION

Definition and significance of embrittlement

The resistance at high temperatures of constructional components with shaped notches is of great practical importance. Numerous tests have been made with a view to explaining the relationships on which these components depend. It has been observed that in many cases embrittlement phenomena occur which are characterised by the properties shown in Table 1.

In time-to-rupture tests, notched specimens sometimes show a lower resistance than unnotched specimens and this is especially important. The lower resistance of notched specimens is caused by several parameters, among which are the following:

1 The geometrical shape of the notch.
2 The chemical composition of the material.
3 The heat treatment.
4 The structural changes that take place during creep.

These parameters are also responsible for the shape of the time-to-rupture curve for unnotched specimens and the elongation at rupture measured in time-to-rupture tests.

It has long been believed, on the basis of time-to-rupture tests on unnotched and notched specimens, that sufficient indications were available to make it possible to predict the embrittlement behaviour of notched machine components. However, model tests on constructional components, as well as cases of damage during service, have shown that the present concepts must be extended.

In the welding of austenitic heat resistant steels, grain boundary fissures that developed under the combined influence of stresses and precipitation effects have very often been observed in the transition zone. The formation of such fissures, however, represents a special case of general embrittlement phenomena in time-to-rupture tests.

The problem

The purpose of this chapter is to discuss the following points in greater detail:

1 Time-to-rupture tests with unnotched specimens.
   a Influence of geometrical stability.
   b Influence of structural stability.
2 Tests with notched specimens.
3 Embrittlement phenomena in constructional components.

SIMPLE MODEL FOR QUANTITATIVE DETERMINATION OF EMBRITTLEMENT PHENOMENA

General

In considering embrittlement phenomena, an early observation was that the concept of a rheological single phase field had obviously to be abandoned. Two rheologically different phases were therefore introduced into the calculations, *i.e.* the crystal and the grain boundary. At that time, however, the results of modern physical investigations on the processes of

TABLE 1

Summary of investigations on the slope of time-to-rupture curves

Kind of tests	Execution of tests	Researchers	Result
Time-to-rupture test at constant temperature on notched and unnotched specimens	Investigation of shape of curves $\log \sigma - \log t_B$ for constant temperature	Thum, A. and K. Richard [7] Wild, M. [8] Sachs, G. and W. F. Brown, Jr. [9]	I, II, III; I, II, III Notched specimens Unnotched specimens $\log \sigma$; $\log t_B$
Formation of rupture	Microscopic investigation of broken specimens	Grant, N. J. and A. G. Bucklin [10] Thum, A. and K. Richard [7]	Field I: transcrystalline ruptures Fields II and III: intercrystalline ruptures
Time-to-rupture tests on unnotched specimens	Time-to-rupture elongation and constriction in dependence of time	Wild, M. [8]	Time-to-rupture elongation and constriction have smallest values at transition of field I to field II
Hardness of broken time-to-rupture specimens	Determination of hardness in dependence of time to rupture	Wild, M. [8]	Strong decrease at transition of field II to field III

Time-to-rupture tests on notched and unnotched specimens at various temperatures	Calculation of enthalpy of activation in dependence of time to rupture	Sachs, G. and W. F. Brown, Jr. [9] Siegfried, W. [11]	Notched specimens: enthalpy of activation independent of temperature. Unnotched specimens: vice versa
Creep tests at different stresses and temperatures	Determination of 1 per cent elongation curves	Wild, M. [8]	Transition of field I to field II can be determined also in 1 per cent elongation curves
Creep tests at different stresses and temperatures	Determination of creep rate as a function of stress	Grant, N. J. and A. G. Bucklin [10]	Break-like transition of field I to field II at same values of stress as in time-to-rupture tests
Determination of notch impact strength of specimens after a preceding creep process	Definition of notch impact strength in dependence of stress and time with and without subsequent initial heat treatment	Thum, A. and K. Richard [7]	Structural changes take place during a creep process which cannot be reversed by subsequent heat treatment: Formation of fissures in grain boundaries
Hot hardness tests	Determination of hot hardness in dependence of stress and temperature	E. C. Underwood [12]	Transition of zone I to zone II at the same value of the Larson–Miller parameter as in time-to-rupture tests

deformation and rupture during creep were not available. Consequently, it was quite natural to generalise certain observations made on specific alloys in which low melting point eutectics were deposited in the grain boundaries, and to assume that pure metals also show a certain 'grain boundary substance' having a lower melting point than that of the crystal, though a higher strength at lower temperatures. On the basis of these reflections, one required a temperature at which grain boundaries and crystals show the same strength, *i.e.* the equicohesive temperature.

This concept, however, could not be confirmed by the latest physical investigations, and thus the equicohesive theory was discarded in the course of time. Certain phenomena observed during embrittlement in time-to-rupture tests and which could be explained by the equicohesive theory have not yet been otherwise explained. It is therefore necessary to find how, in the light of modern theories of flow, the formation of rupture takes place in the grain boundaries, and whether the interaction of the various elementary models developed by physical research could not be approximated rheologically by the deformation of a two phase system.

The physical processes that take place during time-to-rupture tests on unnotched specimens can be divided into different classes as follows.

Geometrical instability

In 1938 Andrade [1] drew attention to the fact that, for certain metals, the third creep stage is subject to necking, thus increasing the effective stress. However, it was also shown that for steel, and especially for long time-to-rupture tests, other processes, in particular the formation of fissures in the grain boundary, are a far more important cause of the third creep stage. For his investigations on lead, Andrade used equipment in which, during the test, the load applied was changed so that the effective stress remained constant. In consequence of the complexity of this apparatus it is impossible, on the basis considered, to carry out time-to-rupture tests on practical alloys on a large scale. Our investigations showed that, with simplifications normally applicable in long time-to-rupture tests, the increase in stress as a result of necking during the creep test can be determined mathematically. A time-to-rupture test with constant load can thus be related to a time-to-rupture test with constant stress.

Deformation of crystals as a result of various mechanisms of dislocation movement

There are numerous models that explain the creep process by means of the theory of dislocations. In practice, it can be assumed that several of these mechanisms interact during a time-to-rupture test. For our investigations, however, we need not consider the details of these models. The fact that all of them provide a mathematical relationship between the stress and the creep rate is sufficient for our purpose.

FORMATION OF FISSURES IN GRAIN BOUNDARIES

In long time-to-rupture tests, rupture normally takes place by formation of intercrystalline fissures. Various mechanisms have been proposed in metal physics to account for the formation of fissures. All the mechanisms have the following points in common:

1 Formation of fissure nuclei by simultaneous step formation in the grain boundaries, as a result of the sliding process in the crystal and sliding of grain boundaries on one another (*see* Fig. 1).
2 Formation of fissure nuclei by sliding of grain boundaries on one another, in which undeformable particles (carbides, etc) are deposited.
3 Formation of fissure nuclei on points at which three grain boundaries intersect subsequent to grain boundary deformation.

The fissure nuclei may grow according to two different processes, *i.e.*:

1 A stress controlled diffusion [2].
2 The sliding of grain boundaries on one another at points at which three grain boundaries intersect as a result of grain boundary deformation (triple-point-fissures).

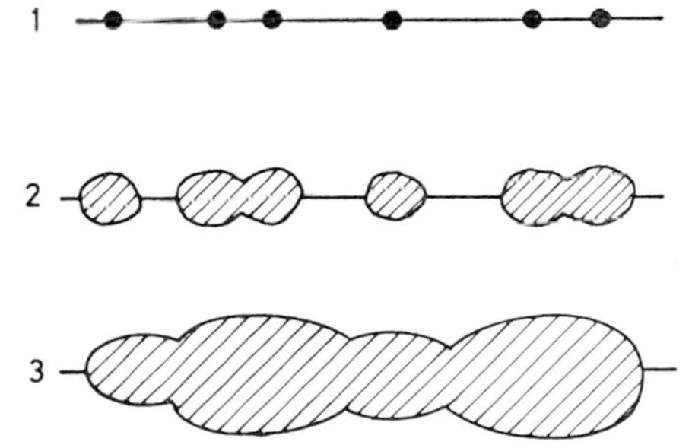

FIG. 1. Growth of fissure nuclei by diffusion (Cottrell [2]).

Model used in the investigations

DEFORMATION OF CRYSTALS AND GRAIN BOUNDARIES

Deformation of crystals, as well as sliding of grain boundaries on one another, takes place in a creep process. When following the extension of a specimen in relation to time, one observes the well known three stages of the creep process, *i.e.* a first stage with decreasing creep rate, a second stage with constant creep rate, and a third stage with increasing creep rate until rupture takes place. When the time-to-rupture is sufficiently long, the duration of the first and the third creep stages is small compared with that of the second stage. It may thus be assumed that there exists a mathematical relationship between stress and creep rate in that the time dependence of the creep rate is ignored and a mean value of the secondary creep rate is adopted. The relationship between stress and the mean value of the secondary creep rate depends on the following factors:

1 Elementary dislocation processes.
2 Ratio of deformation of crystals to that of grain boundaries.

This ratio can be influenced to a large extent by certain structural changes. In many alloys, for example, a zone develops in the neighbourhood of a grain boundary as a result of precipitations in the grain boundaries. This zone is low in alloying elements and therefore shows a lower resistance than the crystals. It can be assumed that in such a structure the contribution of boundary creep is greatly increased. The relationship between stress

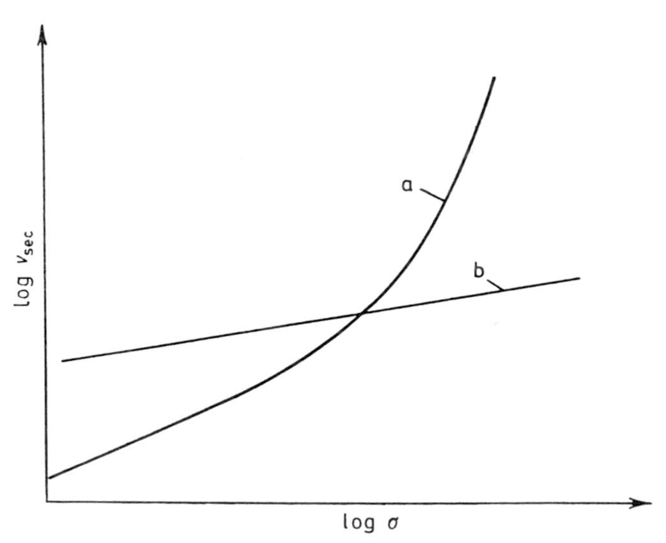

FIG. 2. Relation between stress and mean creep rate with formation of zones low in alloying elements in the grain boundaries.

and creep rate is now strongly dependent on the mutual ratio of the two kinds of creep: creep by deformation of crystals and creep by sliding of grain boundaries on one another. The slope of the creep rate as a function of stress has been plotted in Fig. 2, namely, for a material with a small proportion of grain boundary creep (curve a), and for a material with a large proportion of boundary creep (curve b).

Formation of rupture in the grain boundaries

1 *Formation of fissure nuclei*

All processes of formation of fissure nuclei require a certain deformation by grain boundary creep. The formation of fissure nuclei at points at which three grain boundaries meet has already been investigated by Williams [3], and his results are applied in our calculations. For the formation of fissure nuclei by sliding of grain boundaries when undeformable particles are also present, a relationship between secondary creep rate, time and number of nuclei will be deduced, from the assumption that the fissure nuclei must have a minimum volume before it can grow further.

The formation of fissure nuclei by sliding of grain boundaries over deformation steps will not be considered in isolation, for this kind of formation is also determined qualitatively by the process mentioned above.

2 *Growth of fissure nuclei*

The growth of fissure nuclei has been believed to take place partly according to the process described by Williams [3], and partly by diffusion of vacancies under the influence of stress as indicated by Cottrell [2]. In consequence of the different temperature dependences of these two processes, it is possible to determine their relative importance in the resulting rupture process by evaluating time-to-rupture tests at different temperatures.

Combined effect of the deformation of crystals and grain boundaries

In considering the interaction of the various mechanisms described above, we introduce the following simplifying assumptions:

1 A relationship between stress and mean creep rate for the entire duration of the time-to-rupture test. This assumption presupposes that the changes in structure with time proceed relatively slowly. Under these conditions, the course of the necking is then determined as a function of time.

2 Rupture takes place exclusively by the growth of fissures in the grain boundaries, but the process of deformation is not affected by the growth of fissures.

DETERMINING THE PARAMETERS OF THE MECHANISM OF RUPTURE

Conversions from constant load to constant effective stress, and determining stress dependence of secondary creep rate

TIME-TO-RUPTURE FOR A CONSTANT LOAD TEST

To amplify the reasoning, we assume that the elongation at rupture is constant. The conclusions can easily be extended to the case where the elongation at rupture is a function of the nominal stress. Figure 3 shows

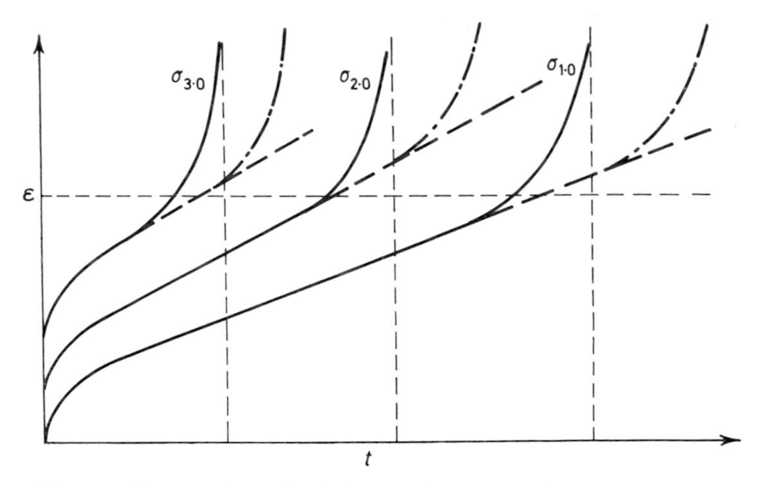

(a) —— Constant nominal stress $\sigma_{3\cdot 0} > \sigma_{2\cdot 0} > \sigma_{1\cdot 0}$

(b) - - - Constant effective stress, formation of rupture by geometric instability

(c) —·— Constant effective stress, formation of rupture by growth of fissures

FIG. 3. Variation of strain with time.

the course of strain as a function of time, for time-to-rupture tests conducted with different nominal stresses $\sigma_{1\cdot 0}$, $\sigma_{2\cdot 0}$, $\sigma_{3\cdot 0}$ at one temperature (curves (a) in Fig. 3). The slope of these curves is decided partly by the fact that the effective stress increases as a result of the formation of a constriction (or neck) in the course of time. If the test had been conducted in equipment that maintained the effective stress constant, the curves would have a slope as in curves (b) of Fig. 3. If rupture were to take place only by formation of a neck, curves (b) would be straight until the elongation at rupture was attained. In practice, however, rupture takes place by formation of fissures in the grain boundaries, and the curves have a slope as in curves (c) of Fig. 3. For times that are substantially less than the time-to-rupture, curves (a) and (b) nearly coincide. Hence, the dependence on stress of the secondary creep rate could be determined from a time-to-rupture test with constant load, provided the elongation was determined as a function of time. As such measurements are rarely available, we have to find whether it is possible to define the relationship between stress and secondary creep rate, in the case where results are known of time-to-rupture tests, whose nominal stress and elongation at rupture have been determined in terms of time.

This problem can be solved mathematically by introducing simplifying assumptions which give sufficient accuracy for time-to-rupture tests of known duration.

Once the relationship between stress and secondary creep rate is established, the time-to-rupture can be calculated. The following assumptions have to be defined mathematically.

1 Presence of a relationship between stress and creep rate

$$\dot{\varepsilon} = \varphi(\sigma) \tag{1}$$

where $\dot{\varepsilon}$ = Creep rate, and σ = Stress.

2 Constancy of volume

$$Fl = F_0 l_0 \tag{2}$$

where F = Cross-sectional area, and l = Length.

Considering natural strain, there results

$$\varepsilon = \ln (l/l_0) = \ln (F_0/F) \tag{3}$$

3 Constancy of load

$$\sigma F = \sigma_0 F_0 \tag{4}$$

Combination of eqns (1)–(4) yields

$$\frac{d\varepsilon}{dt} = \dot{\varepsilon} = \varphi(\sigma_0 \exp \varepsilon) \tag{5}$$

and by integration, the time-to-rupture

$$t = \int_{\varepsilon_0}^{\varepsilon_B} \frac{d\varepsilon}{\varphi(\sigma_0 \exp \varepsilon)} \tag{6}$$

ε_B can be calculated from the reduction of area of broken specimens according to

$$\varepsilon_B = -\ln\left(1 - \frac{\psi_B}{100}\right) \tag{7}$$

where ψ_B = reduction of area in per cent.

Introducing the substitution $\sigma_0 \exp \varepsilon = z$ leads to

$$t = \int_{\sigma_0 \exp \varepsilon_0}^{\sigma_0 \exp \varepsilon} \frac{dz}{z\varphi(z)} \tag{8}$$

Relationship between stress and secondary creep rate

1 *Structurally stable material*

Once the shape of the time-to-rupture curve is obtained graphically or by a mathematical equation, the relationship between stress and creep rate can be derived. To this end, we require the nominal stress, the slope of the tangent line to the time-to-rupture curve, and the elongation at rupture as a function of time-to-rupture. By differentiation of eqn (8), we obtain

$$\frac{1}{m} = \frac{1}{t_B}\left[\frac{1}{[\varphi(\sigma)]_{\sigma_0 \exp \varepsilon_0}} - \frac{1}{[\varphi(\sigma)]_{\sigma_0 \exp \varepsilon_B}}\right] - m = -\frac{d \ln \sigma_0}{d \ln t_B} = n \tag{9}$$

Equation (9) is valid for any integrable function of $\varphi(\sigma)$.

If it is possible in the equation $\dot{\varepsilon} = \varphi(\sigma)$ to use a formula with two parameters, namely

$$\dot{\varepsilon} = v_0 \sinh(\sigma/\sigma_1) \tag{10}$$

where v_0 and σ_1 are constants, or

$$\dot{\varepsilon} = A_1\sigma^{n_1} \tag{11}$$

where A_1 and n_1 are constants, then from eqns (8) and (9) we can determine the parameters v_0 and σ_1 for eqn (10), or A_1 and n_1 for eqn (11), provided that the mathematical relationship between stress and creep rate is accurately reproduced by the formula applied. In most cases, however, such formulae are valid only for a limited range of stress, since transitions can take place from one dislocation mechanism to another. This fact can be taken into account by introducing different values for the parameters v_0 and σ_1 for the various ranges of stress. In the general case, these parameters can be regarded as functions of stress. Equation (10) can therefore be written thus:

$$\dot{\varepsilon} = v_0(\sigma) \sinh [\sigma/\sigma_1(\sigma)]$$

From the mathematical point of view, the introduction of such a formula is suitable because it can be assumed that the parameters v_0 and σ_1 are functions of σ whose change is small compared with that of sinh (σ).

2 *Material that changes structurally during testing*

Should a formula according to eqns (10) and (11) be valid for the entire range of stress, all points on the time-to-rupture curve would have to give the same values for the parameters, *i.e.* the time-to-rupture curve would be determined uniquely by eqn (8). In most cases, however, this is not true, for the following reasons. In long time-to-rupture tests, structural changes nearly always take place with time. Consequently, time, as a variable, still has to be introduced into the relationship between stress and secondary creep rate, and this will complicate the equations to a considerable extent. Therefore, we introduce mean values of time for the parameters for any time-to-rupture test conducted with constant stress. Obviously, a time-to-rupture test of short duration will show mean values of the parameters different from those for a test of long duration, for the latter has a longer time for material changes to take place. Moreover, as in most cases, the elongation at rupture is relatively small and the third creep stage is very short, the change of stress is quite insignificant during most of the time to rupture. To calculate the parameters for the relationship between stress and strain rate, we can thus assume that the relationship between stress and creep rate in a time-to-rupture test is given by

$$\dot{\varepsilon} = v_0(\sigma_0) \sinh [\sigma/\sigma_1(\sigma_0)] \tag{12}$$

where v_0 and σ_1 are functions of σ_0.

Furthermore, we can assume that σ_1 and v_0 change relatively little with variation of σ_0 compared with the change of the hyperbolic sine. By this means, *i.e.* that for parameters v_0 and σ_1 we introduce functions of σ_0 that change only slightly, we are in a position to apply eqns (8) and (9) to any shape of the time-to-rupture curve and thereby to determine the values v_0 and σ_1 in dependence of the nominal stress or of the time-to-rupture.

3 *Determining the mean secondary creep rate*

Applying the simplifications described in the previous paragraph, eqns (8) and (9) can be written as follows:

$$t_B = \int_0^{\varepsilon_B} \frac{d\varepsilon}{v_0 \sinh [(\sigma_0/\sigma_1) \exp \varepsilon]} = \frac{1}{v_0} J\left(\frac{\sigma_0}{\sigma_1}, \varepsilon_B\right) \tag{13}$$

$$\frac{1}{m} = \frac{1}{J[(\sigma_0/\sigma_1), \varepsilon_B]} \left\{\left[\frac{1}{\sinh (\sigma_0/\sigma_1)}\right]_{(\sigma_0/\sigma_1)\exp \varepsilon_B} - \left[\frac{1}{\sinh (\sigma_0/\sigma_1)}\right]_{\sigma_0/\sigma_1}\right. \tag{14}$$

By definition, $$J\left(\frac{\sigma_0}{\sigma_1}, \varepsilon_B\right) = \int_0^{\varepsilon_B} \frac{d\varepsilon}{\sinh [(\sigma_0/\sigma_1) \exp \varepsilon]}$$

Figure 4 shows curves, for different values of ε_B, of $J[(\sigma_0/\sigma_1)\varepsilon_B]$ against σ_0/σ_1.

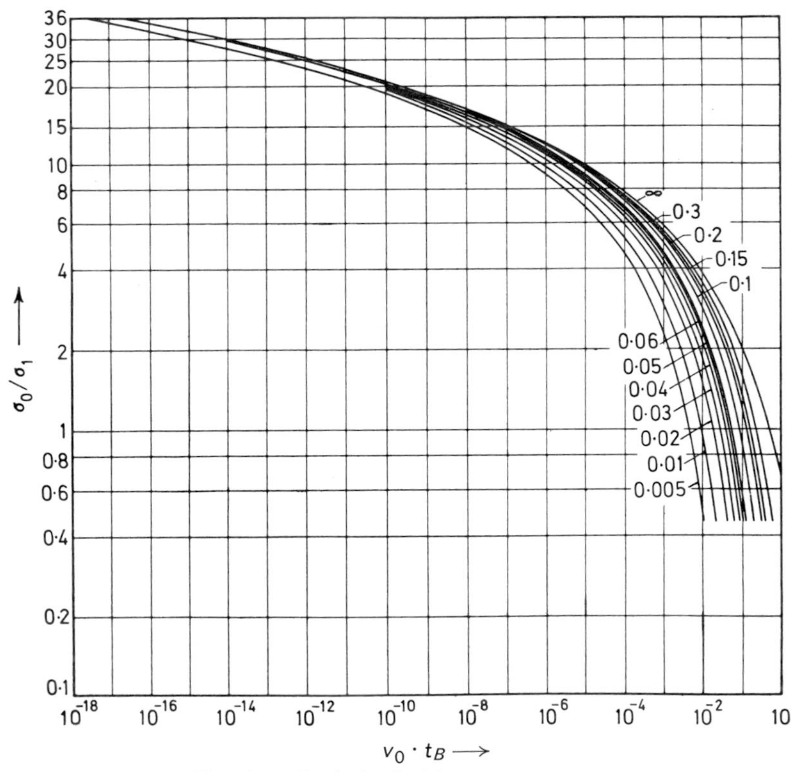

FIG. 4. $J[(\sigma_0/\sigma_1)\varepsilon_B]$ with ε_B as parameter.

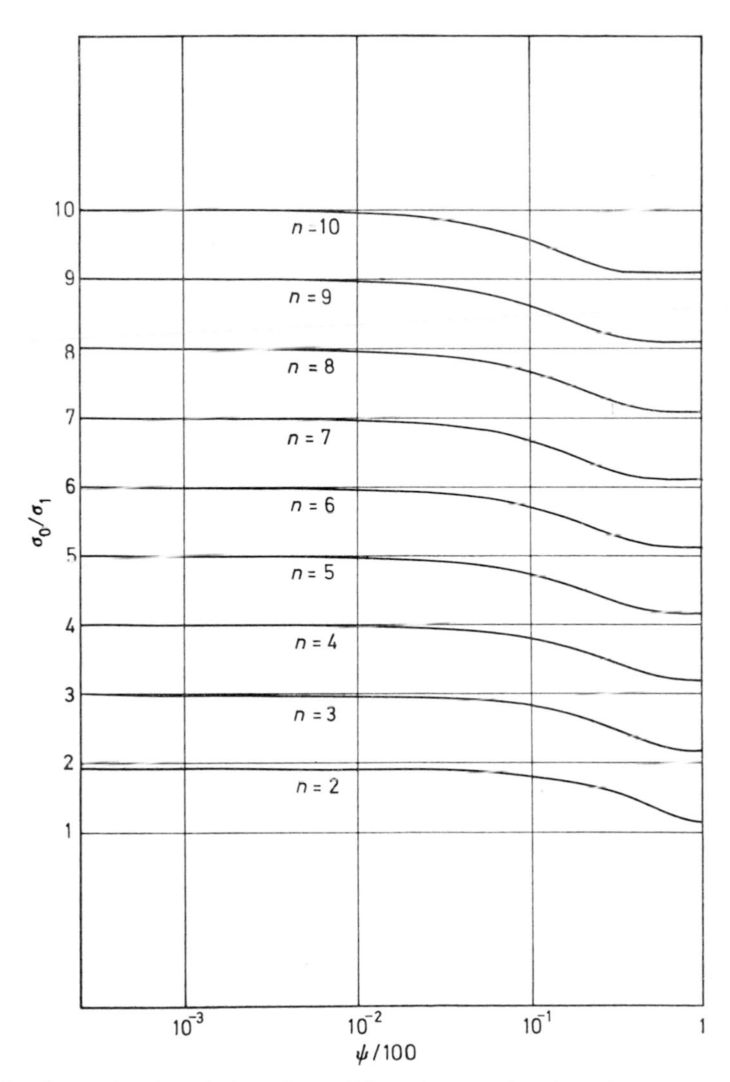

FIG. 5. Curves for the solution of eqn (14), σ_0/σ_1 as a function of $\psi_B/100$ with n as parameter.

As m and ε_B in eqn (14) are known, the only unknown is the ratio σ_0/σ_1. We can, therefore, represent σ_0/σ_1 as a function of $n = -1/m$ (*see* Fig. 5). Once σ_0/σ_1 is known, we also know σ_1, so that, on the basis of eqn (13), v_0 can be calculated by means of eqn (15):

$$v_0 = \frac{1}{t_B} J\left(\frac{\sigma_0}{\sigma_1}, \varepsilon_B\right) \tag{15}$$

$v_0 t_B$ as a function of $\psi/100 = 1 - \exp(\varepsilon_B)$ has been plotted in Fig. 6.

Further, it is interesting to indicate the solution of eqn (14) for small values of ε_B. In this case, eqn (14) becomes

$$n = \coth(\sigma_0/\sigma_1)\sigma_0/\sigma_1 \tag{16}$$

i.e. σ_0/σ_1 is independent of ε_B. Besides, we can put approximately

$$\coth(\sigma_0/\sigma_1) \sim 1 \tag{17}$$

so that

$$\sigma_0/\sigma_1 \approx n \tag{18}$$

4 *Calculation of v_0 for small values of ε_B*

When ε_B is small, eqn (13) becomes

$$v_0 t_B \approx \frac{\varepsilon_B}{\sinh(\sigma_0/\sigma_1)} \tag{19}$$

or

$$v_0 \approx \frac{1}{t_B} \frac{\varepsilon_B}{\sinh(\sigma_0/\sigma_1)} \approx \frac{\varepsilon_B}{t_B \sinh(n)} \tag{20}$$

It can be seen from Figs. 5 and 6 that eqn (20) is valid up to reductions of area of approximately 20 per cent, whereas the error of eqn (18) is relatively insignificant up to reductions of 5 per cent.

By means of Figs. 5 and 6, we are now able to determine the two parameters σ_1 and v_0 as a function of the nominal stress σ_0 for any point of the time-to-rupture curve.

It is therefore possible to indicate also the mean value of the secondary creep rate according to eqn (21) in dependence on stress.

$$\dot{\varepsilon} = v_0 \sinh(\sigma/\sigma_1) \tag{21}$$

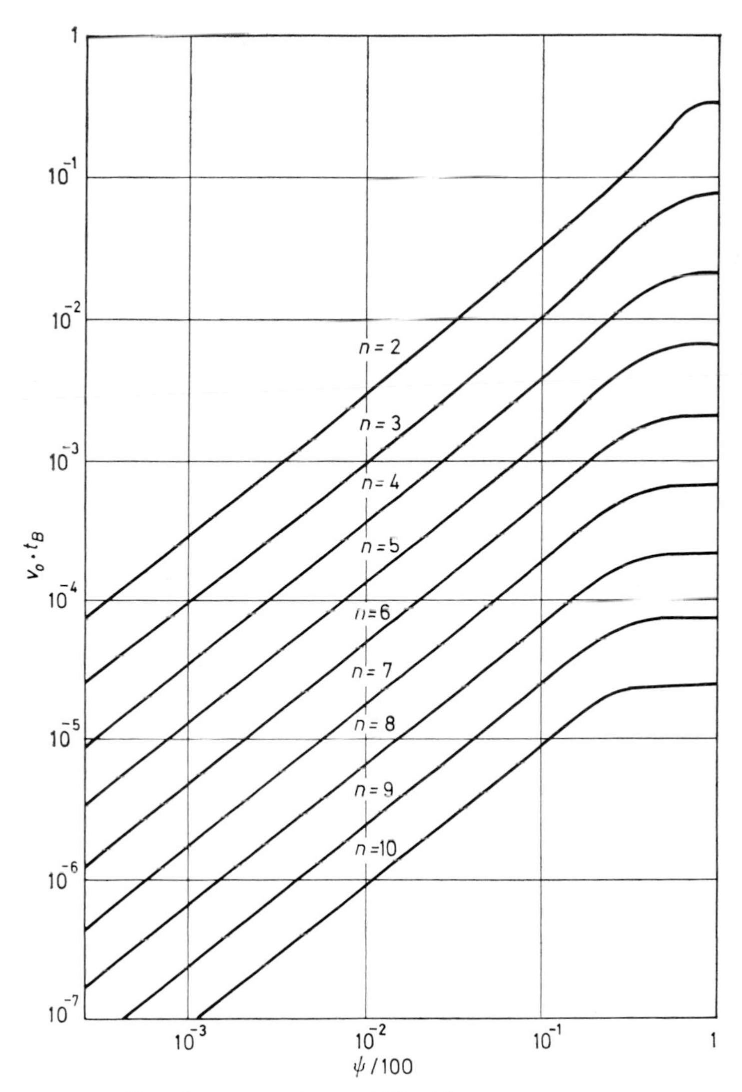

FIG. 6. $v_0 t_B$ against $\psi_B/100$ with n as parameter.

Time-to-rupture for various mechanisms of rupture

CHANGE OF EFFECTIVE STRESS WITH TIME

The calculations given in the preceding section are based on the following mechanism: the test bar lengthened until, at the point of necking, the contraction at rupture ψ_B was reached. This elongation at rupture was determined by measuring the broken specimen. In this case, no further assumptions have been made with regard to the mechanism of formation of rupture.

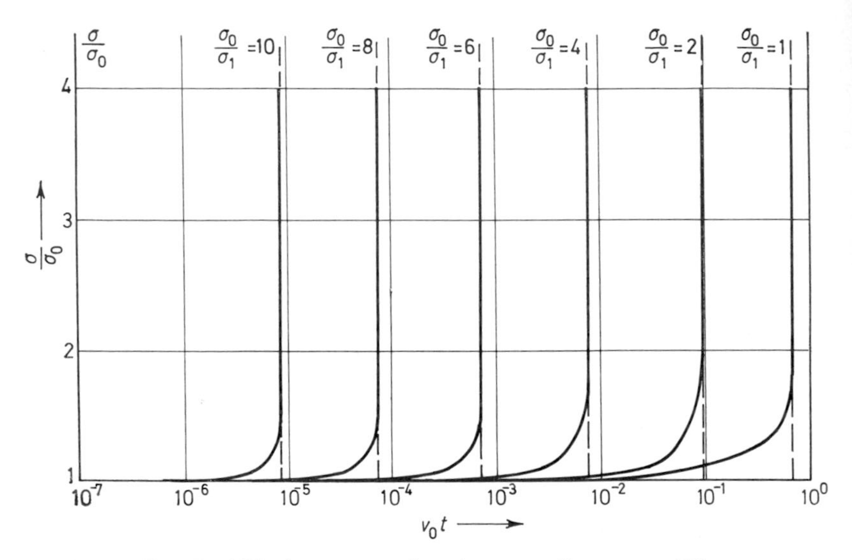

FIG. 7. Effective stress against time according to eqn (22).

Knowing the values v_0 and σ_1 as functions of the nominal stress σ_0, we can indicate the change of the time dependence of the effective stress. Thus, according to eqn (8)

$$t = \int_{\sigma_0}^{\sigma} \frac{dz}{z\varphi(z)} = \frac{1}{v_0} \int_{\sigma_0/\sigma_1}^{\sigma/\sigma_1} \frac{dz}{z \sinh(z)} = \frac{1}{v_0} \left[J\left(\frac{\sigma_0}{\sigma_1}\right) - J\left(\frac{\sigma}{\sigma_1}\right) \right] \tag{22}$$

The value of σ/σ_0 as a function of $v_0 t$ for constant values of σ_0/σ_1 can be determined from Fig. 4 as is shown in Fig. 7.

FORMATION OF RUPTURE BY DIFFUSION OF VACANCIES

Cottrell [2] calculated the time to rupture for the mechanism of diffusion of vacancies at constant stress. He obtained the relation:

$$t_B = \frac{a^3 kT}{24\sigma\Omega D\delta} \tag{23}$$

where

σ = Stress
a = Mean distance between fissure nuclei
Ω = Volume of an atom (or hole)
D = Diffusion coefficient and
δ = Thickness of grain boundary.

In a time-to-rupture test at constant load, the stress changes during the test owing to necking according to eqn (22); moreover, the stress in the metal between the fissures increases owing to their extension, causing a reduction in area. Taking these two factors into account, the change of stress during the test is the following:

$$\sigma' = \sigma_0 \exp \varepsilon \qquad \text{(Formation of a constriction)}$$

$$\sigma = \frac{\sigma'}{1 - (4/3)\pi(r^3/a^3)} \times \frac{\sigma_0 \exp \varepsilon}{1 - (4/3)\pi(r^3/a^3)} \qquad \text{(Increase of stress owing to extension of fissures)} \tag{24}$$

where r = Radius of the circular fissure.

If eqn (24) is introduced into Cottrell's differential equation

$$\frac{dr}{dt} = \frac{3\Omega\delta D\sigma}{arkT} \tag{25}$$

the result is

$$r\frac{dr}{dt} = \frac{3\Omega D\delta}{akT} = \frac{a^3 \exp \varepsilon}{a^3 - (4/3)\pi r^3}\sigma_0 \tag{26}$$

Equation (5) relates t and ε so that ε can be introduced as a variable integral, thus

$$dt = \frac{d\varepsilon}{\varphi(\sigma_0 \exp \varepsilon)} \tag{27}$$

We then obtain

$$p = \frac{a^3kT}{\Omega D\delta} = 25{\cdot}8 \int_{\sigma_0}^{\sigma_0 \exp \varepsilon_B} \frac{du}{\varphi(u)} \tag{28}$$

Defining

$$\chi = \frac{1}{\sigma_0}\int_{\sigma_0}^{\sigma_0 \exp \varepsilon} \frac{du}{\varphi(u)} \tag{29}$$

χ having the dimension of time, yields

$$p = 25{\cdot}8\chi\sigma_0 \tag{30}$$

For the integration, Cottrell made the assumption that a number of uniformly distributed fissure nuclei, whose distance is a, are available at the time $t = 0$.

The integration has to be conducted up to the elongation at rupture. On the basis of metallographic studies, Cottrell presumed that the fissure nuclei develop to spherical cavities. In this case, an extension of the test bar results from the formation of these cavities. This kind of formation of cavities occurs as a rule at very high temperatures, at which the resistance of crystals is low. On the other hand, no spherical cavities form at lower temperatures, particularly in austenitic steels, but the formation of fissures is confined to a very narrow zone within the grain boundary, so that an intercrystalline fissure develops along the grain boundaries. In this case, the part of the elongation at rupture that is attributable to formation of fissures is infinitely small. The elongation at rupture is determined exclusively by deformation of crystals. This is the type of fissure formation that has been observed most frequently in practice and it has led to the so-called equicohesive theory. Our reflections are therefore restricted to this case.

Here, the expression

$$\dot{\varepsilon} = \varphi(\sigma) = v_0(\sigma_0) \sinh [\sigma/\sigma_1(\sigma_0)] \tag{31}$$

for the creep rate can be introduced into eqn (29) and the value χ in eqn (30) calculated as follows:

$$\chi = \frac{2}{(v_0/\sigma_1)\sigma_0} \left\{ \text{arc coth} \left[\exp\left(-\frac{\sigma_0}{\sigma_1} \right) \right] - \text{arc tanh} \left[\exp\left(-\frac{\sigma_0}{\sigma_1} \exp \varepsilon_B \right) \right] \right\} \tag{32}$$

We have shown in the discussion on the relationship between stress and secondary creep rate how the values v_0 and σ_1 can be determined from the slope of the time-to-rupture curve. If these results are introduced into eqn (32), we obtain the slope of χ (Fig. 8) as a function of the reduction of area ψ for the different slopes of the tangent line to the time-to-rupture curve as parameter. We see from Fig. 8 that, for elongations up to approximately 25 per cent, χ differs by only about 10 per cent from the time to rupture, and we can therefore put

$$\chi = \alpha t_B \tag{33}$$

where $\alpha \approx 1$.

Hence, for the value p, we obtain

$$p = \frac{a^3 kT}{\Omega D \delta} = 25{\cdot}8 \alpha t_B \sigma_0 \approx 25{\cdot}8 t_B \sigma_0 \tag{34}$$

Comparing eqn (34) with eqn (23) as has been done by Cottrell for a constant stress, we see that the increase of stress with time, owing to necking and the extension of fissures, has very little influence on the

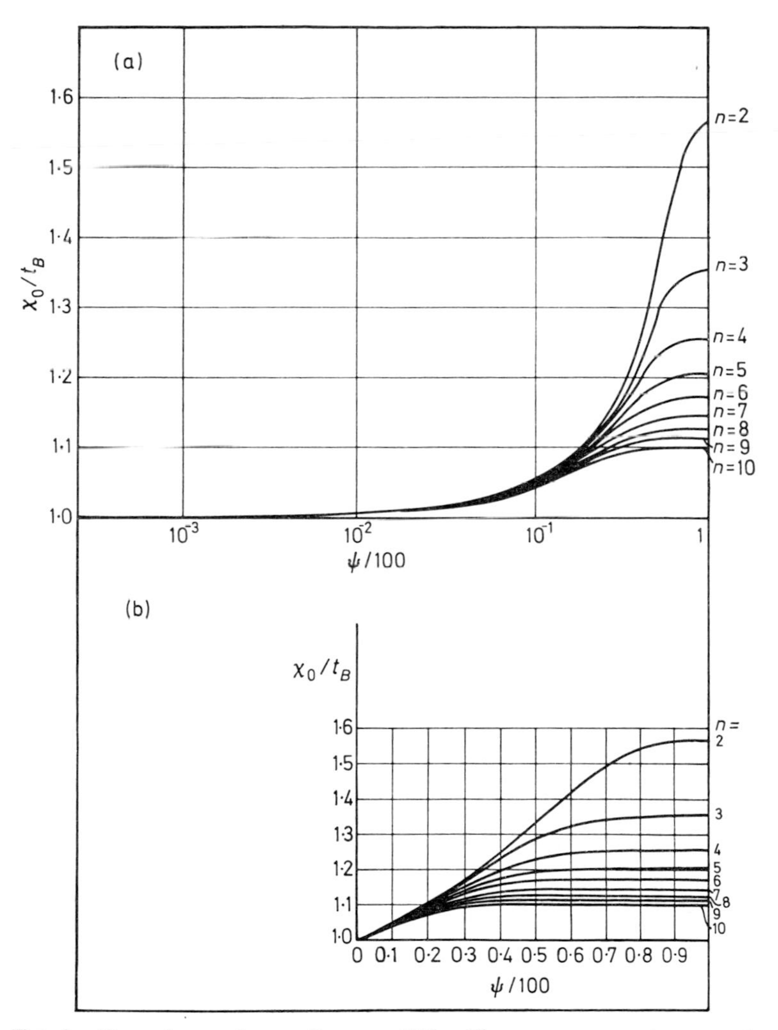

FIG. 8. Dependence of $\alpha = \chi / t_B$ on $\psi_B/100$ with n as parameter (see eqn 33).

time-to-rupture. The value of p, which depends on the distance between given fissure nuclei and the diffusion constants, can then be calculated from eqn (34). When the other values appearing in the expression for p are known, this equation yields the density of fissure nuclei at the time $t = 0$. This relation is valid only if the diffusion of vacancies is the only mechanism for the growth of fissures and no triple-point-fissures develop simultaneously. For the present, we shall discuss a material in which the growth of fissures takes place exclusively by diffusion processes, and we shall return to the growth of triple-point-fissures later.

Nucleation of fissures

1 *Mechanism*

For his calculations, Cottrell assumed that all fissure nuclei exist already at the time $t = 0$ and no others develop during the process of deformation. It is generally supposed that the nucleation of fissures is caused by certain

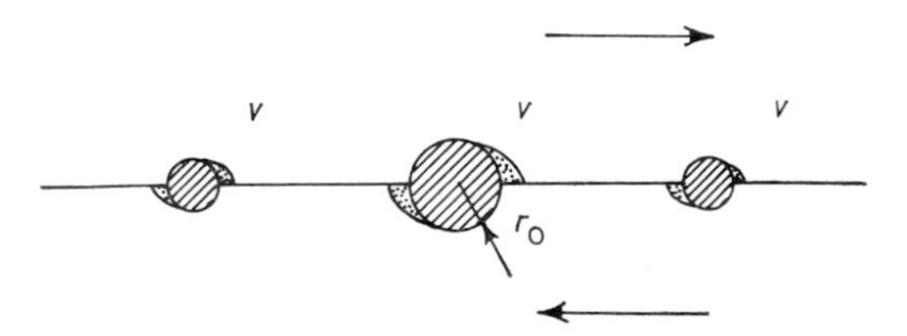

FIG. 9. Fissure nucleus when $v > v_k$.

obstacles situated in the grain boundaries as precipitations and by step formations as a result of sliding planes. These obstacles cause small holes to form by the sliding of grain boundaries on one another; holes, once they have reached a certain size, are nuclei for the formation of fissures. On the basis of a simple assumption we can now deduce some fundamental relations between the shape, size and number of precipitations and the mean creep rate with the number of fissure nuclei. Our derivations are based on the supposition (Fig. 9) that a certain number of precipitations, whose size varies about a statistical mean value, exist in the grain boundary. Then vacancies form by sliding of grain boundaries on one another. We presume that these vacancies appear as fissure nuclei once they have a critical volume, $v = v_k$.

2 *Analysis of the model*

If r_0 is the radius of a particle, according to the previous assumptions a new fissure nucleus develops when

$$\varepsilon_{kg} r_0 = v_k \tag{35}$$

where ε_{kg} = sliding of grain boundaries, or as

$$\varepsilon_{kg} = \dot{\varepsilon}_{kg} t \tag{36}$$

we have

$$\dot{\varepsilon}_{kg} t r_0 = v_k \tag{37}$$

The time for the development of a fissure nucleus is given by

$$t = v_k / \dot{\varepsilon}_{kg} r_0 \tag{38}$$

If, for the quantity of precipitations, we adopt the Gauss error distribution law

$$n(r_0) = \frac{n_0}{\rho(2\pi)^{\frac{1}{2}}} \exp\left[-\frac{1}{2}\frac{(r_0 - r_m)^2}{\rho^2}\right] \tag{39}$$

where

n = Number of particles per centimetre with radius r_0
n_0 = Total number of precipitated particles per centimetre
ρ = Deviation and
r_m = Mean value of r_0,

the number of fissure nuclei at the time t is equal to that of precipitations with a radius greater than

$$r_0(t) > v_k / \dot{\varepsilon}_{kg} t \tag{40}$$

By substitution into eqn (39), we obtain the expression

$$N(t) = \frac{n_0}{\rho} \int_{v_k/\dot{\varepsilon}_{kg} t}^{\infty} \frac{1}{(2\pi)^{\frac{1}{2}}} \exp\left[-\frac{1}{2}\frac{(r_0 - r_m)^2}{\rho^2}\right] \mathrm{d}r_0 \tag{41}$$

Putting

$$Z = \frac{1}{\sqrt{2}} \frac{(v_k/\dot{\varepsilon} t) - r_m}{\rho} = \frac{v_k - r_m \dot{\varepsilon}_{kg} t}{\sqrt{(2)}\rho \dot{\varepsilon}_{kg} t} = Z(t) \tag{42}$$

yields

$$N(t) = \frac{n_0}{2}\{1 - \mathrm{erf}\,[Z(t)]\} = \frac{n_0 \rho}{2} \mathrm{erfc}\,[Z(t)]$$

$$= \frac{n_0}{2} \mathrm{erfc}\left\{\frac{v_k}{\sqrt{(2)}\rho \dot{\varepsilon}_{kg} t} - \frac{r_m}{\rho\sqrt{2}}\right\} = \frac{n_0}{2} \mathrm{erfc}\left(\frac{A}{t} - B\right) \tag{43}$$

where

$$A = \frac{v_k}{\sqrt{(2)}\rho\dot{\varepsilon}_{kg}}$$

$$B = \frac{r_m}{\rho\sqrt{2}}$$

$$\text{erf}\,(x) = \frac{2}{\sqrt{\pi}}\int_{\infty}^{x} \mathrm{e}^{-\xi^2}\,\mathrm{d}\xi = \text{error function}$$

and

$$\text{erfc}\,(x) = 1 - \text{erf}\,(x)$$

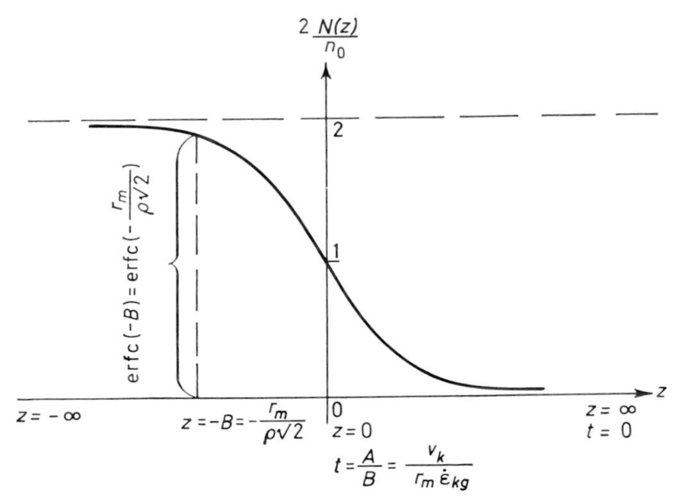

FIG. 10. erfc (z) = erfc $[(A/t) - B]$ against z.

The fundamental curve of erfc (z) is plotted in Fig. 10. We have for

$$\begin{aligned} t = 0: &\quad z = \infty: \quad \text{erfc}\,(z) = 0 \\ t = \frac{A}{B} = \frac{v_k}{r_m\dot{\varepsilon}_{kg}}: &\quad z = 0: \quad \text{erfc}\,(z) = 1 \\ t = \infty: &\quad z = -B = -\frac{r_m}{\rho\sqrt{2}}: \quad \text{erfc}\,(z) = \text{erfc}\left(-\frac{r_m}{\rho\sqrt{2}}\right) \end{aligned} \tag{44}$$

In principle, the curves for $N(t)/n_0$ have the shape plotted in Fig. 11.

These curves increase from $t = 0$ and become asymptotic to the value

$$\frac{N(t_\infty)}{n_0} = \text{erfc}\,(-B) = \text{erfc}\left(-\frac{r_m}{\rho\sqrt{2}}\right) \tag{45}$$

3 *Number of fissure nuclei for various kinds of precipitation*

a Cottrell's original assumptions. In his original formula, Cottrell assumed that all fissure nuclei exist at the time $t = 0$ and have the same size, *i.e.* a disappearingly small radius. In this case, $A = 0$ and $B = \infty$.

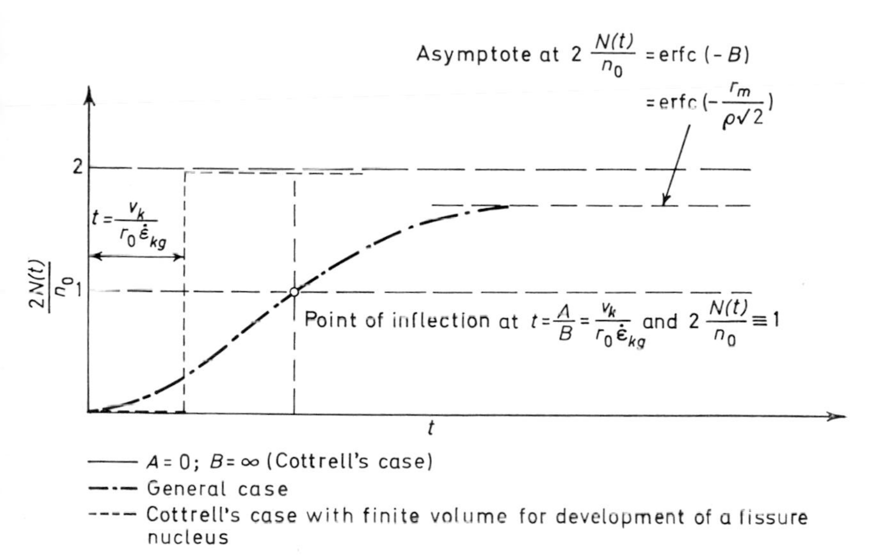

FIG. 11. $2[N(t)/n_0] = \text{erfc}\,[(A/t) - B]$ against t for different values of A and B.

The curve for $N(t)/n_0$, therefore, suddenly increases at the point $t = 0$ to the value 1, and for longer times takes a horizontal slope. In this event,

$$a = \sqrt[3]{(1/n_0)} \tag{46}$$

and, according to eqn (34),

$$p = \frac{a^3 kT}{\Omega D\delta} = \frac{1}{n_0}\frac{kT}{\Omega D\delta} = 25{\cdot}8\alpha\sigma_0 t_B = p_0 \tag{47}$$

where $\alpha \approx 1$. The value p is directly proportional to $1/n_0$, where n_0 is the number of fissure nuclei per centimetre of grain boundary.

b Modification of Cottrell's assumption. We can modify Cottrell's assumption by supposing that a number of particles with a finite radius

$r = r_0$ exists at the time $t = 0$, and that the critical volume of fissures is greater than 0. In this case, the fissure nuclei appear only after the time

$$t = v_k / r_0 \dot{\varepsilon}_{kg} \tag{48}$$

The slope of the curve for $N(t)/n_0$ is given in Fig. 11. A sudden transition from the value 0 to the value 1 takes place at the time $t = v_k / r_0 \dot{\varepsilon}_{kg}$. Integrating Cottrell's differential eqn (25) for this function of distribution of fissure nuclei gives

$$t_B - \frac{v_k}{r_m \dot{\varepsilon}_{kg}} = \frac{1}{n_0} \frac{kT}{\Omega D \delta 25{\cdot}8 \sigma_0} \tag{49}$$

$$25{\cdot}8 \sigma_0 t_B = \frac{1}{n_0} \frac{kT}{\Omega D \delta} + \frac{v_k 25{\cdot}8 \sigma_0}{r_m \dot{\varepsilon}_{kg}} = p_0 + \frac{v_k \text{ const}}{r_m (\dot{\varepsilon}_{kg})^{1-(1/n)}}$$

$$\approx p_0 + \frac{v_k \text{ const}}{r_m \dot{\varepsilon}_{kg}} \tag{50}$$

In the second term on the right hand side of eqn (50) we have $\sigma_0 / \dot{\varepsilon}_{kg}$ and $\dot{\varepsilon}_{kg}$ is a function of σ_0. To evaluate the term, we put $\dot{\varepsilon}_{kg} = A_1 \sigma_0{}^n$. For steels, $n_1 \sim 6/8$. Therefore, we obtain eqn (50).

If $25{\cdot}8 \sigma_0 t_B$ is plotted

$$p_0 = \frac{1}{n_0} \frac{kT}{\Omega D \delta}$$

The greater the value v_k / r_m, the sooner the curve approaches the asymptote.

If $25{\cdot}8 \sigma_0 t_B$ is plotted against $\dot{\varepsilon}_{kg}$ (*see* Fig. 12), we obtain curves that decrease from an initial value to an asymptotic one at

$$p_0 = \frac{1}{n_0} \frac{kT}{\Omega D \delta}$$

c General case of precipitations $N(t) = (n_0/2)$ *erfc* $[(A/t) - B]$. At this point we meet a difficulty in that Cottrell's original equation is no longer valid for the integration. New fissure nuclei develop constantly in the course of time and thus, after a certain period, the diameters of the fissures have different values. However, a certain simplification can be made for evaluating the influence of the precipitation on the product $\alpha \sigma_0 t_B 25{\cdot}8$. As the rate of growth of the fissures is inversely proportional to the radius, the newly developing fissure nuclei will grow much faster than those which exist already. Consequently, a certain equilibrium will be reached after some time, permitting Cottrell's differential equation to be used again. This means that in Fig. 11 the slowly increasing curves can be substituted approximately by curves with a sharp increase at the time $t \simeq v_k/r_m\dot{\varepsilon}_{kg} = A/B$. The case of a general distribution can thus be traced back approximately to that described in the preceding section. On

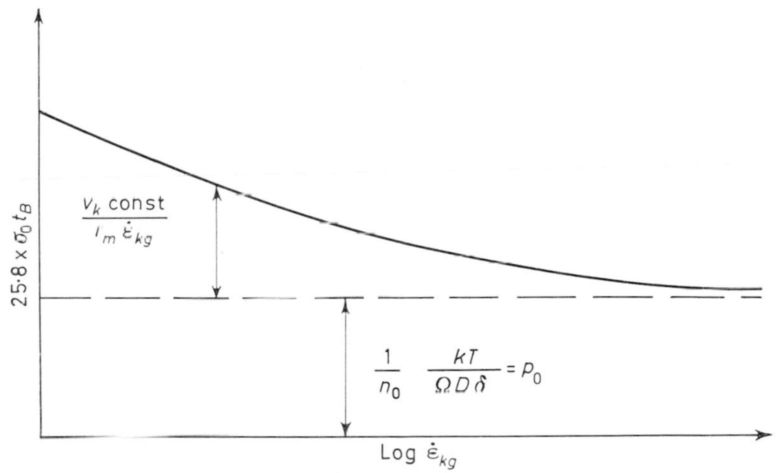

FIG. 12. $25{\cdot}8\sigma_0 t_B$ against $\dot{\varepsilon}_{kg}$ when all precipitates have the same size.

the basis of Fig. 13, we can then discuss the influence of the various parameters that bear on the distribution of precipitations on the time-to-rupture.

i The ratio r_m/ρ

As the ordinate of the asymptote is given by the value erfc $(-r_m/\rho\sqrt{2})$, this ordinate becomes smaller with increasing deviation in the Gauss error distribution law.

ii The value of grain boundary creep $\dot{\varepsilon}_{kg}$

It can be assumed that the sudden transition occurs at the time $t = v_k/r_0\dot{\varepsilon}_{kg}$. Thus, the greater $\dot{\varepsilon}_{kg}$, the shorter is the time for the sudden transition. Integrating Cottrell's differential equation yields

$$25{\cdot}8\sigma_0 t_B = \frac{2}{n_0 \,\text{erfc}\,[-(r_m/\rho\sqrt{2})]} \frac{kT}{\Omega D\delta} + \frac{v_k \text{ const}}{r_m\dot{\varepsilon}_{kg}} \tag{51}$$

(for $\alpha = 1$). When $25{\cdot}8\sigma_0 t_B$ is plotted against $\dot{\varepsilon}_{kg}$ (*see* Fig. 14), we obtain a

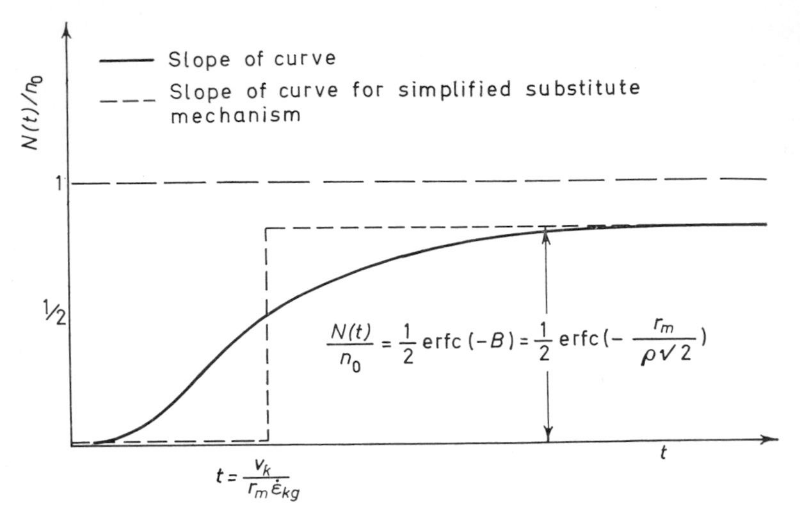

FIG. 13. Number of nuclei as a function of time, and approximation by a Cottrell-mechanism.

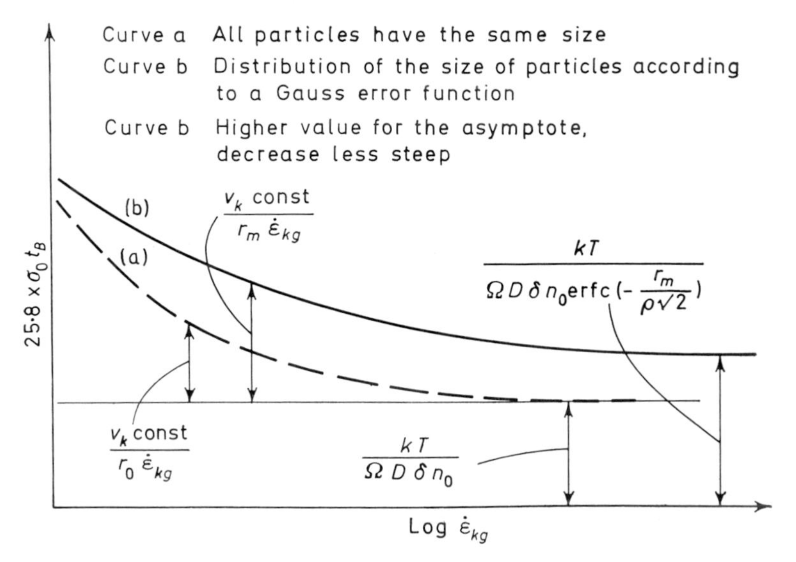

FIG. 14. $25{\cdot}8\sigma_0 t_B$ against $\dot{\varepsilon}_{kg}$ for a constant value and a Gaussian distribution of particle size.

curve showing a horizontal asymptote with the ordinate

$$p_0' = p_0 \frac{1}{\frac{1}{2}\,\mathrm{erfc}\,[-(r_m/\rho\sqrt{2})]} = \frac{kT}{\Omega D\delta n_0}\frac{1}{\frac{1}{2}\,\mathrm{erfc}\,[-(r_m/\rho\sqrt{2})]} \tag{52}$$

For $\dot{\varepsilon}_{kg}$ we have the following ordinate:

$$25{\cdot}8\sigma_0 t_B = p_0' + \frac{v_k\,\mathrm{const}}{r_m\dot{\varepsilon}_{kg}} \tag{53}$$

The influence of the various parameters for the distribution on the shape of the curve can be summarised by stating that the greater the ratio v_k/r_m, the steeper the slope of the curves.

4 *Development of new precipitations during creep*

In the previous discussions, it has been assumed that no new precipitations develop during the deformation process. In most cases, however, this is

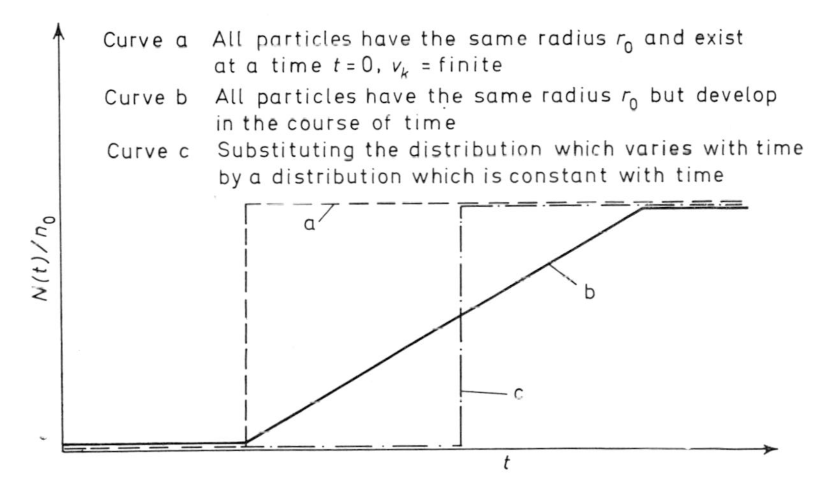

FIG. 15. Fissure nuclei against time at constant particle radius r_0.

not true. Observations show that crystalline transformations and precipitations of new carbides take place on a large scale. On the basis of considerations already discussed, we can now indicate what the curves for $25{\cdot}8\sigma_0 t_B$ against log $\dot{\varepsilon}_{kg}$ will look like. This can be demonstrated best by means of Cottrell's model, in which all precipitations show the same radius, and the critical volume v_k is finite. We have seen in paragraph 3*b* above that the curve $N(t)/n_0$ against time shows a break at the time $t = v_k/r_m\dot{\varepsilon}_{kg}$, plotted as curve (a) in Fig. 15. Supposing that this number of precipitations develops in dependence of t (curve b, Fig. 15), curve (b) can be substituted to a first approximation by curve (c). This has the same effect as an increase of the factor v_k/r_m. Nevertheless, as we see from Fig. 12, an increase of the factor v_k/r_m results in the values of $\sigma_0 t_B$ as a function of log $\dot{\varepsilon}$ decreasing to a larger extent.

5 *Dependence of temperature*

As results from relation

$$25{\cdot}8\sigma_0 t_B = \frac{kT2}{\Omega D n_0 \operatorname{erfc}[-(r_m/\rho\sqrt{2})]} + \frac{v_k \text{ const}}{r_m \dot{\varepsilon}_{kg}} \tag{54}$$

the value $25{\cdot}8\sigma_0 t_B$ depends on the temperature in the same way as the diffusion constant. When the number and the distribution of precipitations do not change with the modification of temperature, Fig. 16 gives the shape of curves for $25{\cdot}8\sigma_0 t_B$ against log $\dot{\varepsilon}_{kg}$ at various temperatures.

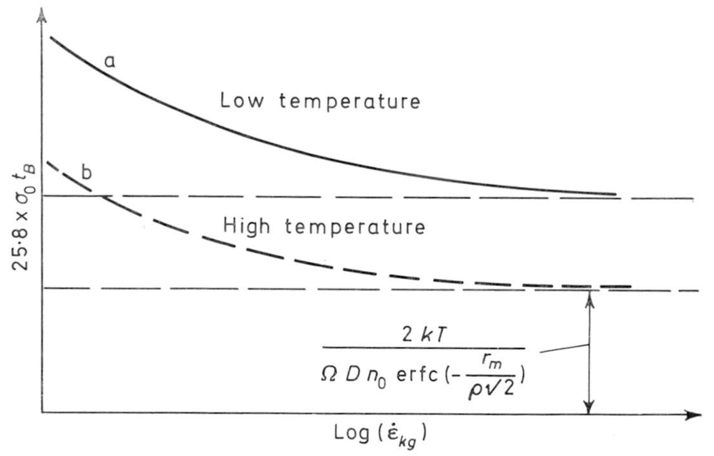

FIG. 16. Influence of temperature on the curves of $25{\cdot}8\sigma_0 t_B$ against $\dot{\varepsilon}_{kg}$ at growth of fissures by diffusion.

6 *Distribution of precipitations varying with temperature*

At the transition from a lower to a higher temperature, the number and size of precipitations are also often changed; for example, an agglomeration of precipitations takes place. In the event of such an agglomeration, the curves for log $25{\cdot}8\sigma_0 t_B$ take a shape in dependence of log $\dot{\varepsilon}_{kg}$ as has been plotted in Fig. 17. In consequence of the increase in temperature, the curve is displaced parallel downwards, whereas the decrease of the number of precipitations means a shifting upwards. The agglomeration of precipitations implies an increase of the radius r_m that causes a flattening, as can be seen from Fig. 17.

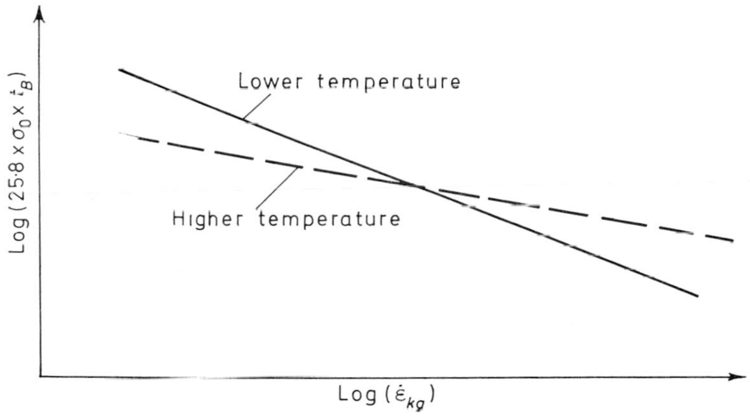

FIG. 17. Shape of curves $25{\cdot}8\sigma_0 t_B = f(\dot{\varepsilon}_{kg})$ when particles agglomerate at high temperatures.

FISSURING BY GROWTH OF TRIPLE-POINT FISSURES

As was shown by Williams [3], the time-to-rupture for this case is given by the relation

$$t_B \sim \frac{2\gamma}{\sigma_0} \frac{1}{L} \frac{1}{\dot{\varepsilon}_{kg}} \tag{55}$$

where γ = Surface stress and L = Mean grain diameter.
Thus,

$$25{\cdot}8\sigma_0 t_B = \frac{2\gamma}{L\dot{\varepsilon}_{kg}} = \frac{\text{const}}{\dot{\varepsilon}_{kg}} \tag{56}$$

If log $25{\cdot}8\sigma_0 t_B$ is plotted as a function of log $\dot{\varepsilon}_{kg}$, we obtain a straight line (*see* Fig. 18). As the constant in eqn (56) is independent of the diffusion, the same straight line is valid for all temperatures.

INTERACTION OF THE TWO FISSURING MECHANISMS

Both the mechanisms of growth of fissures, *i.e.* growth by diffusion of vacancies and formation of triple-point-fissures, act simultaneously in a time-to-rupture test. It has been observed, however, that in most cases each

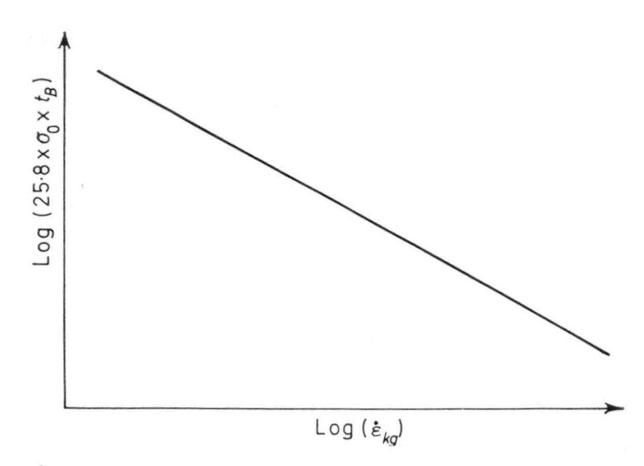

FIG. 18. Curve for $25{\cdot}8\sigma_0 t_B = f(\dot{\varepsilon}_{kg})$ when rupture takes place by growth of triple-point-fissures.

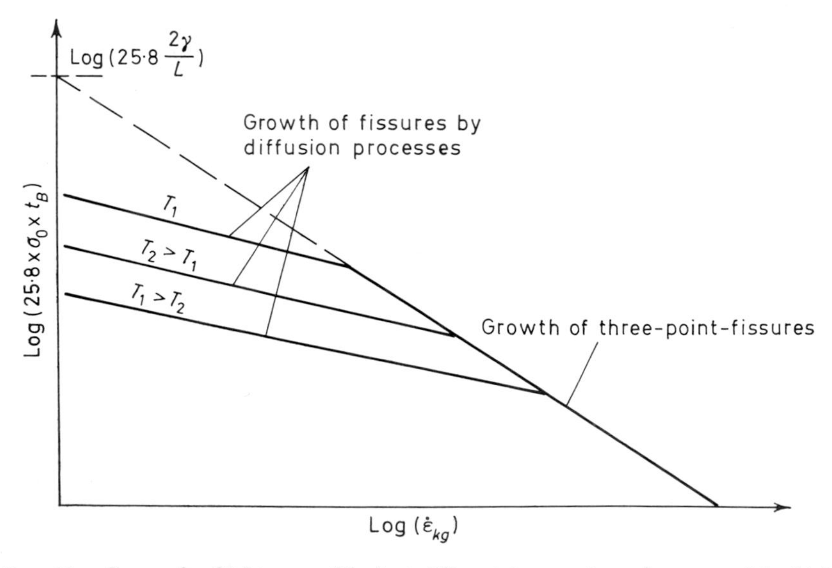

FIG. 19. Curves for $25{\cdot}8\sigma_0 t_B = f(\dot{\varepsilon}_{kg})$ at different temperatures for a material which ruptures at high stresses by growth of triple-point-fissures, and at low stresses by growth of fissures as a result of diffusion.

of the two kinds of fissuring is assigned to certain fields of stress, and that the transition from one kind of fissuring to another takes place at a specific stress. At high stresses and high creep rates, as a rule there is rupture formation by growth of triple-point-fissures, whereas at small stresses and small creep rates, growth of fissures takes place by diffusion processes.

Thus, it is possible to separate the two processes by plotting log $25{\cdot}8\sigma_0 t_B$ against log $\dot{\varepsilon}_{kg}$ for time-to-rupture tests at various temperatures. At small creep rates there is a dependence on temperature according to Fig. 16, whereas at high rates the curves for the different temperatures coincide. On the basis of these reflections, the slope of both curves has been plotted in Fig. 19. We shall see later that this slope can be verified by means of the results of time-to-rupture tests.

EVALUATION OF TIME-TO-RUPTURE TESTS ON A Cr–Ni–Ti STEEL

Description of tests

Time-to-rupture tests extending over very long periods have been made on a type X 10 Cr-Ni-Ti 189 steel at the Max-Planck-Institut fur Eisenforschung in Düsseldorf (Germany) by Krisch [4]. The results of these tests and the reduction of area observed are plotted in Fig. 20. The fact that Krisch indicated mathematical relations for the description of these curves is of great help in an evaluation of our reflections. The time-to-rupture lines can be expressed mathematically by

$$\log \sigma_0 = \alpha_1 + \gamma_1(\log t)^2 \qquad (57)$$

α_1 and γ_1 are dependent on temperature but not on stress; their values are given in Table 2.

TABLE 2

Temperature (°C)	α_1	γ_1
600	1·46	−0·024
650	1·45	−0·037
700	1·33	−0·042
750	1·22	−0·055
800	0·78	−0·029

Evaluation of tests

By substitution of the values for σ and d log t/d log σ, calculated according to eqn (57), into eqns (13), (14) and (32), we obtain σ_0/σ_1, v_0 and α, whence it is possible to calculate the secondary creep rate and the value of $25{\cdot}8\sigma_0 t_B$ according to eqn (21).

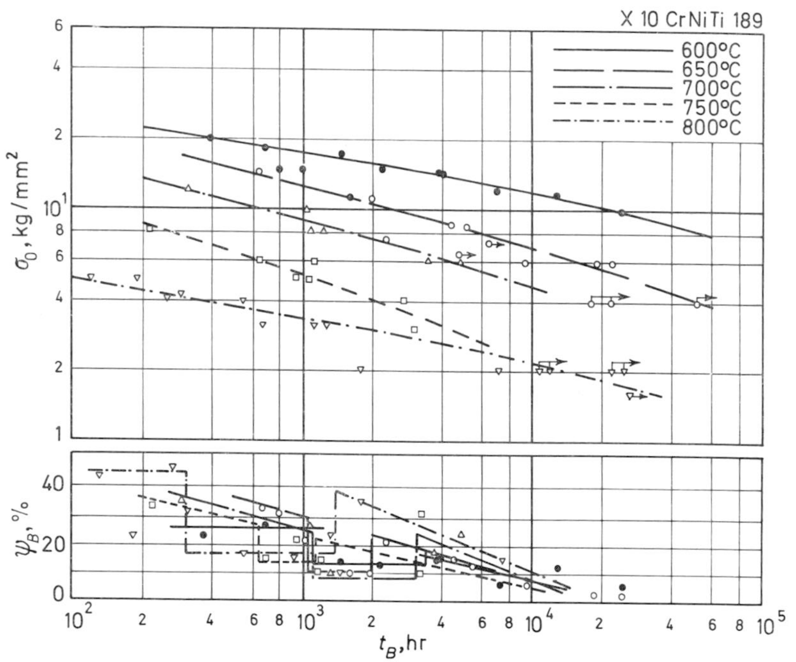

FIG. 20. Results of time-to-rupture tests on steel X 10 Cr–Ni–Ti 189 (Krisch [4]).

Figure 21 shows the stress σ as a function of the logarithm of the secondary creep rate. The value $25{\cdot}8\sigma_0 t_B$ as a function of the secondary creep rate has been plotted on log–log scales in Fig. 22.

Discussion of test results

The curves in Figs. 21 and 22 can be subdivided into three fields (plotted for a series of curves for 600°C in Fig. 21), *i.e.* field I (high stress, high slope), field III (low stress, smaller slope) and, between field I and field III lies a field II with a slope that is approximately equal to that of field I. Moreover, the curve of stress as a function of the logarithm of the secondary creep rate in field II is shifted upwards. With rising temperature, the slope decreases in all three fields.

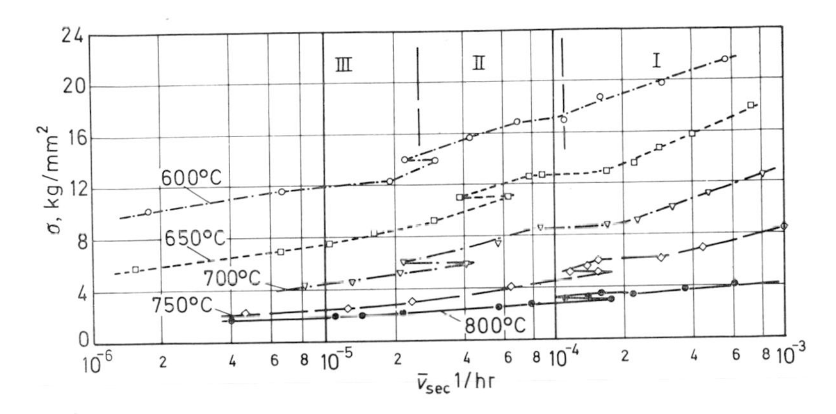

FIG. 21. Mean secondary creep rate against stress for steel X 10 Cr–Ni–Ti 189 calculated from Fig. 20.

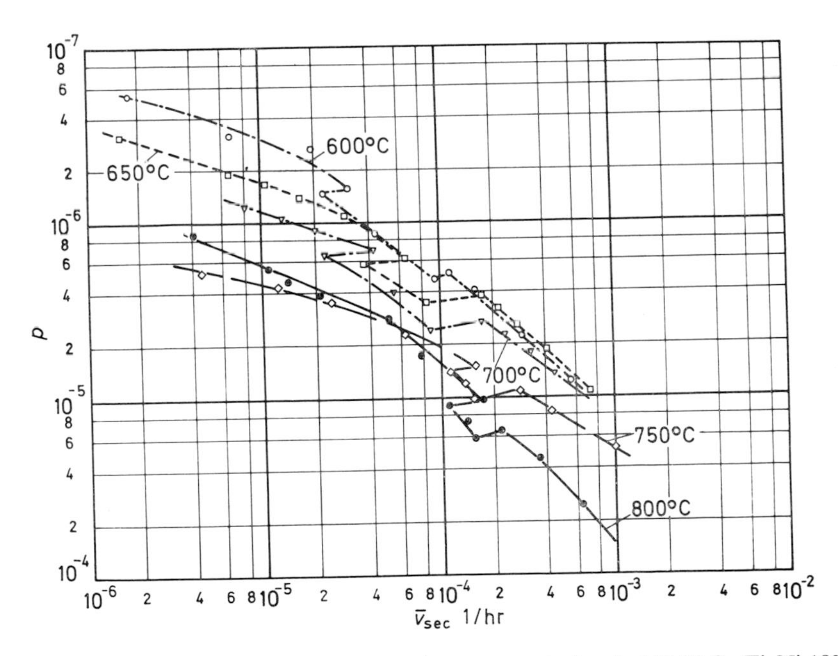

FIG. 22. Parameter p against mean secondary creep rate for steel X 10 Cr–Ti–Ni 189 calculated from Fig. 20.

The curves for $25{\cdot}8\sigma_0 t_B$ as a function of log v_{sec} (*see* Fig. 22) can also be divided into those same three fields.

Comparing Fig. 22 with Fig. 19 and disregarding the translation of curves in field II, we can easily explain the slope of curves for 600°, 650°, 700° and 750°C in Fig. 22 by the slope of curves in Fig. 19. For this reason, the conclusion can be drawn that in field I rupture takes place by formation of triple-point-fissures. The curve for 750°C and that for the formation of triple-point-fissures intersect at a creep rate which is greater than 10^{-3}/hr.

As regards the upward translation of curves in Fig. 21, in case the slope is supposed to be equal to that of field I, a decrease of creep rate has to take place in field II, which can be explained by the fact that the creep process takes place mainly in the grain boundaries. This may be defined by an impoverishment of precipitated particles in the zone adjacent to the grain boundaries. The translation of curves in field II, Fig. 22, can be explained by the fact that it is not the mean value of creep rate but the local creep rate in the grain boundaries which determines whether fissure nuclei are formed. When the grain boundaries are involved to a far larger extent in the creep process, the local creep rate is higher than the mean value. It can therefore be assumed that the curves in Fig. 22 would not show any translation in field II if the local values for the secondary creep rate had been plotted on the abscissa.

EMBRITTLEMENT PHENOMENA

The model of the rupture process during a time-to-rupture test, as described in the preceding paragraphs, also gives the basis for a new way of considering embrittlement phenomena which appear at high temperatures.

Influence of a notch on the state of stress

The state of stress that develops during the creep process on a notched specimen is extremely difficult to calculate. These calculations can be made only for an elastic body. Furthermore, the extent of the stress peak in a purely plastic body for an infinitely sharp notch can be indicated according to Hencky's theory. In this case, the stress peak amounts to the triple value of the nominal stress. A further complication in evaluating the state of stress in a notched specimen, under creep conditions, lies in the geometrical shape of the notch changing in the course of time. The radius of curvature at the notch root markedly increases during a long time-to-rupture test.

The distribution of stress in a notched specimen, consisting of a purely elastic material, has been plotted in Fig. 23. The axial stress shows a strong stress peak in the notch root. As calculations show, the magnitude of this stress peak depends to a large extent on the radius of curvature in the notch root. In the course of time, the radius of curvature increases to a multiple of the initial radius, and thus results in a large reduction of the stress peak. A further reason for the reduction of the stress peak, existing in an elastic body, is that the increase in creep rate with increasing stress is more than linear.

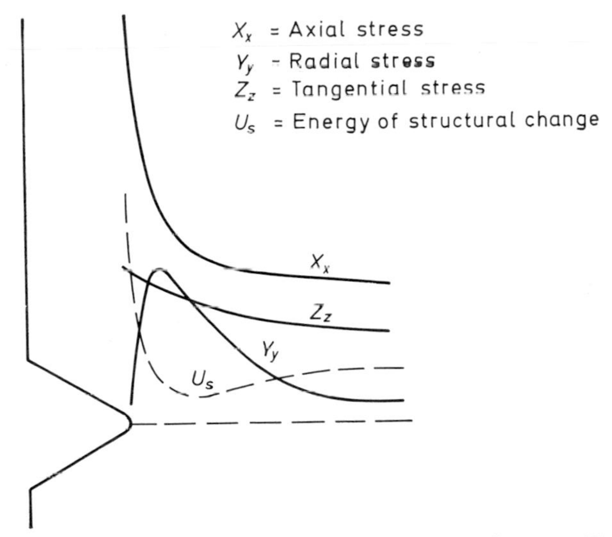

FIG. 23. Stress distribution in a notched specimen under tensile stress (Gensamer).

With respect to the stress gradient in the notched specimen in the course of time, the following considerations apply.

The distribution of stress at the beginning of the time-to-rupture test corresponds approximately to that of the elastic body. The stress peak, however, is very quickly reduced by local flow in the notch root and the consequent increase of the radius of curvature. The more pronounced the stress/creep rate curve, the quicker this reduction takes place. The reduction will thus occur much more rapidly for curve (a) in Fig. 2 than for curve (b). At this concentrated flow, additional fissure nuclei form in the notch root and can lead to premature rupture.

Apart from the formation of a stress peak, there are still radial and tangential components of the state of stress which cause an inhibition of flow. In consequence of the lower flow rate, a decrease in the formation of fissures takes place in comparison with the unnotched specimen.

Influence of a shaped notch on the time-to-rupture strength

On the basis of the above considerations, the influence of a shaped notch on the long time mechanical properties at high temperatures can be described in a general way. The application of a notch causes two opposite tendencies in the formation of fissures (number of fissure nuclei $\sim 1/25{\cdot}8\sigma_0 t_B$):

1 Increase of the number of nuclei as a result of intensified flow in the notch root.

2 Decrease of the secondary creep rate in consequence of the inhibition of flow, and thus also reduced formation of fissure nuclei.

The interaction of these two processes is demonstrated by means of Fig. 24, in which is shown the influence of a shaped notch on a curve of

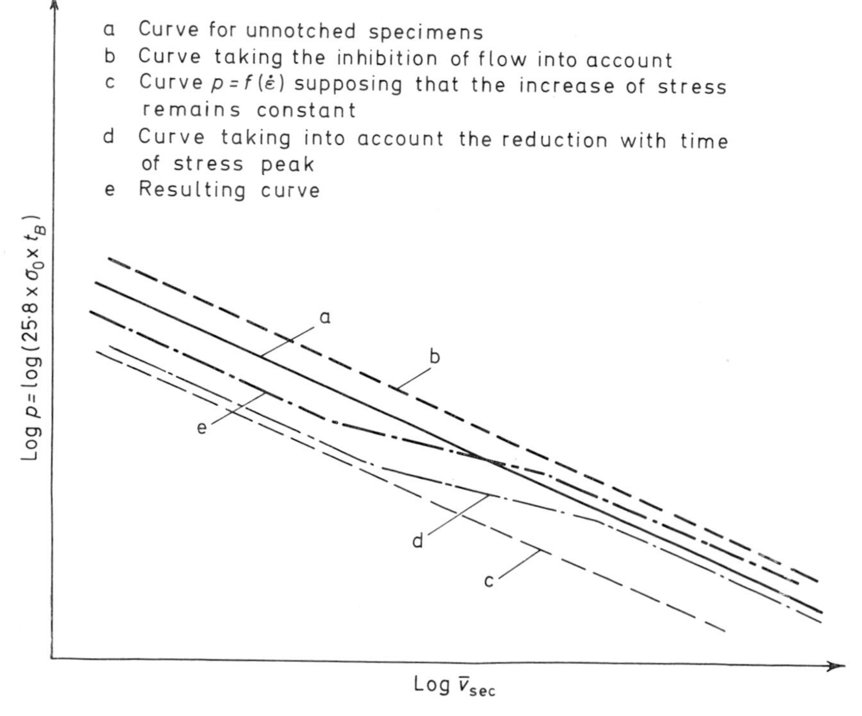

FIG. 24. Influence of shaped notch.

log p/log v_{sec} for the notched specimen curve, which is assumed to be a straight line in double logarithmic scale.

Inhibition of flow

As a result of the inhibition of flow, the mean value of the secondary creep rate of the notched specimen decreases compared with that of the unnotched one. If the value $p(=25{\cdot}8\sigma_0 t_B)$ is plotted against the mean creep rate of the unnotched test bar, the influence of the inhibition of flow results in a displacement of the curve for log p to the right (curve b in Fig. 24). The amount of displacement to the right depends on the degree of inhibition of flow, *i.e.* mainly on the depth of the notch. The inhibition of flow attains its maximum at a mean depth of the notch.

Stress peak

The stress peak causes a local increase of the creep rate which, however, diminishes in the course of time. With reference to the mean creep rate of unnotched specimens, this results in a displacement of the curve of log p/log v_{sec} to the left. Disregarding the reduction with time of the stress

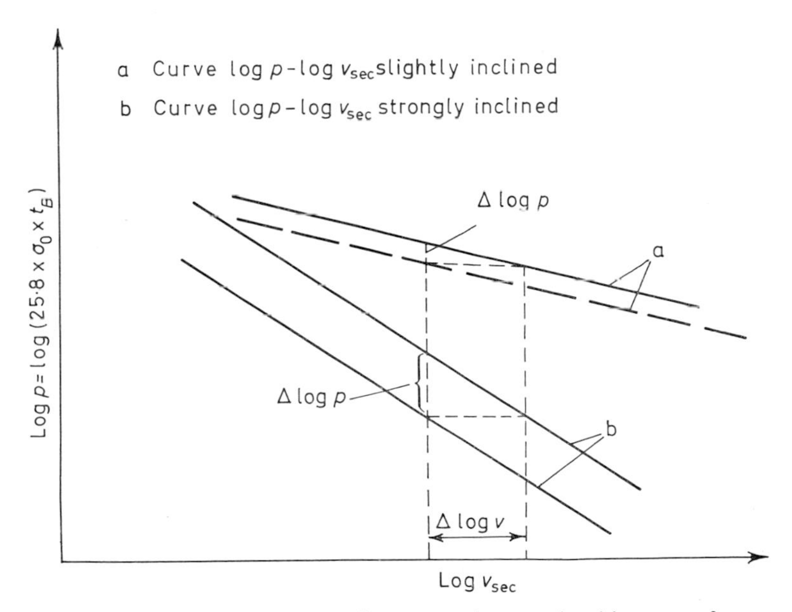

Fig. 25. Displacement of curve log p/log v_{sec} at the same local increase of creep rate above the mean value.

peak, curve (a) in Fig. 24 would be shifted to curve (c). The amount of this displacement to the right depends on the following factors:

1 Amount of local increase of creep rate, given by the magnitude of the stress peak and the shape of the curve $v_{sec} = f(\sigma)$.

2 Slope of the curve of log p/log v_{sec}, *i.e.* of the number, geometrical form and distribution of inclusions in the grain boundaries (Fig. 25). Time-to-rupture tests on notched and unnotched specimens very often show a strong increase of embrittlement for steels with a large number of fissure nuclei.

Considering that the stress peak is reduced in the course of time, curve (c) changes as follows: as mean values with time for the creep rate are plotted in Fig. 24, it can be assumed that the amount of displacement of curve (a) to the left has to be multiplied by the ratio of the time during which the stress peak is effective, to the time to rupture. However, because

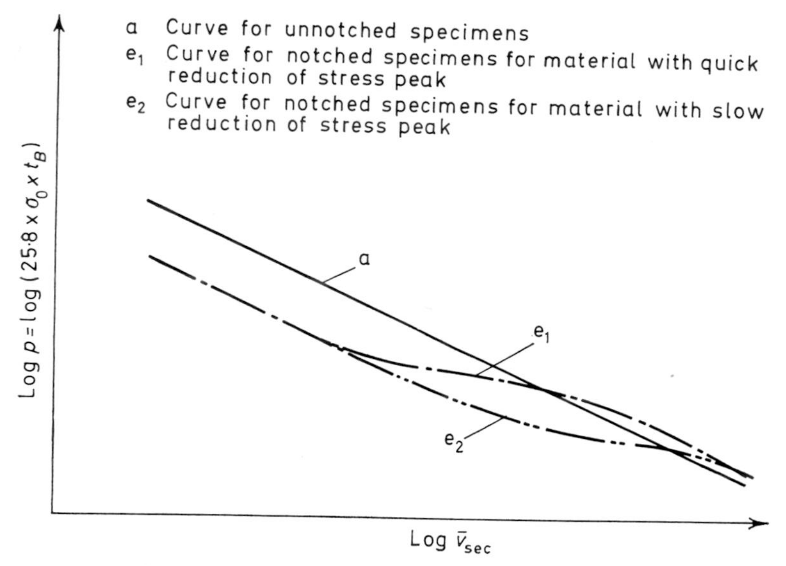

FIG. 26. Influence of zones low in alloying elements.

of the marked increase of the creep rate with the stress, this ratio is significantly smaller at high values for the initial stress than at smaller creep rates. The curve for log p as a function of log v_{sec}, therefore, has the shape of curve (d). The resulting curve for the number of nuclei of the notched specimen is obtained by superimposing curves (b) and (d) (curve e in Fig. 24).

Zones low in alloying elements near grain boundaries

The modification of the curve of log p/log v_{sec} owing to the existence of zones low in alloying elements is represented in Fig. 26. The curve for e_1 corresponds to a notched test bar of a material without formation of grain boundary zones, whereas curve e_2 represents the behaviour of a material showing the formation of such zones. Creep rupture of strength has been plotted against time-to-rupture in Fig. 27 for an unnotched

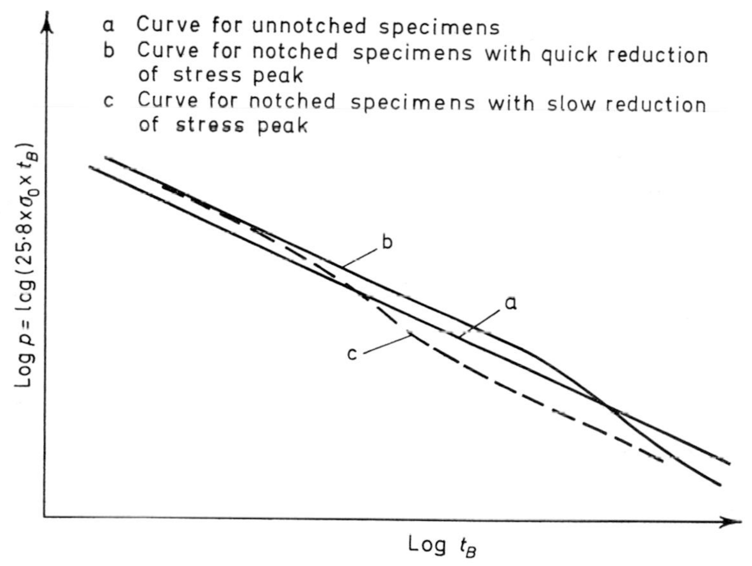

FIG. 27. Creep rupture strength against time-to-rupture.

specimen (curve a), for a notched bar with embrittlement but without formation of grain boundary zones (curve b), and for a notched bar with embrittlement and simultaneous formation of grain boundary zones (curve c). Moreover, the formation of grain boundary zones that are low in alloying elements increases the number of fissure nuclei in most cases with precipitation of carbides and intermetallic compounds in the grain boundaries.

CONCLUSIONS

On the basis of the above analysis, we can conclude that the following factors considerably influence embrittlement:

1 Number, size and shape of precipitations in the grain boundaries.
2 Formation of zones which are low in alloying elements in the neighbourhood of grain boundaries, the gradient in the distribution of the alloying elements being an all important factor.

The influence of these factors in the embrittlement of certain Inconel alloys, and of steam turbine wheels at Tanners Creek power station in USA, is discussed in Appendices 1 and 2, respectively.

REFERENCES

1. ANDRADE, E. N. DE C. and CHALMERS, B. (1938). 'The resistivity of polycrystalline wires in relation to plastic deformation and the mechanism of plastic flow.' *Proc. Roy. Soc. A*, **138,** 348–374.
2. COTTRELL, A. H. (1961). 'Intercrystalline creep fractures.' *Structural Processes in Creep,* pp. 1–18. London: Iron and Steel Institute.
3. WILLIAMS, J. A. (1967). 'Triple-point crack growth in an age-hardened aluminium 20 per cent zinc alloy.' *Acta Metall.*, **15,** 1559–62.
4. KRISCH, A. (1964). 'Creep behaviour, structural changes and cracks observed in creep rupture tests lasting several years' (in German). Nr. 1355 *Forschungs. Landes. Westf.*, 1964. (English translation in *Jnt. Int. Conf. Creep*, New York and London, **1**(1), 81–90.)
5. RAYMOND, E. L. (1967). 'Effect of grain boundary denudation of gamma prime on notch-rupture ductility of Inconel nickel-chromium alloys X-750 and 718.' *Trans. Metall. Soc. AIME*, **239,** 1415–22.
6. RANKIN, A. W. and SEGUIN, B. R. (1955). 'Report of the investigation of the turbine wheel fracture at Tanners Creek.' *ASME Paper No.* 55-A-210. American Society of Mechanical Engineers, New York.
7. THUM, A. and RICHARD, K. (1949). 'Damage line due to creep load in steels' (in German). *Arch. Eisenhüt Wes.*, **20**(7/8), 229–42.
8. WILD, M. (1963). 'Analysis of the creep behaviour of heat-resistant ferritic steels at 450–600°C' (in German). *Arch. Eisenhüt Wes.*, **34,** 934–950.
9. SACHS, G. and BROWN, W. F., JR. (1952). 'Symposium on strength and ductility of metals at elevated temperatures.' *Spec. Tech. Pub.* No. 128, pp. 6–20, American Society for Testing and Materials.
10. GRANT, N. J. and BUCKLIN, A. G. (1950). 'On the extrapolation of short-time stress-rupture data (for Cobalt chromium–nickel alloys).' *Trans. Amer. Soc. Metals*, **42,** 720–761.
11. SIEGFRIED, W. (1963). 'Application of thermodynamics to the evaluation of creep data of steels' (in German). *Arch. Eisenhüt Wes.*, **34,** 713–726.
12. UNDERWOOD, E. C. (1959/60). 'Strength, stability and the equicohesive point.' *J. Inst. Metals*, **88,** 266–271.

Appendix 1

EMBRITTLEMENT OF CERTAIN INCONEL ALLOYS

The influence of heat treatment on the embrittlement of Inconel alloy X-750 and Inconel alloy 718 can be explained by means of these considerations. We refer to a publication by Raymond [5], who investigated the influence of heat treatment on the form of carbides in the grain boundaries and the zone low in alloying elements, as well as on the time-to-rupture strength of notched specimens. The composition of the alloys is given in Table 3.

TABLE A3

Alloy	C	Mn	Fe	S	Si	Cu	Ni	Cr	Al	Ti	Cb + Ta	Mo
X-750	0·03	0·56	6·51	0·007	0·27	0·06	73·19	15·24	0·76	2·50	0·85	—
718	0·05	0·27	16·80	0·007	0·39	0·33	54·24	18·38	0·36	0·81	5·11	3·23

Investigations on the alloy X-750 showed that a triple heat treatment gives the most favourable mechanical properties. This heat treatment consists of the following processes:

1 Solution heat treatment.
2 Intermediate tempering.
3 Final tempering.

Tests with the electron microscope indicated that, by means of the intermediate tempering, the carbides $M_{23}C_6$ precipitated in the grain boundaries have the form of cubes. These precipitations occur in such a way that the carbon stabilises without a steep Cr-gradient appearing in the neighbourhood of the grain boundaries. When the intermediate tempering is left out, a considerable change in the morphology of $M_{23}C_6$ precipitations and a zone that is low in γ-precipitations near the grain boundaries are observed.

Similar observations have been made on the Inconel alloy 718. It appeared that the time-to-rupture of notched specimens could be increased to a large extent once the temperatures of solution and tempering were chosen so as to avoid both a favourable form of Nb-carbides precipitated in the grain boundaries and a zone that is low in precipitations of γ-phases.

Appendix 2

EMBRITTLEMENT PHENOMENA OF THE TURBINE WHEEL FRACTURE AT TANNERS CREEK

GENERAL

The concept that the stress concentration sensitivity in the time-to-rupture test has to be regarded essentially as the result of two oppositely acting tendencies has also been confirmed by investigations made in connection with the turbine wheel fracture at Tanners Creek reported by Rankin and Seguin [6]. The two oppositely acting tendencies are the formation of fissure nuclei in consequence of the local deformation on the one hand, and the reduction of stress peak by modification of the geometrical form on the other hand.

INVESTIGATIONS ON SPECIMENS FROM THE TURBINE OF TANNERS CREEK

According to the investigations made by Rankin and Seguin, the cause of rupture was an intercrystalline flaw starting from the bolt with which

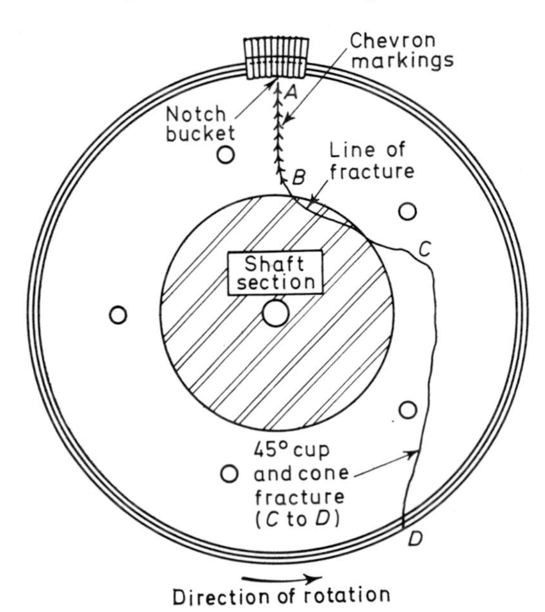

FIG. A28. First stage wheel showing line of fracture (Rankin and Seguin [6]).

the moving blades were fixed on the turbine disc (*see* Fig. A28). The kind of blade fastening can be seen, on an enlarged scale, in Fig. A29. Each blade had been fixed on the disc by means of two cylindrical bolts. To answer the question how the embrittlement phenomena observed in the neighbourhood of the bolts can be explained, time-to-rupture tests have

FIG. A29. Isometric sketch of section through centre of notch opening in turbine wheel rim (Rankin and Seguin [6]).

been made on notched and unnotched specimens of the corresponding steel. Figure A30 represents the results of these investigations. It shows that, even for very sharp notches, no strong decrease in the time-to-rupture is observed, so that the embrittlement during operation cannot be explained at all by means of these test series. Even for very sharp notches, it was impossible to detect embrittlement phenomena. The results of time-to-rupture tests carried out in this connection are plotted in Fig. A31.

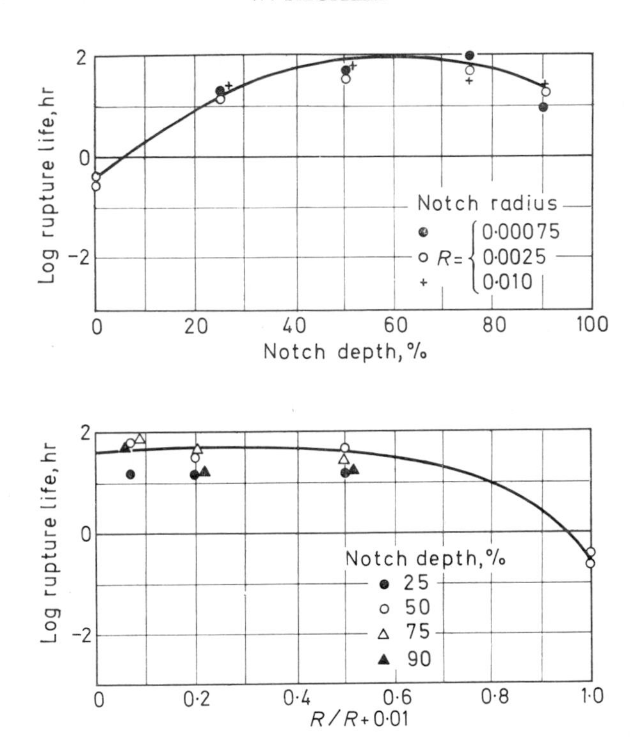

FIG. A30. Effect of notch radius and notch depth on rupture life of Tanners Creek turbine rotor (Rankin and Seguin [6]).

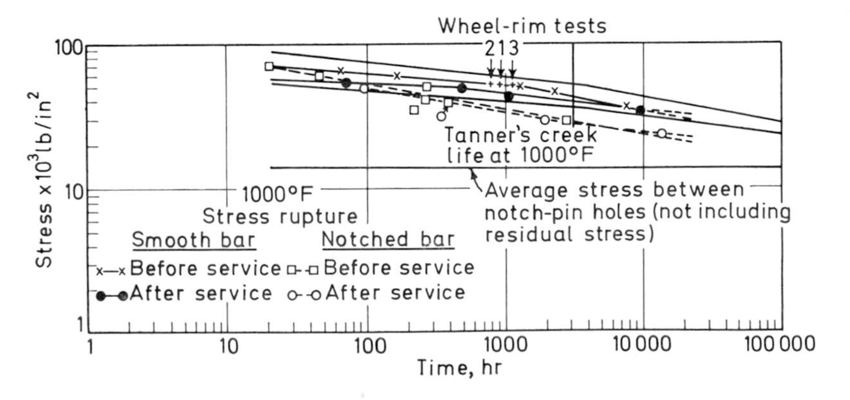

FIG. A31. Smooth bar and notched bar rupture strength of Tanners Creek turbine rotor in the before-service and after-service conditions (Rankin and Seguin [6]).

To find an explanation for this unexpected result, a model has been established which represents the conditions in the turbine much better than tests with notched specimens. An installation has been set up whereby a transverse load could be applied by means of two bolts to a cylindrical bar which is stressed longitudinally. Attention was paid to keeping to the same geometrical conditions as prevailed for the turbine at Tanners Creek. By means of photoelastic tests, the magnitude of the stress peak that is attained in this geometrical system has also been determined.

With this test installation it was possible to simulate the brittle flaws occurring in the turbine. If the time until appearance of these flaws had been plotted in Fig. A31 as a function of the stress peak determined by means of photoelastic tests, the corresponding points (points 1, 2, 3) would fall into the field of scatter of the unnotched specimens. Hence it can be seen that the local stress peak, as has been determined by means of photoelastic tests, has been effective during the whole time of operation. Only in this way is it possible to explain why the embrittlement phenomena of the turbine material appeared at very low nominal stresses and could not be observed on tests with notched specimens.

Rankin and Seguin did not explain why the embrittlement in the wheel-rim test specimen could not be determined by means of time-to-rupture tests on notched and unnotched specimens. On the basis of our conceptions, however, this fact can be explained quite easily.

DIFFERENCE BETWEEN TEST RESULTS ON THE WHEEL-RIM TEST SPECIMEN AT TANNERS CREEK AND ON NOTCHED SPECIMENS

The principal difference between the two kinds of test is that a change of geometrical shape as a result of creep is impossible for the wheel-rim test specimen because of the existing bolts, whereas a very quick change of the radius of curvature in the notch root seems probable for time-to-rupture tests with sharply notched specimens. The conditions for the formation of intercrystalline fissures represented in Figs. 24 and 27 are therefore modified according to Figs. A32 and A33. The value $25{\cdot}8\sigma_0 t_B$ has been plotted on log–log scales in Fig. A32 as a function of the secondary creep rate, for an unnotched bar, a sharply notched one, and for tests with the wheel-rim specimen. As the stress peak is not reduced during the test, the curve e_2 in Fig. 26 does not bend upward, but the corresponding curve in Fig. A32 lies constantly below the curve for the unnotched specimens. If the

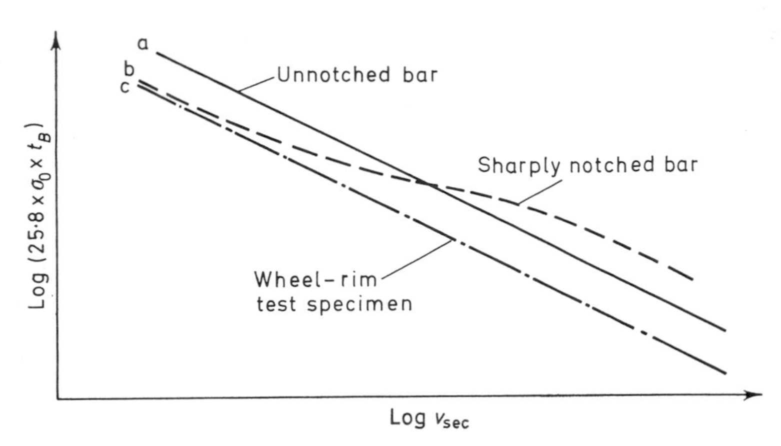

FIG. A32. $25{\cdot}8\sigma_0 t_B = f(v_{sec})$ for different conditions (Rankin and Seguin [6]).

nominal stress is plotted on log–log scales as a function of the time-to-rupture, as can be seen in Fig. A33, the stress for rupture is smaller for the wheel-rim test specimens, even at relatively short times-to-rupture, than for the corresponding test with an unnotched bar. In contrast to Fig. A31, Fig. A33 shows the nominal stress, because this value has to be plotted for time-to-rupture tests with notched specimens since the effective stress changes during the test.

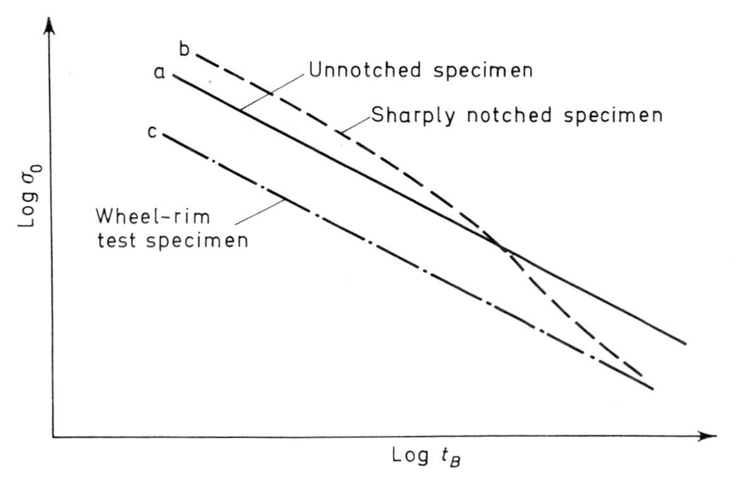

FIG. A33. Log σ_0 as a function of log t_B for different conditions.

Chapter 12

THE ROLE OF CREEP IN HIGH TEMPERATURE LOW CYCLE FATIGUE

S. S. MANSON, G. R. HALFORD AND D. A. SPERA

SUMMARY

The role of creep damage in governing elevated temperature, strain cycling fatigue lives is investigated in this chapter. Experimental and analytical results are presented for two high temperature alloys, Type 316 *stainless steel and the cobalt-base alloy L*-605, *tested in axial strain cycling over a range of frequencies, strain ranges and temperatures. Observed cyclic lives are compared with lives computed on the basis of a linear creep–fatigue damage rule.*

The pure fatigue life for a given cyclic strain range is assumed to be given by the method of universal slopes at the temperature of interest. Fatigue damage is then computed as the ratio of the number of applied cycles to the pure fatigue life.

The pure creep rupture resistance applicable to cyclic stress conditions is evaluated using a cyclic creep rupture test wherein the direction of the rupture stress is reversed each time the creep strain reaches a preset tensile or compressive strain limit. A plot of the cyclic rupture stress against the elapsed rupture time under only the tensile portion of the loading (*corrected for any fatigue damage that may have occurred as a result of cycling*) *serves as the creep rupture curve used in computing creep damage during the ensuing strain cycling fatigue tests. To calculate creep damage in these latter tests, the complete stress history is measured from cycle to cycle and throughout individual cycles as the tests progress. In evaluating creep damage during the axial strain cycling tests, the compressive stresses are assumed to be equally damaging as tensile stresses.*

Creep damage is taken as the ratio of time under a given stress to the time to rupture under the same stress. Since creep damage is based upon a stress–time–failure criterion, an accurate analysis can be made only if the complete stress history is determined accurately—a difficult task to perform in most practical situations outside the laboratory.

Summing the creep and fatigue damage and equating the sum to unity establishes the criterion for failure in accordance with the linear creep–fatigue damage rule. Cyclic lives computed according to this rule are then compared with the experimentally observed lives. Calculated lives based on previously proposed methods of life estimation developed at the NASA laboratory are also presented for comparison.

Although there is generally good agreement between the experimental data and the present analysis, further verification is required before the analysis can be applied to other situations with confidence.

NOTATION

E	Elastic modulus.
D	Ductility $= \ln\left(\dfrac{1}{1 - RA}\right)$.
RA	Reduction of area.
N_f	Number of cycles to pure fatigue failure.
N	A number of cycles.
t	Time.
t_r	Determined from Fig. 2.
$\Delta\varepsilon_t$	Total strain range.
σ_u	Ultimate tensile strength.
φ_{fatigue}	Pure fatigue damage.
φ_{creep}	Pure creep damage.

INTRODUCTION

Many important technological applications involve fatigue at high temperatures, and it is important to assess, in advance of service, the life of machine components subjected to cyclic loading. While reasonably valid procedures have been devised for obtaining estimates of fatigue behaviour at low and intermediate temperatures (for example, the method of universal slopes described in refs. 1 and 2), these procedures are not directly applicable at high temperatures where creep processes begin to interact with fatigue. The significance of creep in influencing fatigue resistance at high temperature is therefore currently receiving considerable attention with the result that our understanding of the details of the process is in a state of continuous advancement. This chapter, which must be viewed as an interim report on the subject, directs attention to the role of creep in low cycle fatigue. It is our intention to investigate the significance of the role that creep can play in governing high temperature cyclic lives. This is accomplished by conducting strain cycling tests on two high temperature alloys and making concurrent measurements of stress, temperature and strain at various frequencies. The results are then analysed in terms of the damage imposed by the creep and fatigue components.

While one purpose of the report is to demonstrate the importance of creep in low cycle fatigue, another is to discuss a promising type of analysis which incorporates basic data from a cyclic creep rupture test.

BACKGROUND

Several approaches have recently been examined at the NASA laboratory to determine how the mechanisms of creep and fatigue, and their interaction processes, may be practically treated. In the first [3], we regarded the influence of the creep mechanism as one of initiating an intercrystalline crack, thereby promoting early fatigue failure by essentially bypassing crack initiation. This assumption brought the computations of fatigue life based on the procedures developed for the sub-creep range into close coincidence with the experimental observations for a number of materials tested in the creep range. However, no attempt was made to account for certain factors, for example the frequency dependence of cyclic life, which has been observed to be of practical importance in some materials.

As a second approach we examined the merit of using a linear summation of creep and fatigue damage [4]: the creep damage being the ratio of the time near peak stress to the conventional rupture time under the same nominal stress and temperature, and the fatigue damage being the ratio of applied cycles to cycles required to cause failure in the absence of a creep effect. The universal slopes method was used to estimate the potential fatigue life in the absence of creep. The summation of the two components was taken to be unity for failure. However, a number of simplifications were still required to render the method tractable. Since it was applied to the problem of strain cycling fatigue, only the strain range was presumed known, whereas the stress required to produce the strain was not known. Nor was the variation of stress throughout each strain cycle known. Simple assumptions were again made, using the universal slopes method, to estimate the stress amplitude of each cycle. The effectiveness of compressive stresses in contributing to creep damage was not directly resolved, although indirectly their effect was allowed for by a single empirical constant that entered into the calculations. While still very approximate, acceptable estimates of high temperature, low cycle fatigue were made for a large number of materials and test conditions. The special merit of the method lay in the fact that no stress measurements were required, and that the estimates of cyclic lives could be obtained for any cyclic strain range using conventional tensile and creep rupture properties.

A closer examination of the role of creep in both thermal and isothermal strain cycling fatigue was made in refs. 5, 6 and 7. Here, again, time ratios and cycle ratios served as the measure of creep and fatigue damage, respectively. The fatigue component was treated in the same manner as in ref. 4, using the method of universal slopes as a basis for estimation of fatigue potential in the absence of creep effect. Rather than estimating the stresses from the universal slopes method, however, they were measured or determined by detailed computation for a number of experimental situations that served as test cases. To determine the creep damage caused by a given stress acting for a given increment of time, it was still necessary to stipulate a creep rupture curve. Two curves were considered, the conventional monotonic creep rupture curve for smooth specimens and the curve for sharply notched specimens. It was reasoned that these curves should provide lower and upper bounds, respectively. In examining a number of experimental results on the basis of this method it was found that in nearly all cases the test data did indeed fall between the computed upper and lower bounds, although the spread in life between the two computed bounds was sometimes large.

BASIS FOR ANALYSIS

To improve the computation of the creep damage, it is necessary to obtain a creep rupture curve that is more relevant to the fatigue application than either the smooth or notched monotonic creep rupture curves. Consideration has thus been given to a cyclic creep rupture curve based on smooth specimen results. Spurious notch effects such as stress concentrations are thereby eliminated, while maintaining an essentially constant cross-sectional area, a zero mean strain, and introducing some of the complexities present in cyclic loading (for example, compressive stresses). This is one approach now under consideration at the NASA laboratory to overcome some of the difficulties we have encountered in the application of previous methods.

As a vehicle for describing the analysis, we shall apply it to the problem of low cycle fatigue at a constant temperature within the creep range. The specimens are subjected to a completely reversed strain about a zero mean value, a type of test that has been commonly discussed in the fatigue literature over the past 10 years. The method of approach, however, is general enough so that any other type of loading can be treated using the same basic principles of analysis.

To apply the analysis to this specific creep–fatigue problem, it is necessary to know at least three important factors: the pure fatigue resistance, the pure creep resistance and how creep and fatigue combine.

Pure fatigue

It is difficult to separate experimentally the fatigue effect from the creep effect at high temperatures. For this reason, and for the reason that in many applications the pure fatigue damage is small compared to the creep damage, we have continued as in refs. 4 to 7 to calculate the pure fatigue life using the relatively simple method of universal slopes [1]:

$$\Delta\varepsilon_t = \frac{3{\cdot}5\sigma_u}{E}(N_f)^{-0{\cdot}12} + D^{0{\cdot}6}(N_f)^{-0{\cdot}6} \tag{1}$$

where

$\Delta\varepsilon_t$ = Total strain range

σ_u = Ultimate tensile strength

E = Elastic modulus

D = Ductility = $\ln\left(\frac{1}{1 - RA}\right)$; RA = reduction of area

N_f = Number of cycles to pure fatigue failure.

For a given strain range, the pure fatigue damage, φ_{fatigue}, done in N cycles, will be taken simply as the cycle ratio:

$$\varphi_{\text{fatigue}} = \frac{N}{N_f} \tag{2}$$

Pure creep

The creep damage is taken as the ratio of time actually spent at a given stress to the time required to cause rupture at that stress, or if the stress is varying, to the appropriate integral representing the same concept. Ideally, the time to rupture should be determined for conditions approaching those associated with the strain cycling tests that are to be analysed.

An intuitively attractive approach is to determine the creep rupture life in a test in which the nominal stress is alternated in a square-wave fashion between a given positive value and negative value of equal magnitude. This general type of test has been discussed by Swindeman [8]. Each stress is permitted to act until a specified strain is achieved (*e.g.* of the order of 0·01) and is then reversed. In this manner, the mean strain is maintained at zero, and hence true stresses remain essentially constant during the entire course of the test. The accumulated times under tension and compression are then noted at failure, and any fatigue damage is accounted for by suitable subtraction, so that the pure cyclic creep rupture resistance can be determined. Such tests, which we will refer to as cyclic creep rupture tests, can be conducted at several different stress levels and strain ranges. Our limited experience indicates that strain ranges between 0·01 and 0·04 are suitable, but that ranges above about 0·05 should be avoided since the peak tensile and compressive true stresses begin to differ greatly, resulting in undesirably biased cycles.

Figure 1 shows a portion of the results of a cyclic creep rupture test at 1300°F (978°K) for the Type 316 stainless steel strained between $\pm 0{\cdot}01$ at a stress level of $\pm 45 \times 10^3$ lb/in^2 (± 31 hbar). It is noted that in any one cycle the compressive creep rate is substantially lower than the corresponding tensile creep rate at the same stress level. This is typical of the behaviour of the two materials studied in this investigation at several levels of stress and strain. Both the tensile and compressive creep rates increased steadily from cycle to cycle for the Type 316 stainless steel in a manner similar to that reported in the literature (for example, ref. 9). The differences in the tensile and compressive creep rates and the increases in rates from cycle to cycle tend to complicate the interpretation of this type of test. Further complications may also develop due to strain hardening, progressive changes in metallurgy, grain boundary rotation [9] and the ability of a cracked specimen to withstand compression better than tension. It is necessary, therefore, to decide whether the time during which damage occurs in such a test should be regarded as the total time or only the time under tension. Then one must decide how to use the creep rupture curve so constructed for the analysis of other test cycles.

While rational considerations could be offered for using each of several possibilities, the problem is too complicated to be treated entirely analytically. Our approach, therefore, has been largely pragmatic: by choosing several reasonable alternatives and making the analysis of the experimental results on the basis of each, we selected the one that gave the best results. The three alternatives considered were:

1 Consider compressive stress to be just as damaging as tensile stress in both the establishment of the cyclic creep rupture curve and in the analysis of the strain cycling tests.
2 Neglect compression damage in both types of tests.
3 Neglect compression damage in the cyclic creep rupture test, but retain it in the subsequent analysis of the strain cycling tests.

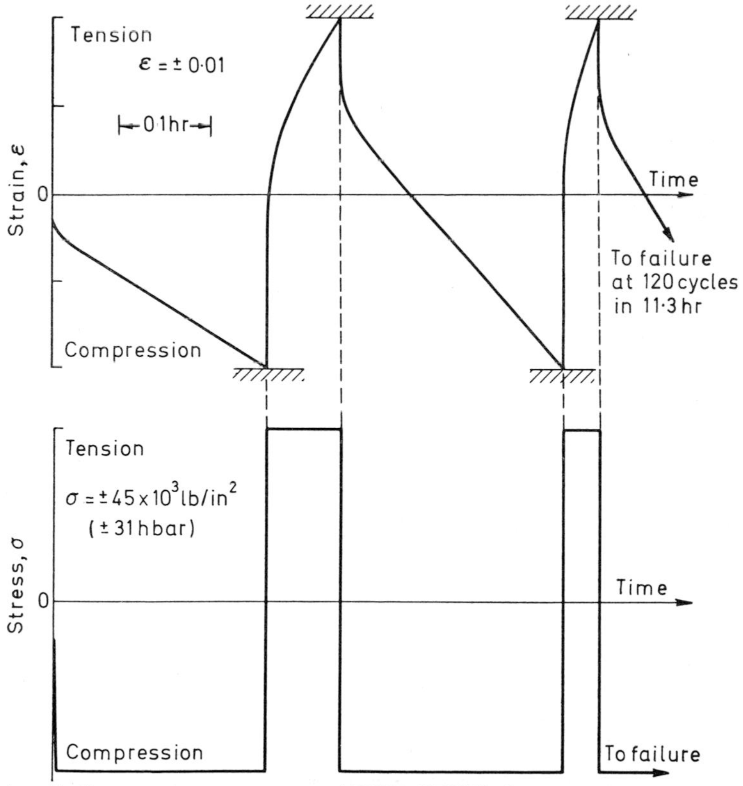

FIG. 1. Cyclic creep rupture test at 1300°F (978°K) for Type 316 stainless steel. Note higher creep resistance in compression and cyclic increases in creep rates in both tension and compression. Results traced from strip chart recordings.

The first alternative is not supported by data in the literature. Several investigators [10–12] have found that periods of compressive creep during monotonic and cyclic tests were not significantly damaging. It is especially noteworthy that one of the first investigators in this field was Dr Johnson, the man whose work is commemorated by this Volume. He found, for example, that compressive creep tests produced no intercrystalline cracking for time intervals at which fracture would have occurred under tensile stresses of the same magnitude.

Alternative (2) is considered in detail in ref. 7 in which a variety of non-isothermal test data from the literature are analysed for creep damage. These data indicated that compressive creep damage should be neglected in some cases but not in others. The following tentative criterion was suggested to determine whether compressive creep damage should be included during a particular time interval. If the product of the stress rate and the total strain rate is greater than zero, compressive creep damage occurs, otherwise it does not. Using this criterion, it was found in ref. 7 that test data which were apparently contradictory were brought into agreement.

In the present investigation, the cyclic rupture tests were conducted with zero stress rate in compression (as well as in tension), so compressive damage was assumed to be negligible in these specific cases, as stated by the above criterion. However, the strain cycling tests produced stress strain hysteresis loops such that the stress rate–strain rate product was always greater than zero. Thus, compressive damage was assumed to be present in the strain cycling tests while it was neglected in the cyclic creep rupture tests. This leads directly to the third alternative listed above.

Regardless of the reasoning that leads to the choice of an alternative, its merits must be based on the accuracy of the calculations it yields. Although the test results to be discussed in this chapter were analysed by each of the three alternatives cited, only the analysis by alternative (3) will be presented because the best results were, indeed, obtained in this way.

Thus, using only the tensile portion of the time in the cyclic creep rupture tests (Fig. 1), isothermal rupture curves were constructed that are assumed to represent the applicable pure creep rupture resistance of the alloys studied. These curves are illustrated in Fig. 2 as solid lines for the cobalt-base alloy L-605 and Type 316 stainless steel. The curves for the total cyclic time and the monotonic curves (shown as a band) used to obtain bounds on life are also shown in Fig. 2 for comparison.

In the subsequent analyses of the strain cycling fatigue tests, the term 'present analysis' will refer to calculated lives derived from the solid,

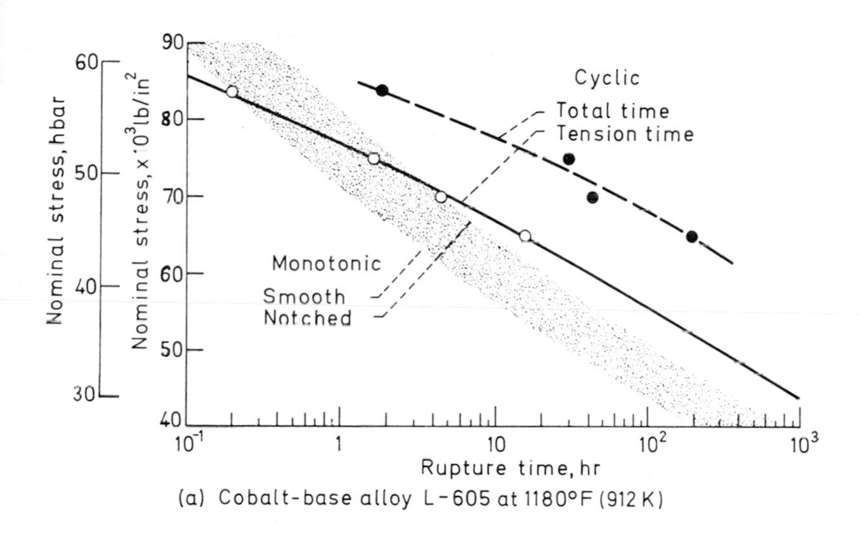

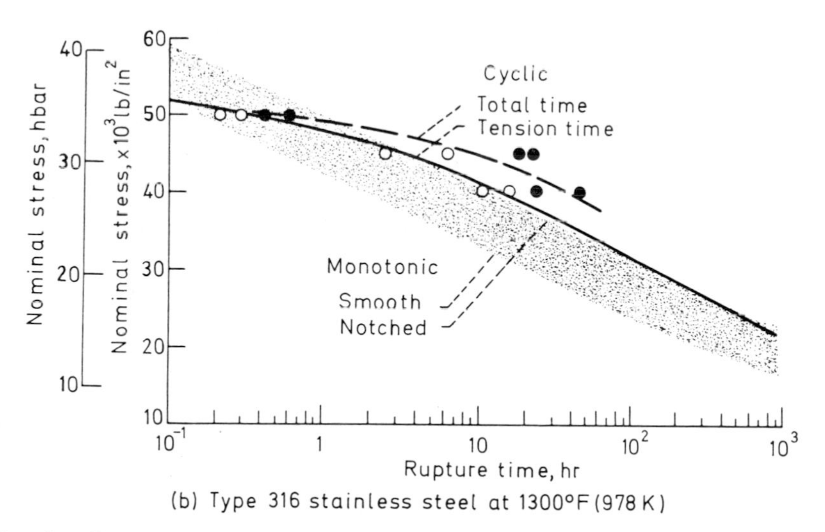

FIG. 2. Comparison of cyclic and monotonic creep rupture strengths of two high temperature alloys.

tension–time curves in these figures, while the term 'method of ref. 5' refers to life calculations based on the limits provided by the monotonic rupture curves for smooth and notched specimens.

In this chapter, creep damage will be taken as a time ratio, or 'life fraction', as it is sometimes referred to. Therefore, the creep damage increment at a given stress during an interval of time dt is given by the time ratio dt/t_r. The value of t_r is determined directly from the solid rupture curve in Fig. 2. Under non-steady stresses, the pure creep damage φ_{creep}, is obtained by integrating:

$$\varphi_{creep} = \int \frac{dt}{t_r} \tag{3}$$

Creep Fatigue Interaction

Creep and fatigue damage may interact in a complex fashion as discussed in ref. 3. However, a simple linear damage rule can serve as an interim treatment that yields reasonable results. This rule is based on the concept that failure occurs when the sum of the fatigue damage and the creep damage is equal to unity:

$$\varphi_{fatigue} + \varphi_{creep} = 1 \text{ (at failure)} \tag{4}$$

The form of eqn (4) is equivalent to that presented by Taira [13].

RESULTS AND DISCUSSION

Reversed strain cycling tests were conducted over a wide range of frequencies on two high temperature alloys: the cobalt-base alloy L-605 at temperatures from room temperature to 1360°F (1010°K) and Type 316 stainless steel at 1300°F (978°K). The behaviour of these two materials differs so markedly that a severe test is provided as to the scope and validity of any analysis purporting to treat them within the same framework.

As seen in Fig. 3, the L-605, when tested at a strain range of 0·009 over a frequency range from $9{\cdot}3 \times 10^{-4}$ to 2·0 Hz, showed a life variation from approximately 30 cycles to 6000 cycles, or a factor of 200. By contrast, the stainless steel, as shown in Fig. 4, exhibited very little frequency dependence within a smaller, but comparable, frequency range for each of several strain ranges investigated. Shown also in Figs. 3 and 4 are the life estimates made according to the method of ref. 4. Extremely good results are seen for the stainless steel, Fig. 4, but the method does not account adequately for the large drop-off in life for the L-605 alloy at the very low frequencies.

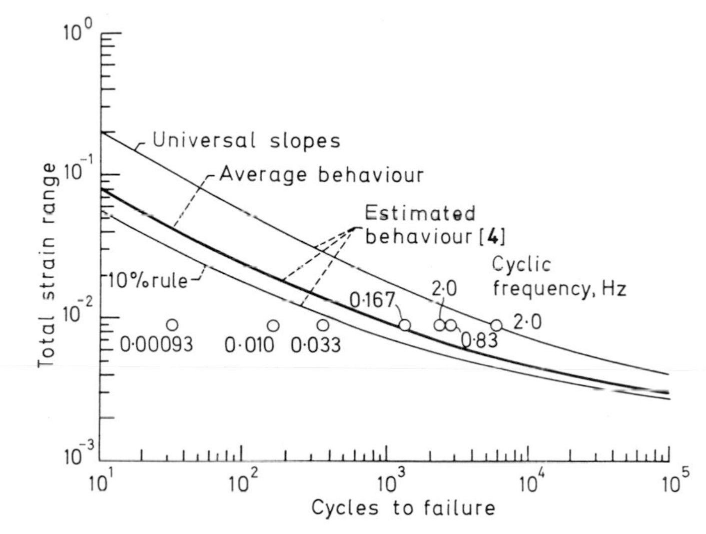

FIG. 3. Cyclic strain resistance of L-605 at 1180°F (912°K). Note strong frequency dependence.

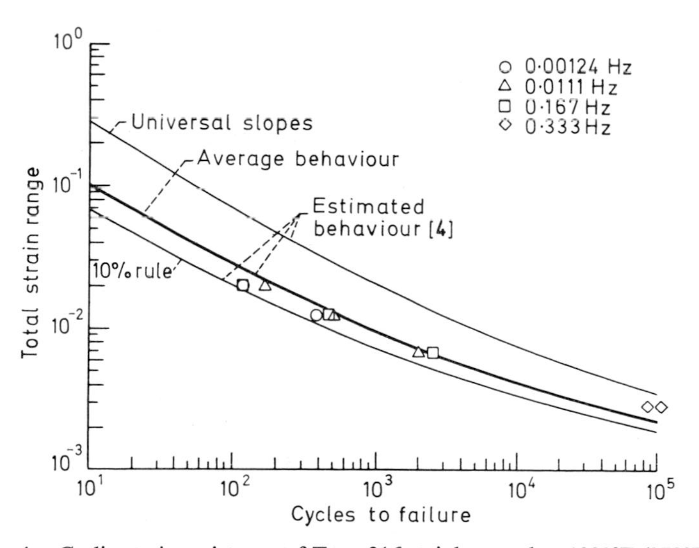

FIG. 4. Cyclic strain resistance of Type 316 stainless steel at 1300°F (978°K).

The results for these two alloys will now be evaluated according to the analysis outlined in this report.

Cobalt-base alloy L-605

The alloy L-605 exhibits a high degree of hardening during the application of strain at elevated temperatures. The cyclic hardening characteristics of this alloy (Fig. 5) show a strong dependence on both the number of applied cycles and the elapsed time. In this figure it can be seen that the longer the time, the higher the stress range required to produce the same strain

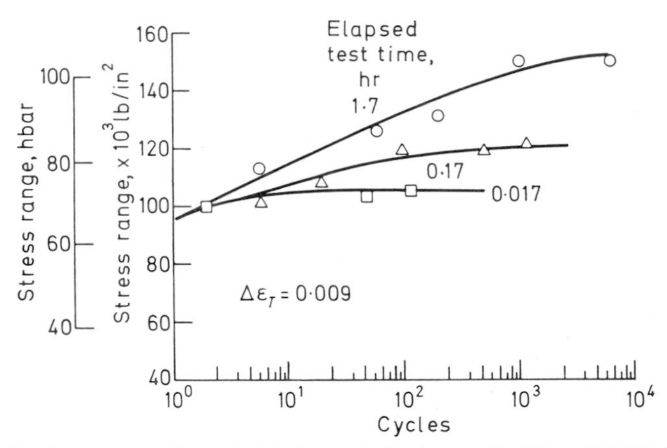

FIG. 5. Isochronous cyclic strain hardening behaviour of L-605 at 1180°F (912°K).

range in the same number of cycles. This complex type of hardening is thought to be caused by a time dependent precipitation of the $M_{23}C_6$ carbide [14] on deformation bands that are generated in a cyclic fashion by the enforced plastic strain.

Electronmicrographs of replicas of sections taken near the fracture zone tend to bear out this explanation of the hardening mechanisms. Figure 6 shows typical structures in a high frequency and a very low frequency test specimen. The essentially featureless structure of the high frequency test specimen (Fig. 6a) results from the application of more than 2300 cycles in only 20 min. This brief exposure at 1180°F (912°K) was insufficient to cause significant precipitation (Fig. 24 of ref. 14), and the micrograph is typical of the as-received alloy [15]. The large particle is presumed to be undissolved M_6C. Such particles were present in the as-received alloy. By contrast, the low frequency test produced the structure shown in Fig. 6b in only 32 cycles which required almost 10 hr. The $M_{23}C_6$ precipitate is seen distributed along series of parallel deformation bands. This microstructure is characteristic of a high strength, low ductility condition for L-605 [16, 17]. Hence, it is also possible that the relatively short fatigue lives of some of the specimens of this alloy might be interpreted as being a consequence of the reduced 1200°F (922°K) ductility brought about by

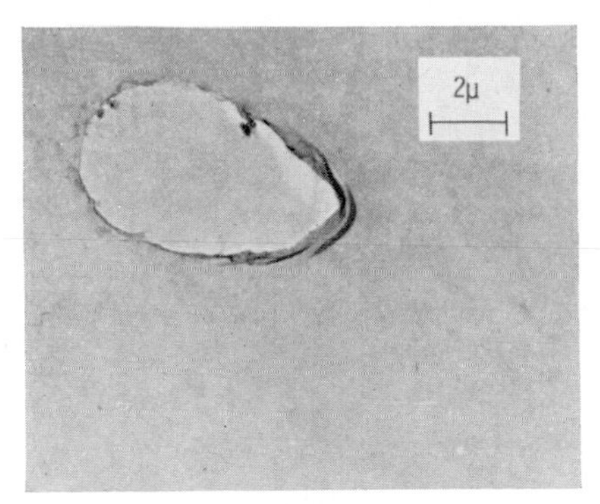

(a) High frequency, 2Hz, $N_f = 2340$ cycles.

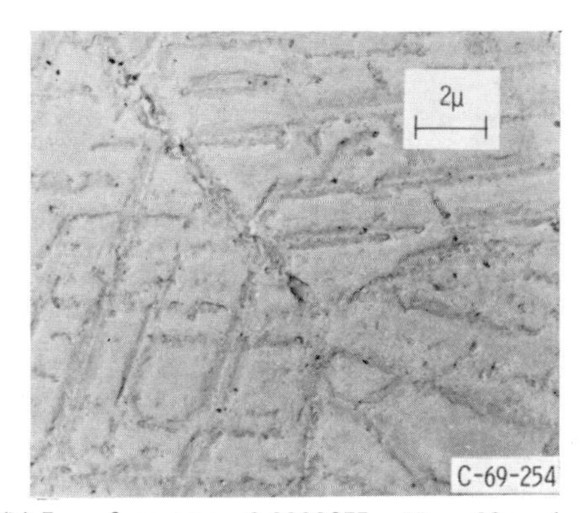

(b) Low frequency, 0·00093Hz, $N_f = 32$ cycles.

FIG. 6. Electronmicrographs of replicas of polished and etched surfaces near fracture zone. L-605 strain cycled at 0·009 total strain range at 1180°F (912°K).

plastic straining [17]. Such reductions in ductility would detrimentally affect both the creep and fatigue behaviour. The low lives might then be equally attributed to the metallurgical changes rather than to the significance of the compressive stresses that we have considered to be damaging in the analysis. This approach, however, has not been pursued in this investigation.

To perform the cyclic life computations according to the linear creep–fatigue damage analysis described above it was first necessary to establish the cyclic creep rupture results already shown in Fig. 2. These were obtained as indicated in Fig. 1 by setting a strain level of $\pm 0{\cdot}01$, and determining the time under tension to failure for a selected number of completely reversed stress levels. Limited testing indicated that approximately the same results are obtained for strain levels other than $\pm 0{\cdot}01$, but this matter

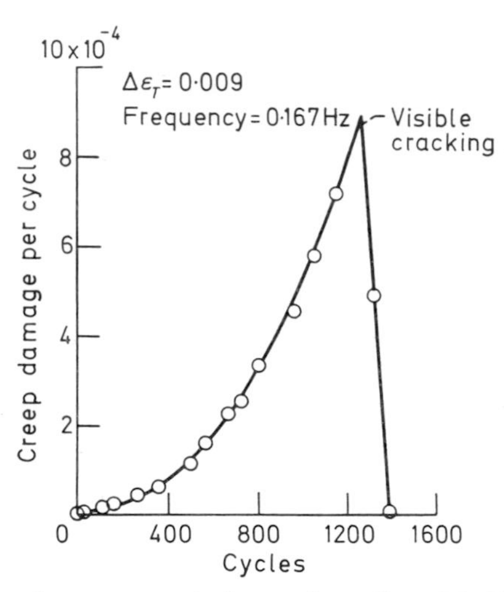

FIG. 7. Typical creep damage rates during strain cycling of L-605 at 1180°F (912°K).

also needs further study. In some cases, notably for the highest stress levels shown in Fig. 2, it was not possible to apply the desired stress level at the very beginning of the test. Rather, a few cycles of a smaller stress level were required to strain and time harden the material so that the desired high stress level could be achieved within the required strain range. However, the computed creep damage and fatigue damage induced by the cycles of lower stress were negligible.

The next step in the analysis of the axial strain cycling tests is the determination of the variation of stress with time throughout the test. The stress varies not only within each cycle but also from cycle to cycle as the alloy hardens (Fig. 5). The complete stress history is then used in conjunction with the cyclic rupture strength from Fig. 2 to determine creep damage per cycle at various stages throughout the life of each specimen. The creep damage per cycle is plotted in Fig. 7 for one illustrative case. This quantity is obtained by integrating the infinitesimal creep damage at each instant

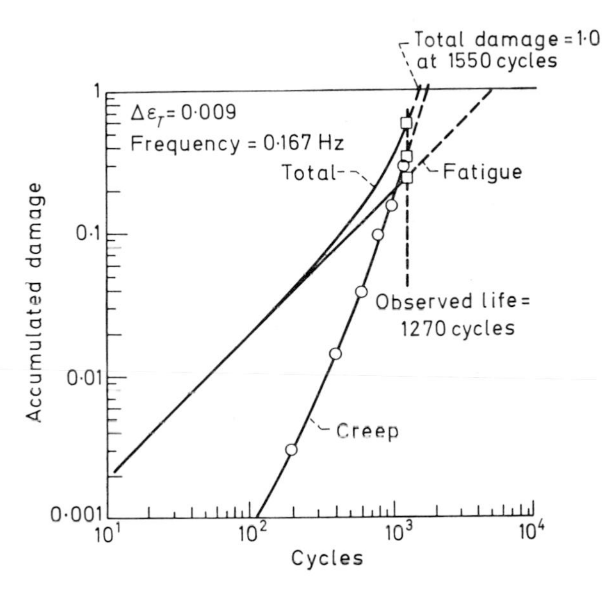

FIG. 8. Progressive increase in creep and fatigue damage during strain cycling of L-605 at 1180°F (912°K).

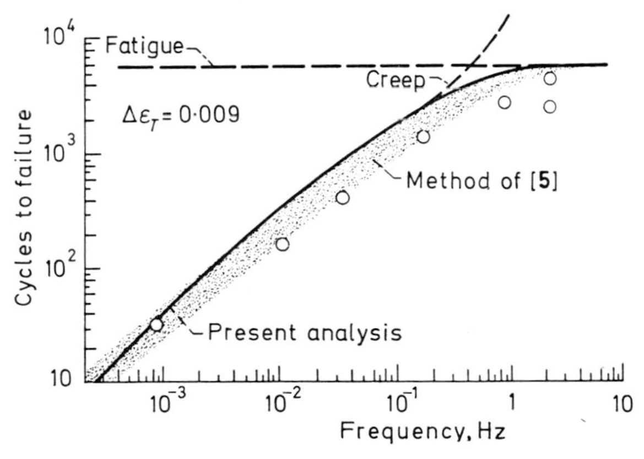

FIG. 9. Comparison of computed and experimental creep fatigue results in low cycle fatigue of L-605 at 1180°F (912°K).

within the cycle according to eqn (3). Note that the creep damage per cycle steadily increases with applied cycles because of the progressively increasing stresses. The accumulated creep damage, as determined by summing the damage per cycle over the number of applied cycles, is shown in Fig. 8. Also shown is the pure fatigue damage, according to eqn (2), and the total damage which is the sum of the fatigue and creep damage. The number of cycles required to accumulate a total damage of unity is taken as the calculated cyclic life, in accordance with eqn (4). Similar calculations for the other frequencies produced the results shown in Fig. 9. The pure fatigue and creep components are shown as dashed curves.

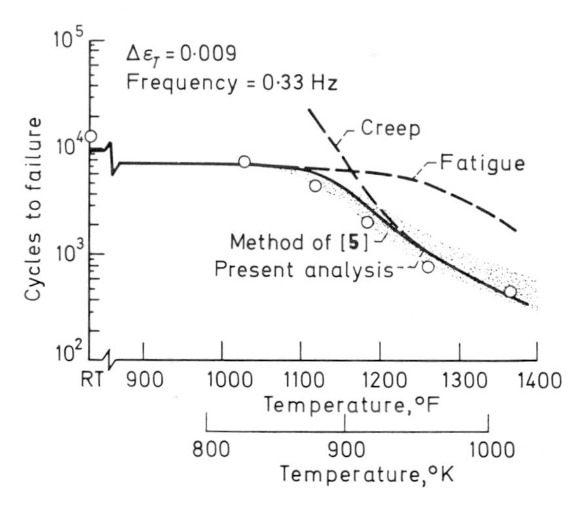

FIG. 10. Creep fatigue results in low cycle fatigue of L-605 over a range of temperatures.

The upper and lower bounds as proposed in ref. 5 are also shown for comparison. It is apparent that at frequencies below about 0·1 Hz the damage imposed is due almost entirely to creep. In the range between 0·1 Hz and approximately 1 Hz both creep and fatigue damage interact and at frequencies above 1 Hz the damage is due almost entirely to fatigue.

Figure 10 shows computed and experimental lives for a strain range of 0·009 and a frequency of 0·33 Hz applied over the temperature range from room temperature to 1360°F (1010°K). The basic procedure here is the same as already discussed. Both the cyclic and monotonic creep rupture curves at temperatures other than 1180°F (912°K) were obtained by assuming that rupture strength variations with temperature are the same as those for sheet stock as analysed in ref. 5. Up to a temperature of about

1150°F (894°K) the life is governed largely by pure fatigue damage, whereas above this temperature, it is controlled by the accumulation of creep damage. For frequencies greater than 0·33 Hz, it is expected that this cross-over temperature would be shifted to higher values.

Stainless steel Type 316

The elevated temperature creep–fatigue behaviour for this alloy is considerably different from that of L-605. Here, instead of progressive metallurgical hardening, there is an early, transient cyclic strain softening, as shown in Fig. 11. Also seen is the fact that the material possesses a

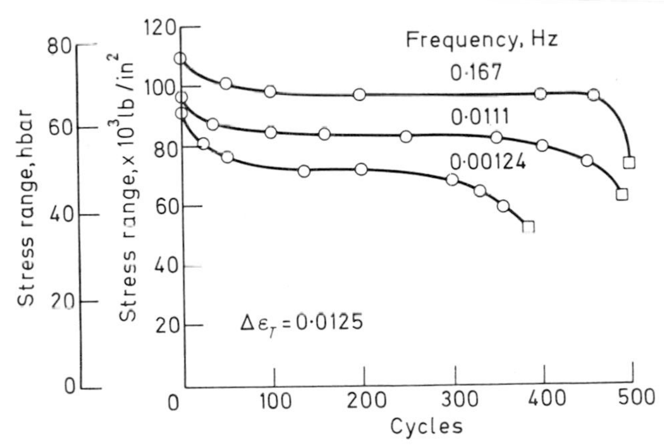

FIG. 11. Frequency dependence of cyclic strain softening behaviour of Type 316 stainless steel at 1300°F (978°K).

positive strain rate sensitivity, since the higher frequencies and hence higher strain rates produce higher stress levels. The strain rate dependence is further illustrated in Fig. 12 where the data are plotted in conventional stress–strain fashion with cyclic strain rate as a parameter. Here it is clearly seen that at a given strain the stress can vary by more than 50 per cent, depending on strain rate.

As the frequency is increased, a given number of cycles requires a shorter time, but during this shorter time stresses are higher so that the accumulated creep damage per cycle remains approximately constant. The result is the relative insensitivity of the cyclic lives to frequency as already seen in Fig. 4.

Figure 13 shows the results of the analysis of the stainless steel at strain ranges of 0·0068, 0·0125 and 0·020 at 1300°F (978°K). It is seen that

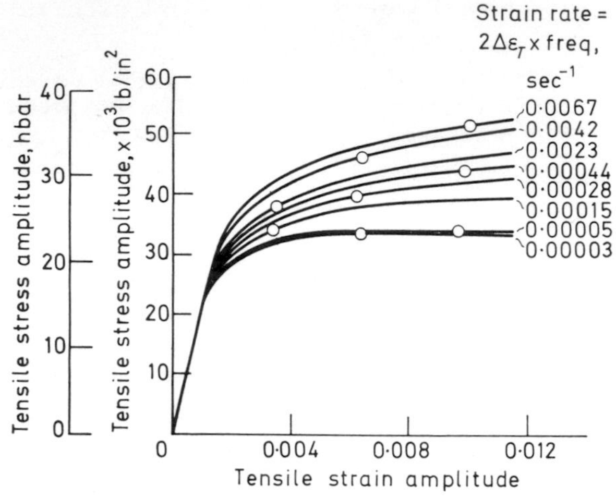

FIG. 12. Strong strain rate dependence of cyclic stress–strain curve of Type 316 stainless steel at 1300°F (978°K).

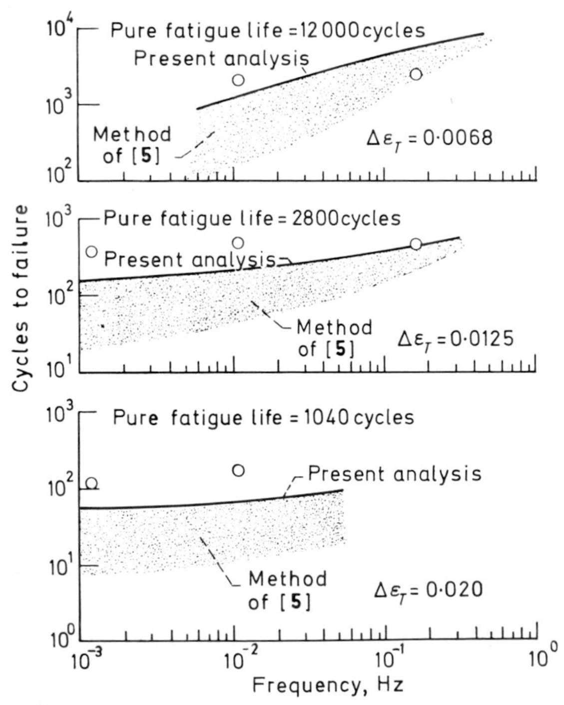

FIG. 13. Comparison of computed and experimental creep fatigue results in low cycle fatigue of Type 316 stainless steel at 1300°F (978°K).

on the basis of the computations, a relatively small frequency dependence of cyclic life is to be expected, and this is verified by the experimental results. A sizable reduction in life would not be anticipated, even if the test frequency were reduced substantially below 10^{-3} Hz.

CONCLUDING REMARKS

The general agreement between computation and experiment obtained for these two materials of divergent hardening characteristics implies considerable promise for this type of analysis. The major conclusion that can be drawn lies in the observation that creep may play an important and sometimes dominant role in high temperature cyclic life. From a practical standpoint, however, a limitation of this analysis should not be overlooked. To perform these studies it was necessary to determine the complete stress history. These stresses were found to be a strong function of strain rate, temperature and individual material metallurgy, and their accurate prediction in advance of actual test might therefore have proved very difficult, if not impossible. Thus, while the analysis described herein appears to be more exact than the ones previously developed at the NASA laboratory and shows promise of being useful in a much wider scope of problems such as thermal fatigue (wherein stress, strain and temperature vary simultaneously), the method of ref. 4 still retains utility when the stresses involved cannot be measured. This would include many applications of practical engineering structural parts for which stress measurements may prove to be prohibitively difficult.

From the results of the calculations made in this chapter, it also becomes clear, even more so than from our previous studies, how important it is to obtain complete information on the temperature and stress history of an engineering part if a meaningful analysis is to be made. The significance of test frequency is also shown and the precaution is suggested to proceed carefully in predicting the effect of very low frequencies from studies in a higher frequency range.

While reasonably good results were obtained in this study by assuming a linear creep–fatigue damage rule, it must be emphasised that such a treatment is at best an approximation to true material behaviour. In other applications, for example long periods of steady load in which alternating stresses are periodically introduced, a linear damage rule may be less applicable. More study is needed for other types of stress patterns as well as for other materials in the mode of loading used for the two materials investigated in this report.

CONCLUSIONS

The following general conclusions are drawn from the results of this investigation.

Creep can play an important and sometimes dominant role in low cycle fatigue at high temperatures. Recognition of this fact may be helpful to materials engineers and designers in seeking to improve high temperature alloys and structures in resisting cyclic loads.

Cyclic tests at a variety of temperatures, frequencies, strain ranges and cycle configurations on the alloy L-605 and 316 stainless steel showed that (a) the simple life-fraction theory was adequate for calculating creep damage when the cyclic creep rupture curve was used as a basis for analysis, (b) the method of universal slopes developed originally for room temperature use was sufficiently accurate at high temperature to be used to calculate pure fatigue damage, and (c) a linear creep–fatigue damage rule explained the observed transitions from one failure mode to the other.

REFERENCES

1. MANSON, S. S. (1965). 'Fatigue: A complex subject. Some simple approximations.' *Exp. Mech.*, **5**(7), 193–226.
2. MANSON, S. S. (1966). *Thermal Stresses and Low-cycle Fatigue*. New York: McGraw-Hill.
3. MANSON, S. S. (1966). 'Interfaces between fatigue, creep and fracture.' *Int. J. Fracture Mech.*, **2**(1), 327–363.
4. MANSON, S. S. and HALFORD, G. R. (1967). 'A method of estimating high-temperature low-cycle fatigue behaviour of materials.' *Thermal and High-strain Fatigue*, pp. 154–170. London: Metals and Metallurgy Trust.
5. SPERA, D. A. (1968). *A linear and creep damage theory for thermal fatigue of materials*. Ph.D. Thesis. University of Wisconsin.
6. SPERA, D. A. (1969). 'The calculation of creep damage during elevated-temperature, low-cycle fatigue.' *NASA TN* D-5317. Washington, D.C.: National Aeronautics and Space Administration.
7. SPERA, D. A. (1969). 'The calculation of thermal fatigue life based on accumulated creep damage.' *NASA TN* D-5489. Washington, D.C.: National Aeronautics and Space Administration.
8. SWINDEMAN, R. W. (1963). 'The interrelation of cyclic and monotonic creep rupture.' *Jt. Int. Conf. on Creep. Proc. Instn Mech. Engrs*, **178**(3A), 3-71–3-76.

9. KITAGAWA, M. (1968). 'Enhanced grain boundary sliding during reversed creep of lead.' *T. & A.M.* Report No. 319. Urbana: University of Illinois.
10. JOHNSON, A. E., HENDERSON, J. and MATHUR, V. D. (1956). 'Combined stress creep fracture of a commercial copper at 25°C.' *Engineer Lond.*, **196,** 261–265.
11. HULL, D. and RIMMER, D. E. (1959). 'The growth of grain-boundary voids under stresses.' *Phil. Mag.*, **4,** Ser. 8 (42), 673–687.
12. KENNEDY, C. R. (1963). 'Effect of stress state on high-temperature low-cycle fatigue.' *ASTM Spec. Tech. Publication* No. 338, 92–104. Philadelphia: American Society for Testing and Materials.
13. TAIRA, S. (1962). 'Lifetime of structures subjected to varying load and temperature.' HOFF, N. J. (Ed.), *Creep in Structures*. New York: Academic Press.
14. MORRAL, F. R., HABRAKEN, L., COUTSOURADIS, D., DRAPIER, J. M. and URBAIN, M. (1968). 'Microstructure of cobalt-base high-temperature alloys.' *Tech. Report* No. 8-21.1. Detroit: American Society for Metals.
15. SANDROCK, G. D., ASHBROOK, R. L. and FRECHE, J. C. (1965). 'Effect of variations in silicon and iron content on embrittlement of a cobalt-base alloy (L-605).' *NASA TN* D-2989. Washington, D.C.: National Aeronautics and Space Administration.
16. SANDROCK, G. D. and LEONARD, L. (1966). 'Cold reduction as a means of reducing embrittlement of a cobalt-base alloy (L-605).' *NASA TN* D-3528. Washington, D.C.: National Aeronautics and Space Administration.
17. NEJEDLIK, J. F. (1966). 'The embrittlement characteristics of a low-silicon modified cobalt-base alloy (L-605) at 1200 and 1600°F.' *Report No.* ER 6870. Cleveland, Ohio: TRW Equipment Laboratories.

Chapter 13

CREEP TESTS ON $2\frac{1}{4}$ PER CENT CHROMIUM 1 PER CENT MOLYBDENUM STEEL IN BAINITIC CONDITION

A. KRISCH

SUMMARY

Creep tests on $2\frac{1}{4}$ *per cent Cr* 1 *per cent Mo steel in a bainitic structural condition were carried out at* 500°–600° *C for* 40 000–70 000 *hr, and showed no significant change in microstructure under the optical microscope. However, in the electron microscope fine precipitations, formed as needles, were visible after only a few hours; the precipitates increased in number and grew in size with time, but disappeared in the tests extending to* 40 000 *hr. A second precipitation process is the formation of similar particles in the bainitic crystals which continued to grow in the longest testing times. Relations between the creep properties and the precipitation processes are discussed.*

INTRODUCTION

In recent years, many long time tests have been undertaken to study the mechanical properties of steels at high temperatures. Extrapolation of the rupture/time curves, obtained in these tests, demands knowledge of the processes occurring in the specimens subjected to long time high temperature stress. These processes may be of a mechanical nature, for instance time dependent (rheological) deformation or the opening and enlarging of microcracks, or they may be internal, involving change of structure.

Former opinion considered that the structure of steel must be stable to obtain good creep resistance, but more recent tests have shown that a stable structure is probably not the best and that a slow precipitation process results in a reduction of slip in the glide planes. Earlier tests [1, 2] on a 1 per cent Cr $\frac{1}{2}$ per cent Mo steel (Type 13 Cr Mo 44) have demonstrated that there is a continuous precipitation of carbides in the ferrite and in the grain boundaries during creep testing, and also that there is an agglomeration of cementite in the pearlite and bainite. These processes had already commenced before testing at 500°C and were still in progress after 50 000 hr. At higher temperatures, up to 600°C, completion of these processes could not be established. It followed that the fine precipitations in the ferrite were essential to maintaining the strength of this steel at high temperature and for long times.

TESTS ON $2\frac{1}{4}$% Cr 1% Mo STEEL

To obtain the bainitic structure, a steel (Type 10 Cr Mo 910) with 0·1 per cent C, 2·43 per cent Cr and 0·98 per cent Mo was cooled from 940°C in air and tempered for 40 min at 750°C; the microstructure is shown in Fig. 1. Creep tests were conducted at 500°, 550°, 575° and 600°C, the

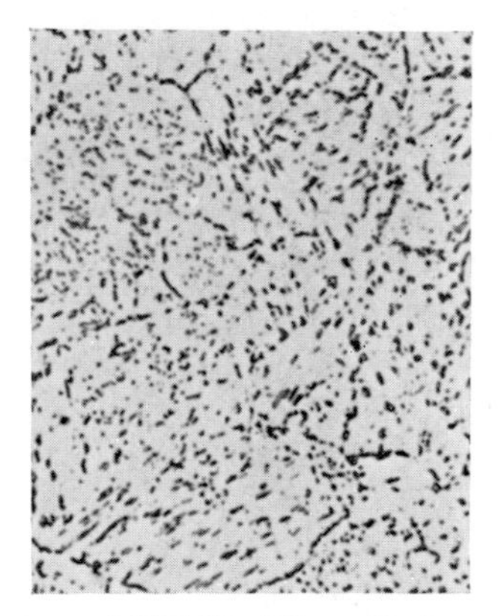

FIG. 1. Microstructure of 10CrMo 910 steel before creep test (×1000).

stresses being chosen so that the rupture/time lines could be determined up to 30 000 hr. The longer tests were continued to 40 000 hr and one at 550°C to 72 000 hr without fracture.

Creep strain/time diagrams for 500°C and 575°C are shown in Fig. 2a and b. They confirmed our knowledge of the mechanical processes in creep tests. On the ordinate scale, the condition of primary creep with decreasing creep rate is difficult to detect, and the condition of tertiary creep with increasing creep rate is not evident. Most tests, terminated before fracture, had not reached the third stage of creep. The creep curves for the specimens which ruptured show a significant increase of creep rate at 3–5 per cent strain. If creep strain/time curves and rupture/time curves are to be extrapolated, this value of strain may be significant. Extrapolation of creep/time curves should be treated with caution and should be made only to 3–5 per cent strain; for theoretical research, this strain gives a more realistic value for the calculation of mean creep rate than does elongation at fracture.

No creep test at 500°C or above showed termination of creep. This was also the case for tests at stresses of less than the 100 000 hr rupture stress. The unbroken specimens, which had been tested for 40 000 hr or longer, had creep rates less than 10^{-6}/hr and most were nearer

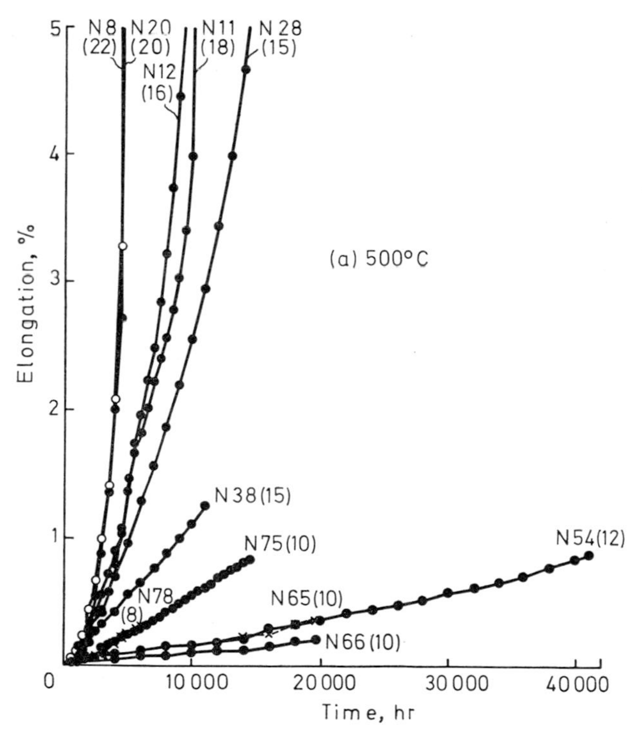

FIG. 2(a). Time elongation curves of creep test at 500°C (stress in kg/mm²).

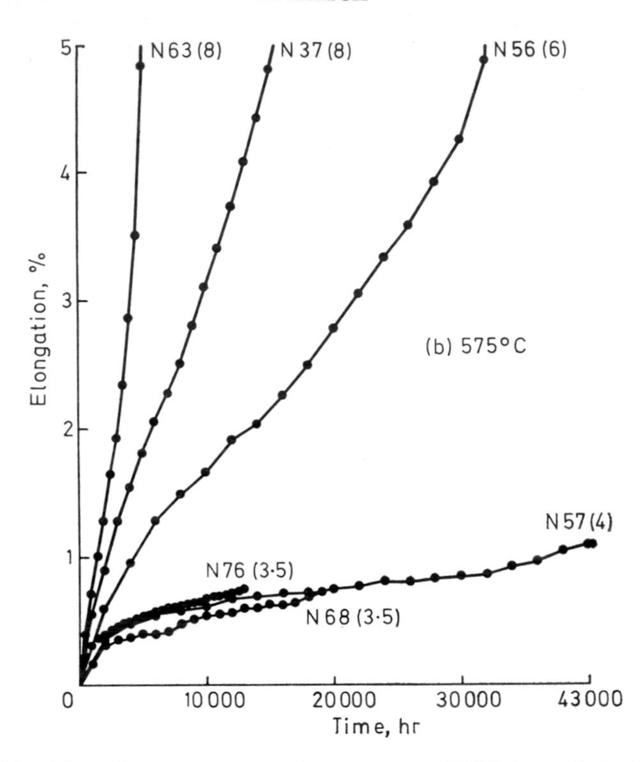

FIG. 2(b). Time elongation curves of creep test at 575°C (stress in kg/mm²).

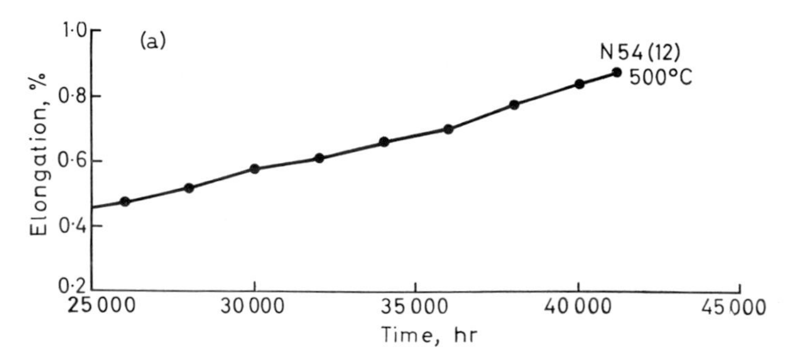

FIG. 3(a). Time elongation curve of creep test at 500°C (stress in kg/mm²).

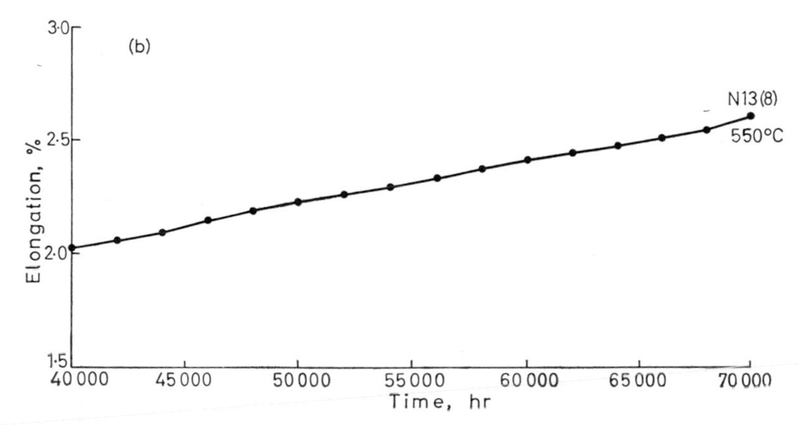

FIG. 3(b). Time elongation curve of creep test at 550°C. Final years of test only.

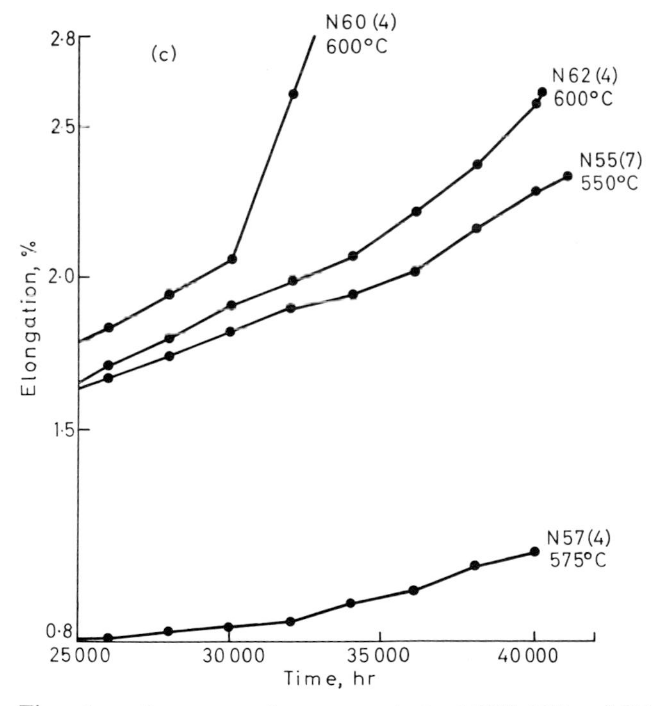

FIG. 3(c). Time elongation curves of some creep tests at 550°, 575° and 600°C. Final years of tests only.

10^{-7}/hr. Specimen No. 54, tested at 500°C and 12 kg/mm^2 (Fig. 3a), showed approximately steady increase of strain up to 35 000 hr and then the creep rate increased to about 3×10^{-7}/hr. This specimen would reach a critical strain of 3 per cent after 80 000–120 000 hr. The longest test, on specimen No. 13, at 550°C and 8 kg/mm^2 (Fig. 3b) showed a low creep rate of less than $2{\cdot}5 \times 10^{-7}$/hr, but in 72 422 hr it had already reached a strain of 2·5 per cent. By linear extrapolation, the strain will be 3 per cent in approximately 100 000 hr and 4 per cent in approximately 150 000 hr; at this strain, accelerated creep should begin.

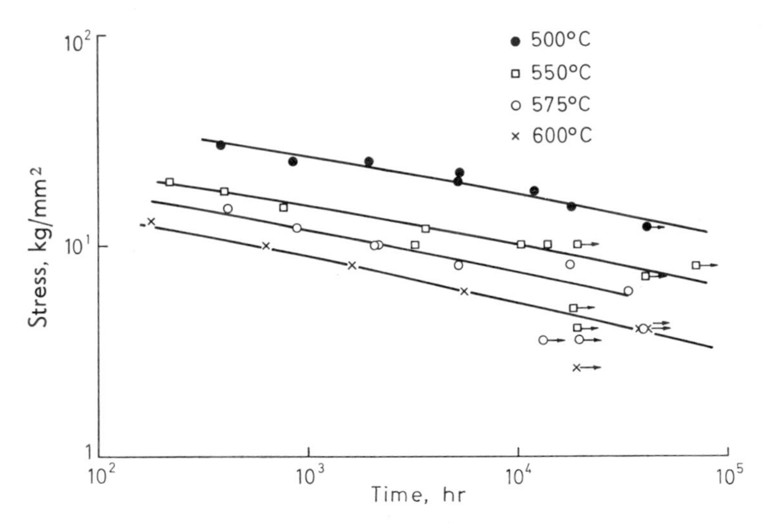

FIG. 4. Time-to-rupture lines of the 10CrMo 910 steel investigated.

Specimen No. 55, tested at 550°C and 7 kg/mm^2 (Fig. 3c) reached the same strain at 35 000 hr as that of specimen No. 13, but had a higher creep rate and should attain accelerated (*i.e.* tertiary) creep at about 50 000 hr. Specimen No. 57, tested at 575°C and 4 kg/mm^2, had a low creep rate of $1\text{–}2 \times 10^{-7}$/hr up to 32 000 hr, but thereafter it had a higher creep rate and the critical strain of 3–4 per cent should be reached before 100 000 hr.

Specimens Nos. 60 and 62, tested at 600°C and 4 kg/mm^2, reached the same strain of 2 per cent at 30 000 hr. Then, however, the creep rate of specimen No. 60 became four times as high, and the test piece broke after 38 284 hr. Metallographic examination of this specimen showed a different structure. Specimen No. 62, tested under the same conditions, should soon rupture, since the creep rate should increase after 35 000 hr.

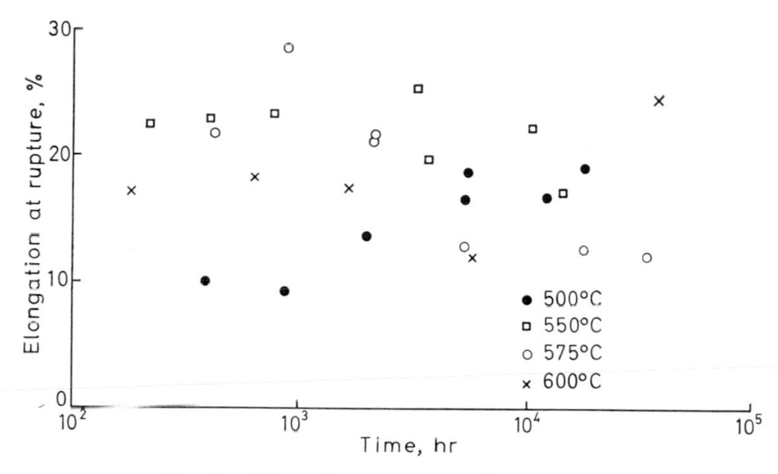

FIG. 5. Elongation at rupture after creep test.

The prediction of a life of several years of a specimen at temperatures of 500°–550°C is possible only if the creep rate is near 2×10^{-7}/hr, as is the case for the Type 13 Cr Mo 44 steel. At 575°C the limiting creep rate would already be exceeded, since the specimen No. 57 tested at 575°C and 4 kg/mm^2 had a creep rate of $1–1{\cdot}5 \times 10^{-7}$/hr from 10 000–30 000 hr, but afterwards the creep rate increased.

Rupture/time curves for temperatures of 550°–600°C are shown in Fig. 4, unbroken specimens being marked by arrows. The extrapolated strengths for 100 000 hr are near the lower limits of the scatter band known to apply for this class of steel. Values of elongation and reduction of area at rupture are relatively good (*see* Figs. 5 and 6). Examination of

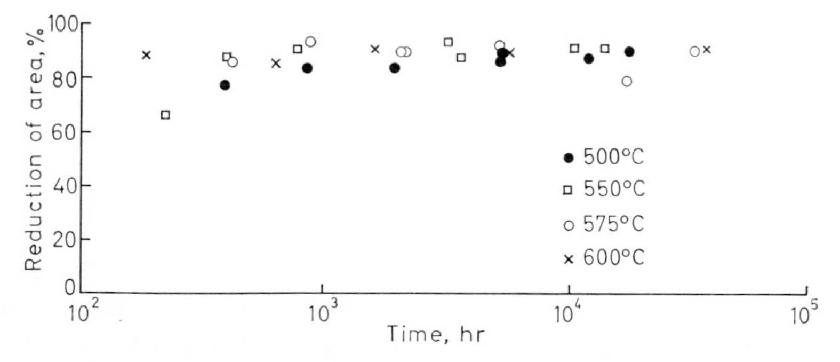

FIG. 6. Reduction of area after creep test.

specimens after creep testing revealed a diminution of hardness compared with the condition after thermal treatment ($H_B = 180$ kg/mm^2). At 500°C the decrease of hardness was about 20 H_B, at 550°C about 35 H_B, and at 600°C about 50 H_B. These hardness values were reached after about 1000 hr; longer tests showed only a very small additional change in hardness.

MICROSTRUCTURAL FEATURES

Although in the ferritic–pearlitic–bainitic structure of 1 per cent Cr $\frac{1}{2}$ per cent Mo steel (Type 13 Cr Mo 44) a significant change of microstructure was visible in the optical microscope [2] at low magnification, this was not observed in the bainitic structure of steel Type 10 Cr Mo 910 with the higher alloy content of $2\frac{1}{4}$ per cent Cr and 1 per cent Mo. The many specimens investigated metallographically, after creep testing, showed practically the same proportion of bainitic structure and did so at the highest magnification (Fig. 7). Some specimens seem to have a slight

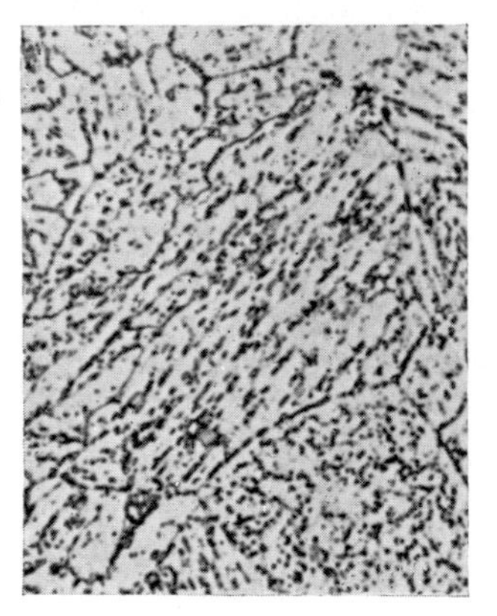

FIG. 7. Microstructure after creep test: 500°C for 41 106 hr at 12 kg/mm^2; elongation 0·9 per cent, not broken (×1000).

enlargement of carbides in the grain boundaries, but it is not certain that this is a real change from the condition, before creep testing, of these specific test pieces.

A definite change, however, occurs within the bainitic structure of Type 10 Cr Mo 910 steel during creep testing. The structure is changed by slow precipitation similar to that in the ferritic structure of Type 13 Cr Mo 44 steel. These precipitated particles are so small that they could be detected only by the electron microscope.

After thermal treatment and before creep testing, the bainitic carbides are visible, in the plastic replicas, as parallelograms nearly 0·7 μm long and 0·15 μm wide (Fig. 8). Finer precipitations are seldom found. After creep testing specimen No. 39 at 500°C for 394 hr, these bainitic carbides were practically unchanged, but some fine precipitation formed as thin plates of 0·1 μm length, and others formed as needles which, on reaching

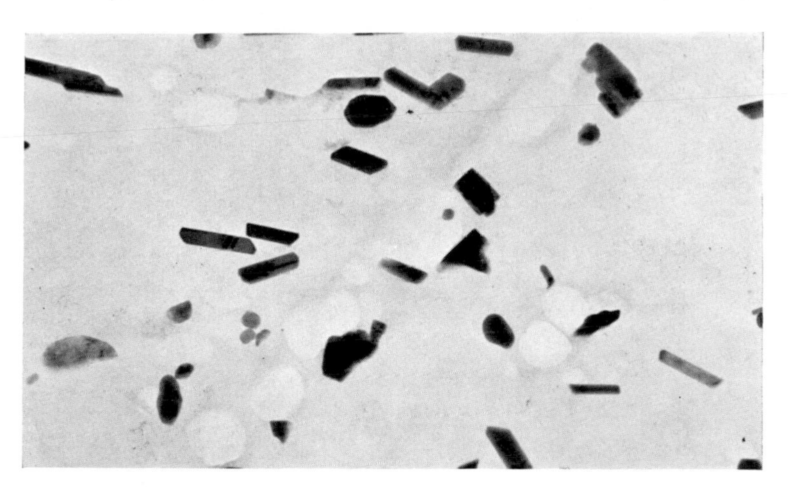

FIG. 8. Extraction replica of precipitations before creep test. (×20 000)

0·4 μm length, were found gathered partly in groups. After 16 310 hr the number of these needles in specimen No. 38 had increased and the distribution was uniform (Fig. 9). Other particles had formed in the bainitic grains. After 41 106 hr at 500°C, the number of finer precipitations in specimen No. 54 again diminished but the number and size of the particles, formed in the bainitic grains, had increased (Fig. 10).

At the higher temperatures the same processes occur: precipitation of fine needles, growth of the needles, and re-solution, accompanied by precipitation of other fine particles within the bainite crystals. The final stage is growth of these precipitated particles. The higher the temperature, the earlier the disappearance of the needles and the growth of the particles.

The bainitic structure of this steel has greater stability in the finely precipitated form. In the preceding tests on the 1 per cent Cr $\frac{1}{2}$ per cent Mo steel [2], the opposite sequence occurred: the agglomeration of pearlitic and bainitic carbides, and stable fine precipitations. If it had been supposed

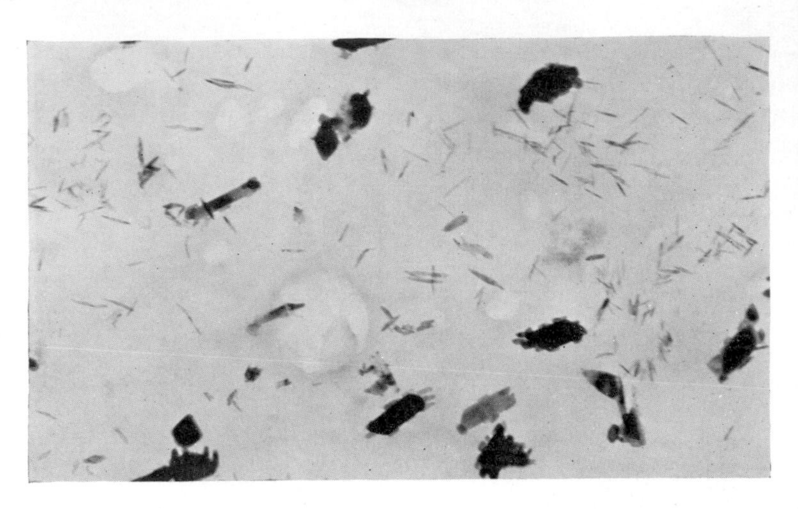

FIG. 9. Extraction replica of precipitation after creep test: 500°C for 16 310 hr at 15 kg/mm^2; elongation 1·4 per cent, not broken (×20 000).

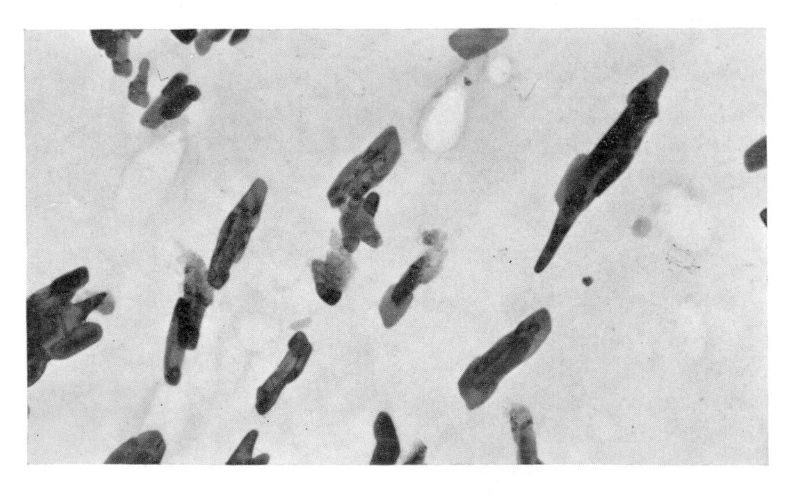

FIG. 10. Extraction replica of precipitation after creep test: 500°C for 41 106 hr at 12 kg/mm^2; elongation 0·9 per cent, not broken (×20 000).

that the strength of the ferritic–pearlitic steel was caused by the high stability of the fine precipitations, the strength of the bainitic steel would be expected to be due also to these fine precipitations. However, after 50 000 hr, the diminished strength is retained only by the coarser bainitic structures—on the assumption that even the finest particles can be detected by electron microscopy.

CONCLUSIONS

Creep tests on $2\frac{1}{4}$ per cent Cr 1 per cent Mo steel showed that transient creep at 500°C ceases after 500 to 1000 hr. At 530°–600°C, creep at decreasing rate was observed over 10 000 to 20 000 hr, but in no test did creep appear to stop completely. The third stage of creep, with increasing creep rate, begins when 3 to 5 per cent strain has been reached; extrapolation of rupture time should therefore be made in two steps, *i.e.* to this strain level and afterwards to rupture. Only if a low creep rate, not more than 2×10^{-7}/hr, has been measured over long times, may lives of several years be predicted in the temperature range 500°–575°C.

The precipitation processes, which are revealed by electron microscopy of the creep specimens, are too slow to be related to the changes in creep strain rate. The relatively high creep rupture strength of this type of steel may not be due only to the fine precipitations; the bainite carbides, and the fine particles associated with them, should increase the creep strength after long times, thus further strengthening the alloyed matrix.

REFERENCES

1. KRISCH, A. (1966). 'Creep rupture tests on a chrome–molybdenum steel' (in German with English abstract). *Arch. Eisenhütt Wes.*, **37** (4), 317–324.
2. KRISCH, A. and NAUMANN, F. K. (1966). 'Structural changes and their significance for creep resistance of a specimen of a chrome–molybdenum steel' (in German with English abstract). *Arch. Eisenhütt Wes.*, **37** (9), 749–757.

Chapter 14

COMBINED STRESS CREEP OF NON-LINEAR VISCOELASTIC MATERIAL

W. N. FINDLEY

SUMMARY

The multiple integral representation for creep behaviour of viscoelastic material is reviewed, together with the modified superposition principle and the product form assumption for the kernels of the multiple integrals. Creep experiments under combined tension and torsion are employed to determine the material constants. Other experiments involving abrupt changes in states of stress and constant rates of stressing are employed to evaluate the usefulness of the theories by comparing the theoretical predictions with the experimental results.

NOTATION

A, B	Coefficients in equations as defined.
f_{ij}	Coefficient of multiple integral relation for constant stress.
$\mathbf{I}$	Kronecker delta $= \delta_{ij}$.
K_j	Kernel function of the time variable.
l	Rate of stressing in torsion.
n	Time exponent.
p	Number of changes in stress.
R, M, N	Coefficients in eqn (20).
t	Current time.
γ	Shear strain.
ε	Creep strain.
$\boldsymbol{\varepsilon}$	Creep strain tensor $= \varepsilon_{ij}$.
ν	Poisson's ratio.

ξ	Arbitrary prior time.
σ	Tensile stress.
$\dot{\sigma}_0$	Rate of stressing.
$\boldsymbol{\sigma}$	Stress tensor $= \sigma_{ij}$.
$\dot{\boldsymbol{\sigma}}$	Stress rate tensor.
τ	Shear or torsion stress.
$\dot{\tau}$	Torsion stress rate.

Subscripts

1, 2	Components on planes at right angles.
I, II	Principal stress directions.

Superscripts

0	Time independent part.
+	Time dependent part.

Note: The bar $^{-}$ indicates the trace of a tensor.

INTRODUCTION

Many materials are non-linear in their mechanical behaviour over some range of stress and temperature. In the range of the variables where creep is important most materials will exhibit non-linear characteristics, especially in the time dependent strain. For very critical designs where creep is involved the approximation of the actual non-linear behaviour by linear relations may not be sufficiently reliable, especially under combined stress where synergistic effects not found in linear theory may account for a large fraction of the observed time dependent strain.

This chapter will not attempt a complete review of the various relationships (*see* [1–7] for example) which have been employed to describe the time dependence and non-linear creep behaviour. Instead, it will describe one approach to the problem and give examples of its application to creep under different states of combined stress.

The method to be described is based on a multiple integral representation proposed for stress relaxation by Green and Rivlin [2]. In this approach, the strain is taken to be a function of all past history of strain (or stress) and the non-linearity is introduced by employing a series of multiple integrals containing kernel functions describing the viscoelastic behaviour of the material. Onaran and Findley [6] presented equivalent multiple integral functions for creep, evaluated the kernel functions required for constant stress creep and demonstrated that a reasonably accurate representation of the experimental results of constant stress creep of plastics could be obtained by this method.

Several investigators [7, 8] have outlined a systematic set of experiments for determining the general form of the kernel functions required to describe creep behaviour under varying stress, but without presenting experimental results. In a recent work by Onaran and Findley [9] experiments were performed to evaluate some of the kernel functions involving the mixed time parameters required to describe varying stress. The results of this investigation showed that the accuracy required of the experimental results to yield a satisfactory determination of these kernel functions was very difficult, if not impossible, to achieve.

As an alternative to direct determination of the kernel functions a possible form of the functions may be chosen and the predictions of this form compared with suitable experiments. The product form of kernel functions proposed by Nakada [10] was compared with results of constant rate of stressing tests and step loading experiments [11] with good results.

Another alternative described by Lai and Findley [12], called the modified superposition method, was also found to yield good agreement with experiments. Nolte and Findley [13] showed that the modified superposition method could be derived from the multiple integral relations by limiting the memory of the material to recent events. This limitation yielded a series of single integrals of exactly the same form as the modified superposition method.

MULTIPLE INTEGRAL METHOD

Experiments have shown [4, 6] that three integrals of the infinite series of multiple integrals is sufficient to describe the behaviour of many plastics. Retaining only the first three terms, it is possible to express the multiple integral relation as follows for creep:

$$
\begin{aligned}
\boldsymbol{\varepsilon}(t) = & \int_0^t (\mathbf{I} K_1 \bar{\dot{\boldsymbol{\sigma}}}_1 + K_2 \dot{\boldsymbol{\sigma}}_1)\, \mathrm{d}\xi \\
& + \int_0^t \int_0^t [\mathbf{I}(K_3 \bar{\dot{\boldsymbol{\sigma}}}_1 \bar{\dot{\boldsymbol{\sigma}}}_2 + K_4 \overline{\dot{\boldsymbol{\sigma}}_1 \dot{\boldsymbol{\sigma}}_2}) + K_5 \bar{\dot{\boldsymbol{\sigma}}}_1 \dot{\boldsymbol{\sigma}}_2 + K_6 \dot{\boldsymbol{\sigma}}_1 \dot{\boldsymbol{\sigma}}_2]\, \mathrm{d}\xi_1\, \mathrm{d}\xi_2 \\
& + \int_0^t \int_0^t \int_0^t [\mathbf{I}(K_7 \overline{\dot{\boldsymbol{\sigma}}_1 \dot{\boldsymbol{\sigma}}_2 \dot{\boldsymbol{\sigma}}_3} + K_8 \bar{\dot{\boldsymbol{\sigma}}}_1 \overline{\dot{\boldsymbol{\sigma}}_2 \dot{\boldsymbol{\sigma}}_3}) + K_9 \bar{\dot{\boldsymbol{\sigma}}}_1 \bar{\dot{\boldsymbol{\sigma}}}_2 \dot{\boldsymbol{\sigma}}_3 \\
& + K_{10} \overline{\dot{\boldsymbol{\sigma}}_1 \dot{\boldsymbol{\sigma}}_2} \dot{\boldsymbol{\sigma}}_3 + K_{11} \bar{\dot{\boldsymbol{\sigma}}}_1 \dot{\boldsymbol{\sigma}}_2 \dot{\boldsymbol{\sigma}}_3 + K_{12} \dot{\boldsymbol{\sigma}}_1 \dot{\boldsymbol{\sigma}}_2 \dot{\boldsymbol{\sigma}}_3]\, \mathrm{d}\xi_1\, \mathrm{d}\xi_2\, \mathrm{d}\xi_3 \qquad (1a)
\end{aligned}
$$

where $\boldsymbol{\varepsilon} = \varepsilon_{ij}$ is the strain tensor, $\mathbf{I} = \delta_{ij}$ is the Kronecker delta which has the value of unity when $i = j$ and zero when $i \neq j$. K_1 to K_{12} are the kernel functions of the time variables as follows:

$$K_j = K_j(t - \xi_1), \qquad j = 1, 2$$

$$K_j = K_j(t - \xi_1, t - \xi_2), \qquad j = 3, 4, 5, 6$$

$$K_j = K_j(t - \xi_1, t - \xi_2, t - \xi_3), \qquad j = 7, \ldots, 12 \tag{1b}$$

t is current time, ξ is an arbitrary prior time, $\boldsymbol{\sigma} = \sigma_{ij}$ is the stress tensor $\dot{\boldsymbol{\sigma}}$ is the stress rate tensor and the bar ($\bar{\ }$) denotes the trace of the stress rate tensor or product of tensors as indicated in the following.

The trace of a tensor is the sum of the diagonal terms of the matrix. Thus, in the tensor,

$$\sigma_{ij} = \begin{vmatrix} \sigma_{11} & \sigma_{12} & \sigma_{13} \\ \sigma_{21} & \sigma_{22} & \sigma_{23} \\ \sigma_{31} & \sigma_{32} & \sigma_{33} \end{vmatrix}$$

the trace $\bar{\boldsymbol{\sigma}} = \operatorname{tr} \sigma_{ij} = \sigma_{11} + \sigma_{22} + \sigma_{33}$. It may be shown that the three invariants of stress may be expressed as functions of the traces of the stress or products of stress tensors. Thus, the terms in eqn (1) having bars over them are invariants of stress. In eqn (1)

$$\bar{\dot{\boldsymbol{\sigma}}}_1 = \operatorname{tr} \dot{\boldsymbol{\sigma}}(\xi_1) = \dot{\sigma}_{ii}(\xi_1)\dagger = \dot{\sigma}_{11}(\xi_1) + \dot{\sigma}_{22}(\xi_1) + \dot{\sigma}_{33}(\xi_1)$$

is the trace of the stress rate tensor,

$$\bar{\dot{\boldsymbol{\sigma}}}_1\bar{\dot{\boldsymbol{\sigma}}}_2 = \operatorname{tr} \dot{\sigma}_{ij}(\xi_1) \operatorname{tr} \dot{\sigma}_{ij}(\xi_2) = \dot{\sigma}_{ii}(\xi_1)\dot{\sigma}_{jj}(\xi_2), \qquad i, j = 1, 2, 3$$

$$\overline{\dot{\boldsymbol{\sigma}}_1\dot{\boldsymbol{\sigma}}_2} = \operatorname{tr} [\dot{\sigma}_{ik}(\xi_1)\dot{\sigma}_{kl}(\xi_2)]$$

$$\overline{\dot{\boldsymbol{\sigma}}_1\dot{\boldsymbol{\sigma}}_2\dot{\boldsymbol{\sigma}}_3} = \operatorname{tr} [\dot{\sigma}_{ik}(\xi_1)\dot{\sigma}_{kj}(\xi_2)\dot{\sigma}_{jl}(\xi_3)]$$

Equation (1) for triaxial stressing has been reduced from 13 kernel functions to 12 by employing the Hamilton–Cayley equation relating the stress tensor and stress invariants [14].

† Repeated subscripts in a single term indicate that the sum is to be taken of the three terms found by substituting the three number subscripts in succession.

A further reduction results from the Hamilton–Cayley equation for biaxial stress with the result that

$$
\begin{aligned}
\varepsilon(t) = & \int_0^t (\mathbf{I}K_1\bar{\dot{\sigma}}_1 + K_2\dot{\sigma}_1)\,\mathrm{d}\xi_1 \\
& + \int_0^t\int_0^t [\mathbf{I}(K_3'\bar{\dot{\sigma}}_1\bar{\dot{\sigma}}_2 + K_4'\overline{\dot{\sigma}_1\dot{\sigma}_2}) + K_5'\bar{\dot{\sigma}}_1\dot{\sigma}_2]\,\mathrm{d}\xi_1\,\mathrm{d}\xi_2 \\
& + \int_0^t\int_0^t\int_0^t [\mathbf{I}(K_7'\bar{\dot{\sigma}}_1\bar{\dot{\sigma}}_2\bar{\dot{\sigma}}_3 + K_8'\bar{\dot{\sigma}}_1\overline{\dot{\sigma}_2\dot{\sigma}_3}) \\
& \qquad + K_9'\bar{\dot{\sigma}}_1\bar{\dot{\sigma}}_2\dot{\sigma}_3 + K_{10}\overline{\dot{\sigma}_1\dot{\sigma}_2}\dot{\sigma}_3]\,\mathrm{d}\xi_1\,\mathrm{d}\xi_2\,\mathrm{d}\xi_3
\end{aligned}
\tag{2a}
$$

where

$$
\begin{aligned}
K_3' &= K_3 - K_6/2 \\
K_4' &= K_4 + K_6/2 \\
K_5' &= K_5 + K_6 \\
K_7' &= -K_7/2 - K_{11} - K_{12}/2 \\
K_8' &= K_8 + 3K_7/2 + K_{11}/2 + K_{12}/2 \\
K_9' &= K_9 + K_{11} + K_{12}/2 \\
K_{10}' &= K_{10} + K_{12}/2
\end{aligned}
\tag{2b}
$$

CREEP AT CONSTANT STRESS

If a state of stress is applied to a previously unstressed material at time $t = 0$ and held constant then all the time functions $t - \xi_1, t - \xi_2, t - \xi_3$ are the same. The integrals in eqn (1a) then may be evaluated by means of the Dirac delta function for which, in this case, $\dot{\sigma}(\xi) = \sigma\delta(\xi)$ with the result:

$$
\begin{aligned}
\varepsilon(t) = & \mathbf{I}(K_1\bar{\sigma} + K_3\bar{\sigma}\,\bar{\sigma} + K_4\overline{\sigma\,\sigma} + K_7\overline{\sigma\,\sigma\,\sigma} + K_8\bar{\sigma}\,\overline{\sigma\,\sigma}) \\
& + \sigma(K_2 + K_5\bar{\sigma} + K_9\bar{\sigma}\,\bar{\sigma} + K_{10}\overline{\sigma\,\sigma}) + \sigma^2(K_6 + K_{11}\bar{\sigma}) + \sigma^3(K_{12})
\end{aligned}
\tag{3a}
$$

where $K_j = K_j(t)$ are functions of time t. Equation (3a) describes the strain for a constant state of triaxial stress. For constant biaxial stress, eqn (2a) reduces to

$$
\begin{aligned}
\varepsilon(t) = & \mathbf{I}(K_1\bar{\sigma} + K_3'\bar{\sigma}\,\bar{\sigma} + K_4'\overline{\sigma\,\sigma} + K_7'\bar{\sigma}\,\bar{\sigma}\,\bar{\sigma} + K_8'\bar{\sigma}\,\overline{\sigma\,\sigma}) \\
& + \sigma(K_2 + K_5'\bar{\sigma} + K_9'\bar{\sigma}\,\bar{\sigma} + K_{10}\overline{\sigma\,\sigma})
\end{aligned}
\tag{3b}
$$

The nine time functions in eqn (3b) may be determined from experiments in combined tension and torsion using three pure tension tests, two pure torsion tests and two combined tension and torsion tests. All tests except one pure tension and one pure torsion should be in the non-linear stress range. These functions may be determined numerically [6] or graphically [15].

Time functions

The creep versus time relationship in the primary stage of creep is a curve of continuously decreasing slope as shown in Fig. 1. For structural plastics in the range of stress of technical importance this stage of creep appears to continue indefinitely. Thus creep curves derived from the behaviour

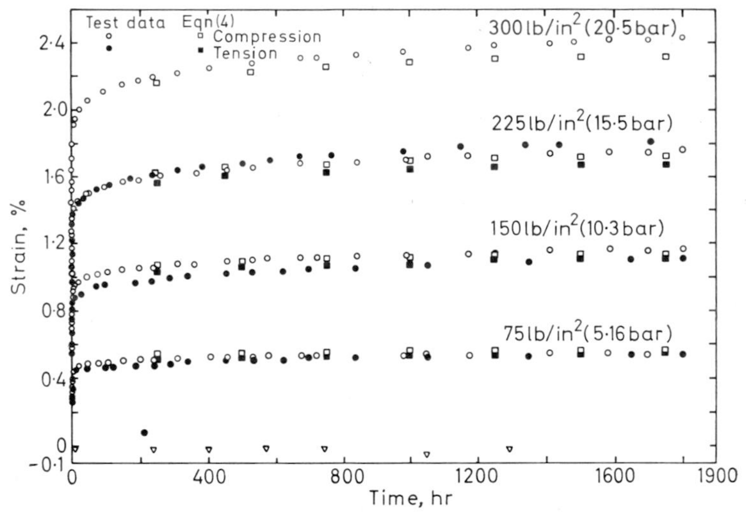

FIG. 1. Strain versus time for tension and compression of polyethylene at 75°F and 50 per cent rh. (From Ref. 20, courtesy ASME.)

of simple mechanical models composed of linear springs and dashpots are not a good approximation of actual behaviour over a long time span, since they tend asymptotically either to a constant strain or a constant strain rate. A power function of time, however, is capable of describing the creep behaviour adequately over a much longer time span for viscoelastic materials, as shown in Fig. 2 [16]. The creep data for the first 2000 hr of

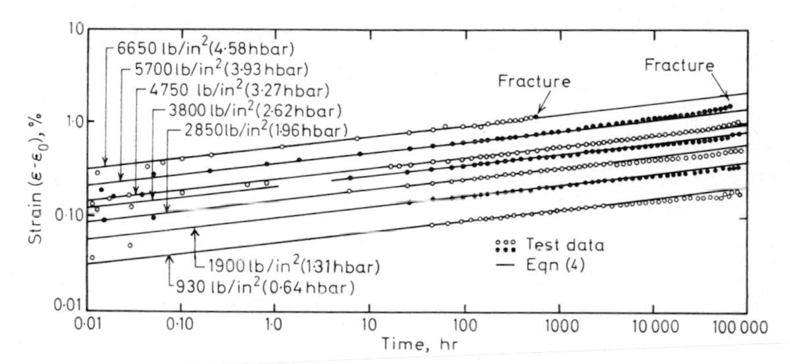

FIG. 2. Log–log plot of strain versus time for tension creep tests of grade C canvas laminate at 77°F and 50 per cent rh (strains corrected for shrinkage). (From Ref. 16, courtesy ASTM.)

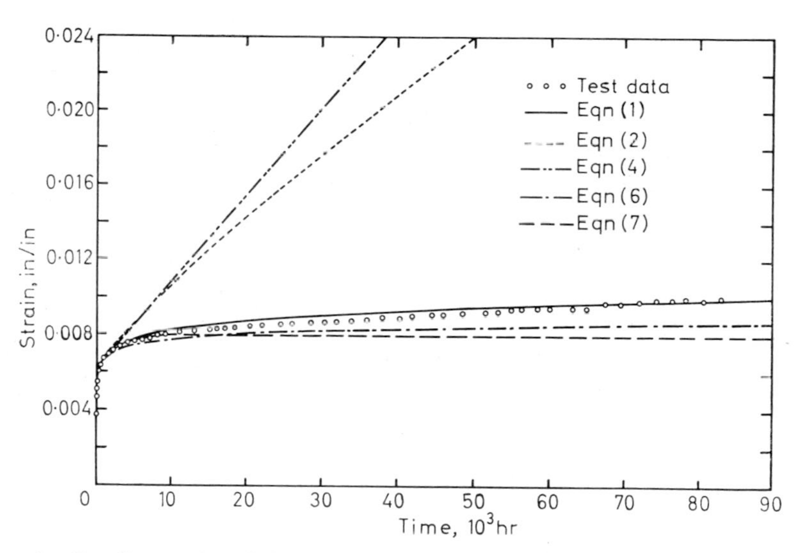

FIG. 3. Rectilinear plot of the creep test of grade C canvas laminate at 3800 lb/in² (2·62 hbar) 70°F and 50% rh together with the prediction of several theories based on the best representation for the first 2000 hr. *Note:* the eqns referred to are; (1) same as eqn (4) in text; (2) $\varepsilon = \varepsilon_0 + A \log t + Bt$; (4) $\varepsilon = \varepsilon_0 + A(1 - e^{-ct}) + Bt$; (6) $\varepsilon = \varepsilon_0 + A \ln t$; (7) $\varepsilon = \varepsilon_0 + A(1 - e^{-ct})$. (From Ref. 16, courtesy ASTM.)

creep of the canvas laminate shown in Fig. 2 were fitted by the power function

$$\varepsilon = \varepsilon^0 + \varepsilon^+ t^n \tag{4}$$

where ε is strain, t time, n a constant independent of stress, ε^0 is the time independent strain and ε^+ is the coefficient of the time dependent term. ε^0 and ε^+ are functions of stress. Then, the creep tests were continued to nearly 100 000 hr as also shown in Fig. 2. The equation of the form (4) derived from the first 2000 hr was extrapolated from 2000 hr to 100 000 hr with the excellent agreement shown in Fig. 2.

The same data for 2000 hr of creep were also described as well as the equations would allow by the other relationships, listed in Fig. 3. These equations were also extrapolated to 100 000 hr for one stress with the

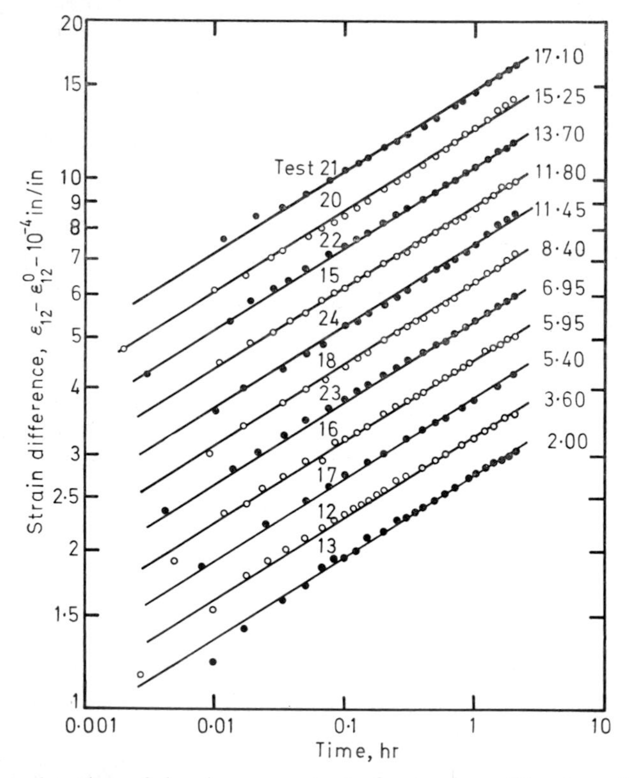

FIG. 4. Log–log plots of shearing component of creep for solid polyurethane under combined tension and torsion. (From Ref. 15, courtesy ASME.) *Note:* numbers on right are the ordinate at 1 hr, which equals ε^+ in eqn (5).

results shown in Fig. 3. The prediction from eqn (4) (the same as equation 1 in Fig. 3) was reasonably good, whereas the prediction from the other time functions shown was far less satisfactory.

It has been found from creep experiments on several plastics in tension, torsion, compression and combined tension and torsion that the creep–time curve may be represented reasonably well by eqn (4) with a constant value of n for a given material independent of stress or state of combined stress. This is shown in Fig. 4 for the shearing (torsion) component of strains from creep of polyurethane (full density) at different stresses and different combinations of tension and torsion. Equation (4) may be rewritten by rearranging and taking logarithms as follows

$$\log(\varepsilon \quad \varepsilon^0) = \log \varepsilon^+ + n \log t \tag{5}$$

Thus, if $\log(\varepsilon - \varepsilon^0)$ is plotted versus $\log t$ as in Fig. 4, eqn (5) predicts a straight line of slope n. Since all curves in Fig. 4 are parallel, n is independent of stress and stress state.

In view of the above the form of each of the kernel functions K_j for constant stress will be taken to be similar in form to eqn (4) in what follows.

General biaxial stress at constant stress

For general biaxial stress, the state of stress is represented by normal stress components σ_1, σ_2 on two planes at right angles plus shear stresses τ on the same planes. Thus,

$$\boldsymbol{\sigma} = \begin{vmatrix} \sigma_1 & \tau & 0 \\ \tau & \sigma_2 & 0 \\ 0 & 0 & 0 \end{vmatrix} \tag{6a}$$

and

$$\bar{\sigma} = \sigma_1 + \sigma_2, \qquad \overline{\sigma\sigma} = \sigma_1^2 + \sigma_2^2 + 2\tau^2 \tag{6b}$$

Substituting eqn (6) in eqn (3b) yields the following six strain components for the general biaxial stress case. For the normal strains $\mathbf{I} = 1$, for the shear strains $\mathbf{I} = 0$:

$$\varepsilon_{11}(t) = A + \sigma_1 B \tag{7a}$$

$$\varepsilon_{22}(t) = A + \sigma_2 B \tag{7b}$$

$$\varepsilon_{33}(t) = A \tag{7c}$$

$$\varepsilon_{12}(t) = \gamma(t)/2 = \tau B \tag{7d}$$

$$\varepsilon_{13} = \varepsilon_{23} = 0 \tag{7e}$$

where

$$A = K_1(\sigma_1 + \sigma_2) + K_3'(\sigma_1 + \sigma_2)^2 + K_4'(\sigma_1{}^2 + \sigma_2{}^2 + 2\tau^2) + K_7'(\sigma_1 + \sigma_2)^3 + K_8'(\sigma_1 + \sigma_2)(\sigma_1{}^2 + \sigma_2{}^2 + 2\tau^2)$$

$$B = K_2 + K_5'(\sigma_1 + \sigma_2) + K_9'(\sigma_1 + \sigma_2)^2 + K_{10}(\sigma_1{}^2 + \sigma_2{}^2 + 2\tau^2)$$

Combined tension and torsion at constant stress

The state of stress represented by combined tension (or compression) σ and torsion τ is given by

$$\boldsymbol{\sigma} = \begin{vmatrix} \sigma & \tau & 0 \\ \tau & 0 & 0 \\ 0 & 0 & 0 \end{vmatrix} \tag{8a}$$

and

$$\bar{\boldsymbol{\sigma}} = \sigma, \qquad \overline{\boldsymbol{\sigma}\,\boldsymbol{\sigma}} = \sigma^2 + 2\tau^2 \tag{8b}$$

The strain components may be found by substituting eqn (8b) in eqn (3b) or by taking $\sigma_1 = \sigma$, $\sigma_2 = 0$ in eqn (7) with the result:

$$\begin{aligned} \varepsilon_{11}(t) &= A' + \sigma B' \\ &= (K_1 + K_2)\sigma + (K_3' + K_4' + K_5')\sigma^2 + (K_7' + K_8' \\ &\quad + K_9' + K_{10})\sigma^3 + 2K_4'\tau^2 + 2(K_8' + K_{10})\sigma\tau^2 \end{aligned} \tag{9a}$$

$$\begin{aligned} \varepsilon_{22}(t) &= \varepsilon_{33}(t) = A' \\ &= K_1\sigma + (K_3' + K_4')\sigma^2 + (K_7' + K_8')\sigma^3 + 2K_4'\tau^2 + 2K_8'\sigma\tau^2 \end{aligned} \tag{9b}$$

$$\begin{aligned} \varepsilon_{12}(t) &= \gamma(t)/2 = \tau B' \\ &= K_2\tau + 2K_{10}\tau^3 + K_5'\sigma\tau + (K_9' + K_{10})\sigma^2\tau \end{aligned} \tag{9c}$$

$$\varepsilon_{13} = \varepsilon_{23} = 0 \tag{9d}$$

where

$$A' = K_1\sigma + K_3'\sigma^2 + K_4'(\sigma^2 + 2\tau^2) + K_7'\sigma^3 + K_8'\sigma(\sigma^2 + 2\tau^2)$$

$$B' = K_2 + K_5'\sigma + K_9'\sigma^2 + K_{10}(\sigma^2 + 2\tau^2)$$

Biaxial normal stress at constant stress

Biaxial tension–tension, compression–compression or tension–compression may be described in terms of the principal stress σ_{I}, σ_{II} as follows:

$$\boldsymbol{\sigma} = \begin{vmatrix} \sigma_{\mathrm{I}} & 0 & 0 \\ 0 & \sigma_{\mathrm{II}} & 0 \\ 0 & 0 & 0 \end{vmatrix} \tag{10a}$$

$$\bar{\boldsymbol{\sigma}} = \sigma_{\mathrm{I}} + \sigma_{\mathrm{II}}, \qquad \overline{\boldsymbol{\sigma}\,\boldsymbol{\sigma}} = \sigma_{\mathrm{I}}^{2} + \sigma_{\mathrm{II}}^{2} \tag{10b}$$

Similarly the strain components are found by substituting eqn (10) in eqn (3b) or taking $\sigma_1 = \sigma_{\mathrm{I}}$, $\sigma_2 = \sigma_{\mathrm{II}}$ and $\tau = 0$ in eqn (7) with the result:

$$\varepsilon_{11}(t) = A'' + \sigma_{\mathrm{I}} B'' \tag{11a}$$

$$\varepsilon_{22}(t) = A'' + \sigma_{\mathrm{II}} B'' \tag{11b}$$

$$\varepsilon_{33}(t) = A'' \tag{11c}$$

$$\varepsilon_{12} = \gamma(t)/2 = \varepsilon_{23} = \varepsilon_{13} = 0 \tag{11d}$$

where

$$A'' = K_1(\sigma_{\mathrm{I}} + \sigma_{\mathrm{II}}) + K_3'(\sigma_{\mathrm{I}} + \sigma_{\mathrm{II}})^2 + K_4'(\sigma_{\mathrm{I}}^2 + \sigma_{\mathrm{II}}^2) + K_7'(\sigma_{\mathrm{I}} + \sigma_{\mathrm{II}})^3 + K_8'(\sigma_{\mathrm{I}} + \sigma_{\mathrm{II}})(\sigma_{\mathrm{I}}^2 + \sigma_{\mathrm{II}}^2)$$

$$B'' = K_2 + K_5'(\sigma_{\mathrm{I}} + \sigma_{\mathrm{II}}) + K_9'(\sigma_{\mathrm{I}} + \sigma_{\mathrm{II}})^2 + K_{10}(\sigma_{\mathrm{I}}^2 + \sigma_{\mathrm{II}}^2)$$

Pure tension or compression at constant stress

For this uniaxial state of stress

$$\boldsymbol{\sigma} = \begin{vmatrix} \sigma & 0 & 0 \\ 0 & 0 & 0 \\ 0 & 0 & 0 \end{vmatrix} \tag{12a}$$

$$\bar{\boldsymbol{\sigma}} = \sigma, \qquad \overline{\boldsymbol{\sigma}\,\boldsymbol{\sigma}} = \sigma^2 \tag{12b}$$

The strain components are found by substituting eqn (12b) in eqn (3b) or by taking all stress components except $\sigma_1 = \sigma_I$ equal to zero in eqns (7), (9) or (11) with the result:

$$\varepsilon_{11}(t) = A''' + \sigma B''' \tag{13a}$$

$$= (K_1 + K_2)\sigma + (K_3' + K_4' + K_5')\sigma^2 + (K_7' + K_8' + K_9' + K_{10})\sigma^3$$

$$\varepsilon_{22}(t) = \varepsilon_{33}(t) = A''' \tag{13b}$$

$$\varepsilon_{12} = \varepsilon_{13} = \varepsilon_{23} = 0 \tag{13c}$$

where

$$A''' = K_1\sigma + (K_3' + K_4')\sigma^2 + (K_7' + K_8')\sigma^3$$

$$B''' = K_2 + K_5'\sigma + (K_9' + K_{10})\sigma^2$$

Pure shear (torsion) at constant stress

This state of stress is given by

$$\boldsymbol{\sigma} = \begin{vmatrix} 0 & \tau & 0 \\ \tau & 0 & 0 \\ 0 & 0 & 0 \end{vmatrix} \tag{14a}$$

$$\overline{\boldsymbol{\sigma}} = 0, \qquad \overline{\boldsymbol{\sigma}\,\boldsymbol{\sigma}} = 2\tau^2 \tag{14b}$$

The corresponding strain components are obtained by substituting eqn (14b) in eqn (3b) or taking $\sigma_1 = \sigma_I = 0, \sigma_2 = \sigma_{II} = 0$ in eqns (7) or (9) as follows:

$$\varepsilon_{11}(t) = \varepsilon_{22}(t) = \varepsilon_{33}(t) = 2K_4'\tau^2 \tag{15a}$$

$$\varepsilon_{12}(t) = \gamma(t)/^2 = \tau(K_2 + 2K_{10}\tau^2) = K_2\tau + 2K_{10}\tau^3 \tag{15b}$$

$$\varepsilon_{13} = \varepsilon_{23} = 0 \tag{15c}$$

Note that eqn (15a) predicts a time dependent normal strain the same in all directions (a time dependent change in volume) unless K_4' is either zero or a time independent constant.

Poisson's ratio

Poisson's ratio ν is given by $\varepsilon_{22}/\varepsilon_{11}$ under a pure tension stress $\sigma_{11} = \sigma_{\text{I}}$. Thus, taking the ratio of eqn (13b) to (13a), yields

$$\nu = \frac{A'''}{A''' + \sigma B'''} = 1/[1 + \sigma(B'''/A''')] \tag{16}$$

Since B''' and A''' are time and stress dependent functions it is clear that Poisson's ratio for a viscoelastic material is a constant of the material only in the unlikely event that $B''' = 0$. If Poisson's ratio is redefined in terms of the time independent strains only, rather than total strains, then it is a time independent function of stress as defined by eqn (16) using only the time independent parts of K_j^0 of K_j. If, as is nearly true, the time independent portion of the strain is taken to be linear then the redefined Poisson's ratio is a constant of the material $K_1^0/(K_1^0 + K_2^0)$.

DETERMINATION OF THE TIME FUNCTIONS

The nine time functions K_j given in eqn (3b) for constant stress may be determined from seven creep tests, three tension (or two tension and one compression), two torsion and two combined tension and torsion. The stresses employed should be distributed over the non-linear range except for one tension and one torsion which should be in the approximately linear range.

If each strain component of the combined stress creep experiments can be represented by eqn (4), with n a constant for the material, then each K_j may be represented by

$$K_j = K_j^0 + K_j^+ t^n \tag{17}$$

and the time independent K_j^0 and time dependent K_j^+ terms may be determined independently from corresponding experimental values.

For example, consider the time dependent coefficients. From three tension creep tests at stresses $\sigma_1 < \sigma_2 < \sigma_3$ the time dependent coefficients ε_1^+, ε_2^+, ε_3^+ may be determined. Substituting these values successively for the left hand side of eqn (9a) and using the corresponding value of σ on the right hand side with $\tau = 0$, three simultaneous equations result.

From these the time dependent part of the coefficients of the σ, σ^2 and σ^3 terms in eqn (9a) may be determined.

From eqn (9c) and the values of $\varepsilon_{12}{}^+$ from a pair of pure torsion creep tests, $K_2{}^+$ and $K_{11}{}^+$ may be determined. From a pure torsion creep test the axial strain introduced on the left of eqn (9a) with $\sigma = 0$ yields $K_4{}'^+$. From the shear strains resulting from two combined tension σ and torsion

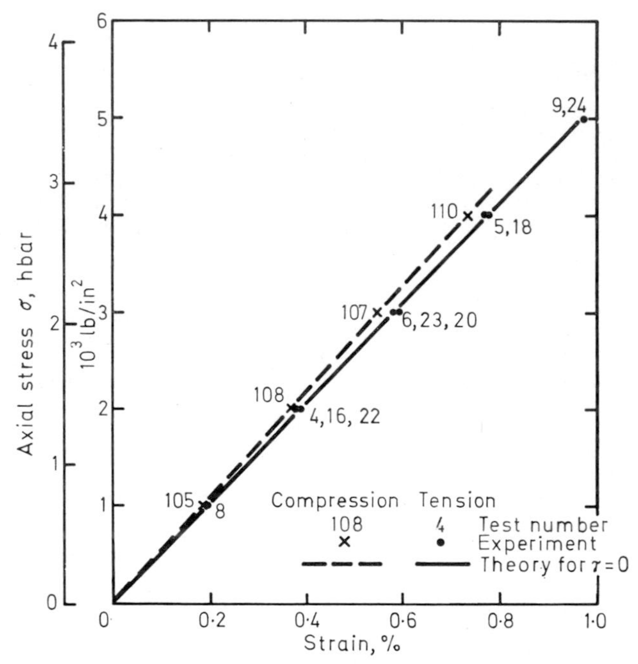

FIG. 5. Axial component of time independent strain, $\varepsilon_{11}{}^0$ versus stress for solid polyurethane at 750°F and 50% rh. (From Ref. 15, courtesy ASME.)

τ creep tests eqn (9c) yields $K_5{}'^+$ and $K_{10}{}'^+$ (since K_{11} has been determined). Finally, from the axial strain for one combined tension and torsion creep test eqn (9a) yields $K_8{}'^+$ since all other terms are known. In a similar manner the time independent K_j's may be determined.

The effect of stress on the time independent strain and time dependent coefficient are shown for a typical plastic in Figs. 5 and 6, respectively. It will be observed that the time independent term is nearly a linear function of stress, whereas the time dependent coefficient is non-linear. Thus, a simplification may be achieved by taking the time independent terms as inear if desired.

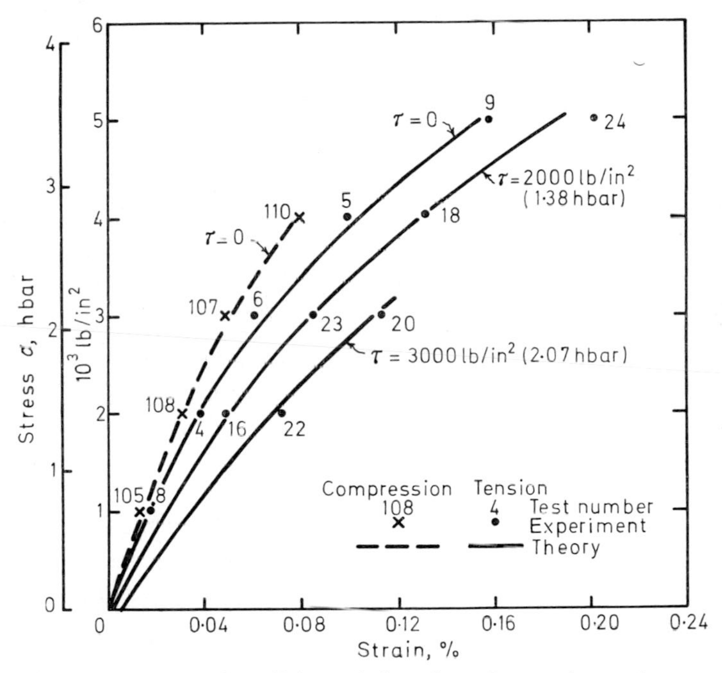

FIG. 6. Axial component of coefficient of time dependent strain, ε_{11}^{+} versus stress for solid polyurethane under combined tension and torsion at 75°F and 50% rh. *Note:* the effect of shear stress. (From Ref. 15, courtesy ASME.)

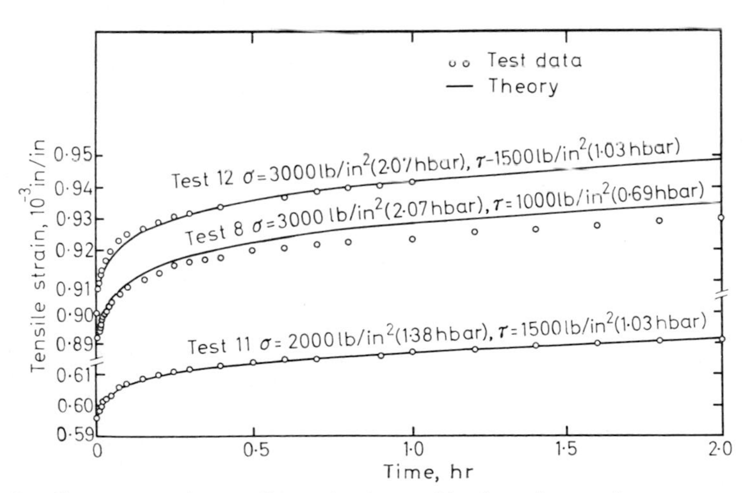

FIG. 7. Creep curves for tensile strains in combined tension–torsion creep tests on polycarbonate at 75°F and 50% rh. (From Ref. 17, courtesy *Polym. Engng. Sci.*)

That the creep behaviour of at least several plastics can be well represented for combined tension and torsion by eqn (9) has been shown by experiments [6, 15, 17]. The comparison between experiments and eqn (9) is shown for combined tension and torsion creep of polycarbonate in Figs. 7 and 8.

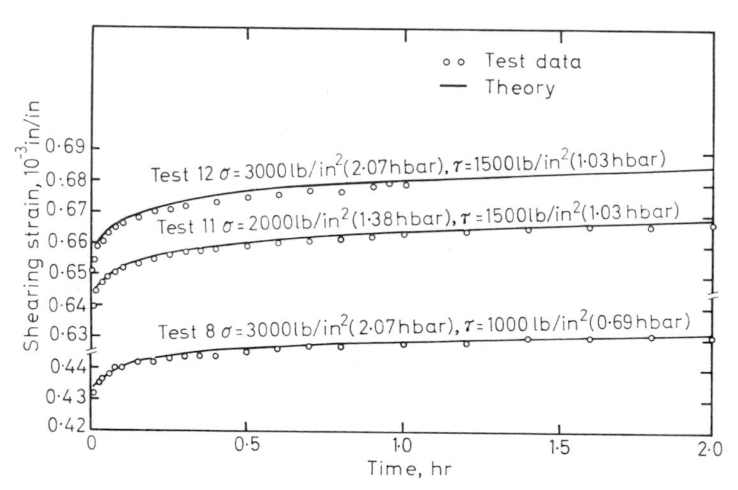

FIG. 8. Creep curves for torsion strains in combined tension–torsion creep tests on polycarbonate at 75°F and 50% rh. (From Ref. 17, courtesy *Polym. Engng. Sci.*)

CREEP UNDER VARIABLE STRESS

Complete evaluation of the kernel functions K_j in eqn (1) when more than one time parameter $(t - \xi_i)$ is involved as in $K_j, j = 3, \ldots, 12$, is very involved [7, 9] and the experimental precision necessary, to say nothing of the hundreds of experiments required, is very difficult to obtain. Accordingly, for varying stress, simpler procedures are desirable.

Product form

One possibility is to assume a form for the multiple time functions eqn (1b). One such assumption is that they are products [10, 11] as follows:

$$K_j(t - \xi_1, t - \xi_2) = K_j(t - \xi_1)K_j(t - \xi_2), \qquad j = 3 \ldots 6 \tag{18a}$$

$$K_j(t - \xi_1, t - \xi_2, t - \xi_3) = K_j(t - \xi_1)K_j(t - \xi_2)K_j(t - \xi_3), \qquad j = 7 \ldots 12 \tag{18b}$$

Employing time functions of the form eqn (4) in eqn (18) yields for the time dependent parts

$$K_j^+(t-\xi_1)^{n/2}K_j^+(t-\xi_2)^{n/2} \tag{19a}$$

$$K_j^+(t-\xi_1)^{n/3}K_j^+(t-\xi_2)^{n/3}K_j^+(t-\xi_3)^{n/3} \tag{19b}$$

Equation (19) may be employed in the multiple integral form eqn (1a), or its equivalent for biaxial stress eqn (2a), to calculate creep under varying stress. For example, inserting eqn (19) into eqn (2) for uniaxial stress yields the following for constant rate of stressing $\dot{\sigma}_0$:

$$\begin{aligned}\varepsilon_{11}(t) &= \int_0^t [R^0 + R^+(t-\xi)^n]\dot{\sigma}_0\,\mathrm{d}\xi \\ &+ \int_0^t\int_0^t [M^0 + M^+(t-\xi_1)^{n/2}(t-\xi_2)^{n/2}]\dot{\sigma}_0^{\,2}\,\mathrm{d}\xi_1\,\mathrm{d}\xi_2 \\ &+ \int_0^t\int_0^t\int_0^t [N^0 + N^+(t-\xi_1)^{n/3}(t-\xi_2)^{n/3}(t-\xi_3)^{n/3}]\dot{\sigma}_0^{\,3}\,\mathrm{d}\xi_1\,\mathrm{d}\xi_2\,\mathrm{d}\xi_3 \\ &= R^0\dot{\sigma}_0 t + R^+\dot{\sigma}_0(n+1)^{-1}t^{n+1} + M^0\dot{\sigma}_0^{\,2}t^2 \\ &+ M^+\dot{\sigma}_0^{\,2}[(n/2)+1]^{-2}t^{n+2} + N^0\dot{\sigma}_0^{\,3}t^3 \\ &\qquad + N^+\dot{\sigma}_0^{\,3}[(n/3)+1]^{-3}t^{n+3}\end{aligned} \tag{20}$$

Results from the product form are in reasonable agreement with experimental results (for the time dependence at least) as shown by Findley and Onaran [11]; *see* Fig. 9. The theoretical values shown in Fig. 9 were computed using results of creep of the same material under constant values of combined tension and torsion.

Modified superposition principle

A simpler procedure, especially for abrupt changes in stress, is called the modified superposition principle [12]. It may be described as follows.

At a change in stress or state of stress at time t_1 it is considered that the strain is the sum of: (a) the strain that would have occurred had the stress not been changed, (b) minus the strain that would have occurred had the prior stress been applied at the time t_1 to a previously unstressed specimen, (c) plus the strain that would have occurred had the current stress been applied at time t_1 to a previously unstressed specimen.

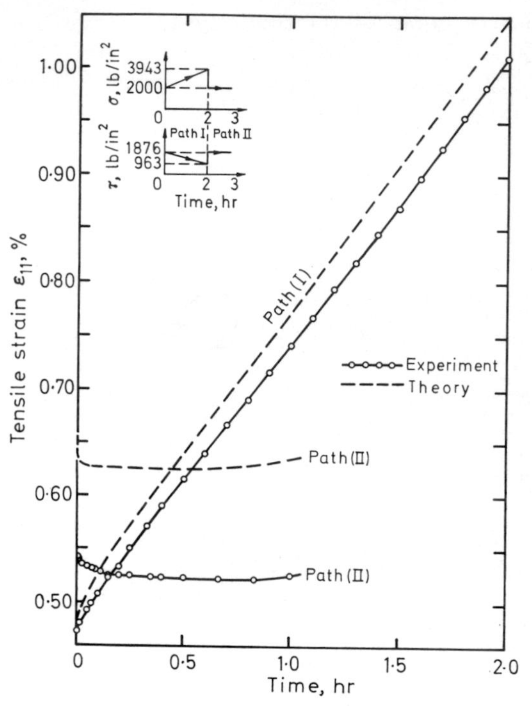

FIG. 9. Tensile strain versus time for combined tension and torsion test of polyvinylchloride at constant rate of stressing at 75°F and 50% rh. (From Ref. 11, courtesy Soc. for Rheology.)

Thus, the strain after an abrupt change in state of stress from $\sigma_{kl}^{(1)}$ to $\sigma_{kl}^{(2)}$ is as follows for time functions of the form eqn (17):

$$\varepsilon_{ij} = f_{ij}^{0}(\sigma_{kl}^{(1)}) + [f_{ij}^{0}(\sigma_{kl}^{(2)}) - f_{ij}^{0}(\sigma_{kl}^{(1)})] + f_{ij}^{+}(\sigma_{kl}^{(1)})t^{n} + [f_{ij}^{+}(\sigma_{kl}^{(2)}) - f_{ij}^{+}(\sigma_{kl}^{(1)})](t - t_1)^{n} \quad (21)$$

For p changes in stress, the strain following the p-th change in stress is

$$\varepsilon_{ij} = f_{ij}^{0}(\sigma_{kl}^{(p)}) + \sum_{p=1}^{p} [f_{ij}^{+}(\sigma_{kl}^{(p)}) - f_{ij}^{+}(\sigma_{kl}^{(p-1)})][t - t_{(p-1)}]^{n}$$

$$p + 1 \geq t > p \quad (22)$$

where f_{ij}^{0} and f_{ij}^{+} stand for the time independent and time dependent coefficients, respectively, of the multiple integral relation for constant stress in eqns (3a), (3b), (7), (9), (11), (13) or (15).

For example, consider the sequence of tension σ and torsion τ stresses shown in Fig. 10. The resulting strain during the different time periods shown may be computed as follows from combined tension and torsion eqn (9) and the modified superposition principle eqn (21). Let $f_{11}(\sigma, \tau) =$ eqn (9a) for tensile strains and $f_{12}(\sigma, \tau) =$ eqn (9c) for shearing strains. By substituting the values of K_j^0, K_j^+ for the given material and the indicated stress components compute

$$f_{11}(\sigma, 0) = A_1^0 + A_1^+ t^n$$

$$f_{11}(\sigma, \tau) = A_2^0 + A_2^+ t^n$$

$$f_{11}(0, \tau) = A_3^0 + A_3^+ t^n$$

$$f_{12}(0, \tau) = B_1^0 + B_1^+ t^n$$

$$f_{12}(\sigma, \tau) = B_2^0 + B_2^+ t^n$$

$$f_{12}(\sigma, 0) = 0$$

where the A's and B's are pure numbers.

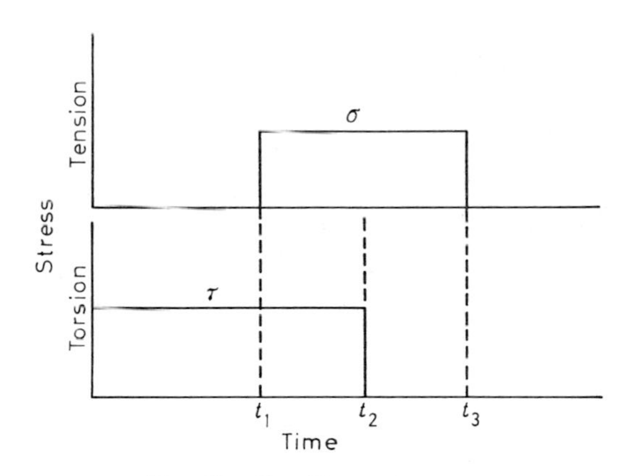

FIG. 10. Loading programme.

For $0 < t < t_1$, $\sigma = 0$, $\tau = \tau$

$$\varepsilon_{11} = A_3^0 + A_3^+ t^n$$

$$\varepsilon_{12} = B_1^0 + B_1^+ t^n$$

Note: ε_{11} is not zero here as the second-last term of eqn (9a) is not zero.

For $t_1 < t < t_2$, $\sigma = \sigma$, $\tau = \tau$

$$\varepsilon_{11} = A_3{}^0 + A_3{}^+ t^n - A_3{}^0 - A_3{}^+(t - t_1)^n + A_2{}^0 + A_2{}^+(t - t_1)^n$$

$$\varepsilon_{12} = B_1{}^0 + B_1{}^+ t^n - B_1{}^0 - B_1{}^+(t - t_1)^n + B_2{}^0 + B_2{}^+(t - t_1)^n$$

For $t_2 < t < t_3$, $\sigma = \sigma$, $\tau = 0$

$$\varepsilon_{11} = A_3{}^+ t^n - A_3{}^+(t - t_1)^n + A_2{}^0 + A_2{}^+(t - t_1)^n - A_2{}^0 - A_2{}^+(t - t_2)^n + A_1{}^0 + A_1{}^+(t - t_2)^n$$

$$\varepsilon_{12} = B_1{}^+ t^n - B_1{}^+(t - t_1)^n + B_2{}^0 + B_2{}^+(t - t_1)^n - B_2{}^0 - B_2{}^+(t - t_2)^n$$

For $t_3 < t$, $\sigma = \tau = 0$

$$\varepsilon_{11} = A_3{}^+ t^n - A_3{}^+(t - t_1)^n + A_2{}^+(t - t_1)^n - A_2{}^+(t - t_2)^n + A_1{}^0 + A_1{}^+(t - t_2)^n - A_1{}^0 - A_1{}^+(t - t_3)^n$$

$$\varepsilon_{12} = B_1{}^+ t^n - B_1{}^+(t - t_1)^n + B_2{}^+(t - t_1)^n - B_2{}^+(t - t_2)^n$$

Note that the time independent strain corresponds to the state of stress at the current time, whereas the time dependent strain is a function of the entire past history of stressing.

It may be observed that ε_{12} is the same from $t = t_2$ on. Thus, the modified superposition principle predicts no change in the creep recovery in torsion when tension is removed during recovery in torsion. Actually experiments show a small effect (*see* Fig. 16b of ref. 18). An effect was also predicted by the product form. Thus, while the modified superposition principle describes creep behaviour under many complex stress histories quite well, there are some features of actual behaviour which are not described by this method.

In Fig. 11 are shown the shearing strains resulting from combined tension and torsion creep tests in which abrupt changes were made in the tension and torsion stresses as indicated in Fig. 11. Also shown by solid lines are the predictions for the second, third and fourth steps of the loading as determined from the modified superposition principle as described above. These computations were based on constant stress creep tests only. The agreement between the theory and experiments shown in Fig. 11 is generally quite satisfactory.

Equation (22) was written in integral form by Pipkin and Rogers [19] and the integral form of eqn (22) was derived from the complete multiple

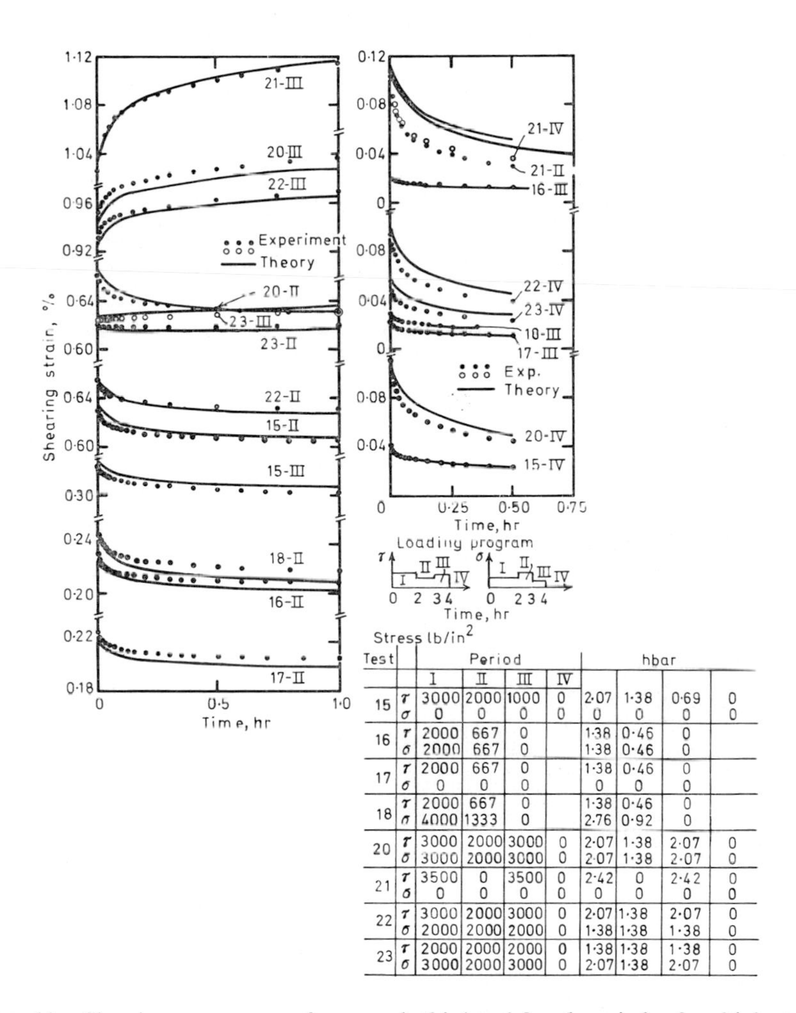

Stress lb/in²

Test		Period I	Period II	Period III	Period IV	hbar			
15	τ	3000	2000	1000	0	2·07	1·38	0·69	0
	σ	0	0	0	0	0	0	0	0
16	τ	2000	667	0		1·38	0·46	0	
	σ	2000	667	0		1·38	0·46	0	
17	τ	2000	667	0		1·38	0·46	0	
	σ	0	0	0		0	0	0	
18	τ	2000	667	0		1·38	0·46	0	
	σ	4000	1333	0		2·76	0·92	0	
20	τ	3000	2000	3000	0	2·07	1·38	2·07	0
	σ	3000	2000	3000	0	2·07	1·38	2·07	0
21	τ	3500	0	3500	0	2·42	0	2·42	0
	σ	0	0	0	0	0	0	0	0
22	τ	3000	2000	3000	0	2·07	1·38	2·07	0
	σ	2000	2000	2000	0	1·38	1·38	1·38	0
23	τ	2000	2000	2000	0	1·38	1·38	1·38	0
	σ	3000	2000	3000	0	2·07	1·38	2·07	0

FIG. 11. Shearing creep curves for second, third and fourth periods of multiple step loading for solid polyurethane under combined tension, σ, and torsion, τ, at 75°F and 50% rh. (From Ref. 15, courtesy ASME.)

integral form eqn (1) by limiting the memory of the material to recent events by Nolte and Findley [13].

The integral form of eqn (22) corresponding to three terms of the multiple integral equation for biaxial stress eqn (2a) yields the following series of single integrals:

$$\boldsymbol{\varepsilon}(t) = \int_0^t [\mathbf{I}(K_1\bar{\dot{\boldsymbol{\sigma}}} + K_3'\bar{\dot{\boldsymbol{\sigma}}}\,\bar{\dot{\boldsymbol{\sigma}}} + K_4'\overline{\dot{\boldsymbol{\sigma}}\,\dot{\boldsymbol{\sigma}}} + K_7'\bar{\dot{\boldsymbol{\sigma}}}\,\bar{\dot{\boldsymbol{\sigma}}}\,\bar{\dot{\boldsymbol{\sigma}}} + K_8'\bar{\dot{\boldsymbol{\sigma}}}\,\overline{\dot{\boldsymbol{\sigma}}\,\dot{\boldsymbol{\sigma}}}$$

$$+ K_2\dot{\boldsymbol{\sigma}} + K_5'\bar{\dot{\boldsymbol{\sigma}}}\,\dot{\boldsymbol{\sigma}} + K_9'\bar{\dot{\boldsymbol{\sigma}}}\,\bar{\dot{\boldsymbol{\sigma}}}\,\dot{\boldsymbol{\sigma}} + K_{10}\overline{\dot{\boldsymbol{\sigma}}\,\dot{\boldsymbol{\sigma}}}\,\dot{\boldsymbol{\sigma}}]\,\mathrm{d}\xi \qquad (23)$$

Equation (23) may be employed to predict creep strain under varying stress from knowledge of creep behaviour under constant stress alone.

For example, consider pure torsion applied at a constant rate of stressing l to a material having a time function $K_j(t) = K_j{}^0 + K_j{}^+t^n$. Thus, $\dot{\sigma} = \dot{\tau} = l$. For pure torsion, eqn (23) reduces to:

$$\varepsilon_{12}(t) = \int_0^t [K_2(t - \xi)\dot{\tau} + 2K_{10}(t - \xi)\dot{\tau}^3]\,\mathrm{d}\xi \qquad (24)$$

Introducing the time function and stress rate and integrating yields:

$$\varepsilon_{12}(t) = K_2{}^0lt + 2K_{10}{}^0l^3t + K_2{}^+(n + 1)^{-1}t^{n+1}$$

$$+ 2K_{10}{}^+l^3(n + 1)^{-1}t^{n+1} \qquad (25)$$

CONCLUSIONS

The results of experiments on creep of plastics suggest the following:

1 The time dependent part of creep strain may be expressed with good accuracy as a power function of time with a stress independent power of the same value for all combinations of tension and torsion.

2 The stress dependence in the range of non-linear stress may be described by a multiple integral function.

3 For variable stressing the behaviour may be expressed reasonably well either by a product form for the kernel functions of the integrals involving multiple time parameters or by a modified superposition principle which amounts to a limited memory form of the multiple integral representation. The latter is easier to apply to step load changes and is somewhat closer to the experimental results, but does not describe all features of creep behaviour.

REFERENCES

1. Markovitz, H. (1957). 'Normal stress effect in polyisobutylene solutions. II Classification and application of rheological theories.' *Trans. Soc. Rheology*, **1,** 25–36.
2. Green, A. E. and Rivlin, R. S. (1957/60). 'The mechanics of non-linear materials with memory.' *Arch. for Rational Mechanics and Analysis*, **1** (Pt I); **3** (Pt II); **4,** (Pt IV).
3. Finnie, I. and Heller, W. R. (1959). *Creep of Engineering Materials*, New York: McGraw-Hill.
4. Ward, I. M. and Onat, E. T. (1965). 'Non-linear mechanical behavior of oriented polypropylene.' *J. Mech. and Phys. of Solids*, **11,** 217.
5. Leaderman, H., McCrackin, F. and Nakada, O. (1963). 'Large longitudinal retarded elastic deformation of rubberlike network polymers. II Application of a general formulation of nonlinear response.' *Trans. Soc. Rheology*, **7,** 111–123.
6. Findley, W. N. and Onaran, K. (1965). 'Combined stress creep experiments on a non-linear viscoelastic material to determine the kernel functions for a multiple integral representation of creep.' *Trans. Soc. Rheology*, **9** (2), 299–327.
7. Lockett, F. J. (1965). 'Creep and stress-relaxation experiments for non-linear materials.' *Int. J. Engng Sci.*, **3,** 59–75.
8. Dong, R. (1964). 'Studies in mechanics of nonlinear solids.' University of California Report UCRL-12039.
9. Onaran, K. and Findley, W. N. (1968). 'Experimental determination of some kernel functions in the multiple integral method for nonlinear creep of polyvinyl chloride.' *Tech. Rep.* No. 8, Purchase Order No. 4627903, Lawrence Radiation Laboratory, University of California, and *Techn. Rep.* No. 4, Grant No. NGR 40-002-027, National Aeronautics and Space Administration. *J. Appl. Mech.* (in press).
10. Nakada, O. (1961). 'Theory of non-linear viscoelasticity. II Analysis of non-linear creep of plastics.' *Reports on Progress in Polymer Physics in Japan*, 4.
11. Findley, W. N. and Onaran, K. (1968). 'Product form of kernel functions for nonlinear viscoelasticity of PVC plastic under constant rate stressing.' *Trans. Soc. Rheology*, **12** (2), 217–242.
12. Findley, W. N. and Lai, J. S. Y. (1967). 'A modified superposition principle applied to creep of nonlinear viscoelastic material under abrupt changes in state of combined stress.' *Trans. Soc. Rheology*, **11** (3), 361–380.
13. Nolte, K. G. and Findley, W. N. (1968). 'Multiple step, nonlinear creep of polyurethane predicted from constant stress creep by three integral representations.' *Techn. Rep.* No. 6, Grant No. NGR 40-002-027, National Aeronautics and Space Administration. *Trans. Soc. Rheology* (in press).
14. Prager, W. (1961). *Introduction to the Mechanics of Continua*, Boston: Ginn and Co.
15. Nolte, K. G. and Findley, W. N. (1970). 'Relationship between the creep of solid and foam polyurethane resulting from combined stresses.' *J. bas. Engng*, **92** D (1), 105–114.

16. FINDLEY, W. N. and PETERSON, D. B. (1958). 'Prediction of long-time creep with ten-year creep data on four plastic laminates.' *Proc. Amer. Soc. for Testing Materials*, **58,** 841–861.
17. LAI, J. S. Y. and FINDLEY, W. N. (1969). 'Combined tension–torsion creep experiments on polycarbonate in the nonlinear range.' *Polym. Engng Sci.*, **9** (5), 378–382.
18. FINDLEY, W. N. and STANLEY, C. A. 'Combined stress creep experiments on rigid polyurethane foam in the nonlinear region with application to multiple integral and modified superposition theory.' *ASTM J. Mater.* (in press).
19. PIPKIN, A. C. and ROGERS, T. G. (1968). 'A nonlinear integral representation for viscoelastic behavior.' *J. Mech. and Phys. Solids*, **16,** 59–72.
20. O'CONNOR, D. G. and FINDLEY, W. N. (1962). 'Influence of normal stress on creep in tension and compression of polyethylene and rigid polyvinyl chloride copolymers'. *J. Engng for Industry*, **84** (2), 237–247.

SECTION III

CREEP AND DESIGN OF COMPONENTS

Chapter 15

CREEP OF TUBULAR SPECIMENS UNDER COMBINED STRESS

S. TAIRA AND R. OHTANI

SUMMARY

Creep tests of pressurised tubes of four steels indicate that strains follow the von Mises effective stress rule. For rupture, three materials follow von Mises, but the 'brittle' material tends to follow the maximum tensile stress criterion. For prediction of the rupture life of tubes, the mean diameter formula appears satisfactory.

NOTATION

b, n, α	Material constants.
C, K	Defined in eqn (8).
D, d	Outside and inside diameters of tubes.
I	Defined in eqn (11).
P	Axial load.
p	Internal pressure.
r	Radius of tubes.
t	Time and wall thickness of tube.
t_r	Time to rupture.
y	Constant in eqn (16).
ε_c	Creep strain.
ε^*	Effective strain.
ε_c^*	Effective creep strain.
σ	Stress.
σ^*	Effective stress.

σ_a Additional axial stress.
σ_f Simple tensile stress giving fracture in same time as another stress.
σ_m Hydrostatic component of stress.

Subscripts

i, o Values at inner and outer radii respectively.
r Radial component.
rup Rupture value.
t Tangential component.
z Axial component.

INTRODUCTION

There are several ways to approach the study of problems concerning the creep of metallic materials under combined stress systems. A number of theoretical and experimental works have hitherto been published on applications to practical service [1–30]. The results of the investigations in this field indicate that the von Mises criterion is valid for the creep deformation of isotropic materials. On the other hand, for creep fracture, the von Mises criterion is not always the best criterion for the prediction of rupture life and the maximum principal stress criterion is superior in some cases. The recent trend of investigation of the creep rupture of internally pressurised tubes at elevated temperatures has led to the tentative conclusion that the so-called mean diameter formula shows the best correlation with the results of uniaxial tensile creep testing [31, 32]. However, this cannot be simply interpreted in terms of either the von Mises or the maximum principal tensile stress criterion, because both predict smaller values of rupture stress than the mean diameter formula. These complications may engender reservations about applying any simple criterion to predict rupture under multiaxial stress.

With these findings in mind, the stress criterion for creep rupture of tubes under internal pressure was considered theoretically, and the validity of the mean diameter formula for the design of pressure vessels and boiler tubes was discussed. For this purpose, creep and creep rupture tests at elevated temperature were conducted on cylindrical specimens of several steels under internal pressure, and under combined axial load and internal pressure. The measured residual stress in the creep strained cylinders was used to derive experimentally the stress distributions in the creep condition.

MATERIALS TESTED AND TEST PIECES

Four kinds of steel were tested. Materials A and B are low carbon steels and material C is a $2\frac{1}{4}$ per cent Cr-1 per cent Mo steel, in the form of commercial boiler tubes of 50 mm outside diameter and wall thickness of about 12·5 mm. Material D is an 18 per cent Cr, 8 per cent Ni, Nb austenitic stainless steel (type AISI 347), in the form of 25 mm diameter hot rolled bars. Chemical compositions, conditions of heat treatment and mechanical properties at room temperature of the materials are listed in Table 1.

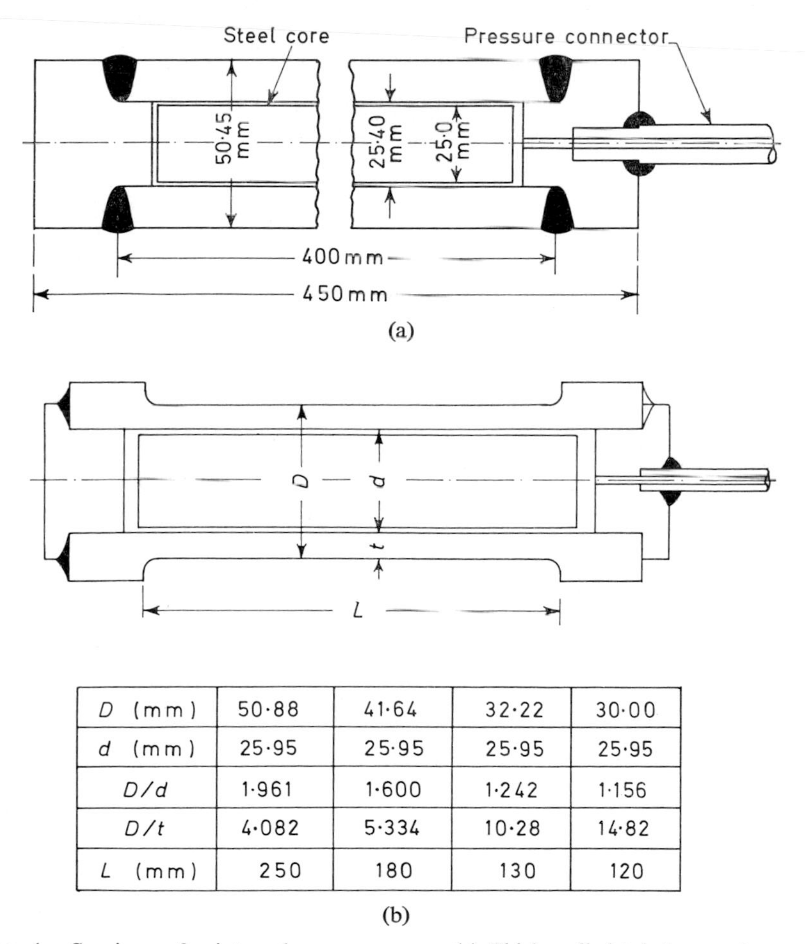

D (mm)	50·88	41·64	32·22	30·00
d (mm)	25·95	25·95	25·95	25·95
D/d	1·961	1·600	1·242	1·156
D/t	4·082	5·334	10·28	14·82
L (mm)	250	180	130	120

FIG. 1. Specimens for internal pressure creep. (a) Thick walled tubular specimen of material B. (b) Tubular specimen of material C.

TABLE 1

Chemical compositions, conditions of heat treatment and mechanical properties at room temperature

Material	Compositions (per cent)										Heat treatment
	C	Si	Mn	P	S	Cu	Ni	Cr	Mo	Nb+Ta	
A	0·14	0·28	0·48	0·014	0·010						920°C × 1 hr
B	0·19	0·28	0·55	0·017	0·008	0·12	0·07	0·07	0·02		920°C × 1 hr
C	0·11	0·38	0·48	0·010	0·008	0·11	0·10	2·16	0·93		920°C × 1 hr
D	0·05	0·50	1·54	0·018	0·018	0·07	12·65	17·83		0·86	1100°C × 1 hr → W.Q.

Material	Yield point		Tensile strength		Elongation (per cent)	Reduction of area (per cent)
	(kg/mm^2)	(hbar)	(kg/mm^2)	(hbar)		
A	28·8	(28·2)	46·4	(45·5)	28·0	64·7
B	30·4	(29·8)	46·2	(45·3)	32·7	69·8
C	29·2	(28·6)	50·4	(49·4)	27·2	77·0
D	28·8	(28·2)	61·4	(60·2)	64·5	69·2

Figure 1 shows the tubular specimens made of the materials B and C used for the creep test under internal pressure. Figure 2 shows the test pieces of materials A and D used for the combined stress creep test. Both ends of the specimens are closed by welding and solid steel cores are inserted into them to reduce the internal vapour volume. The specimens shown in Fig. 2 have end screws for the application of additional axial load.

As detailed descriptions of the creep testing apparatus have already been published [28, 33, 34], they will not be further described here.

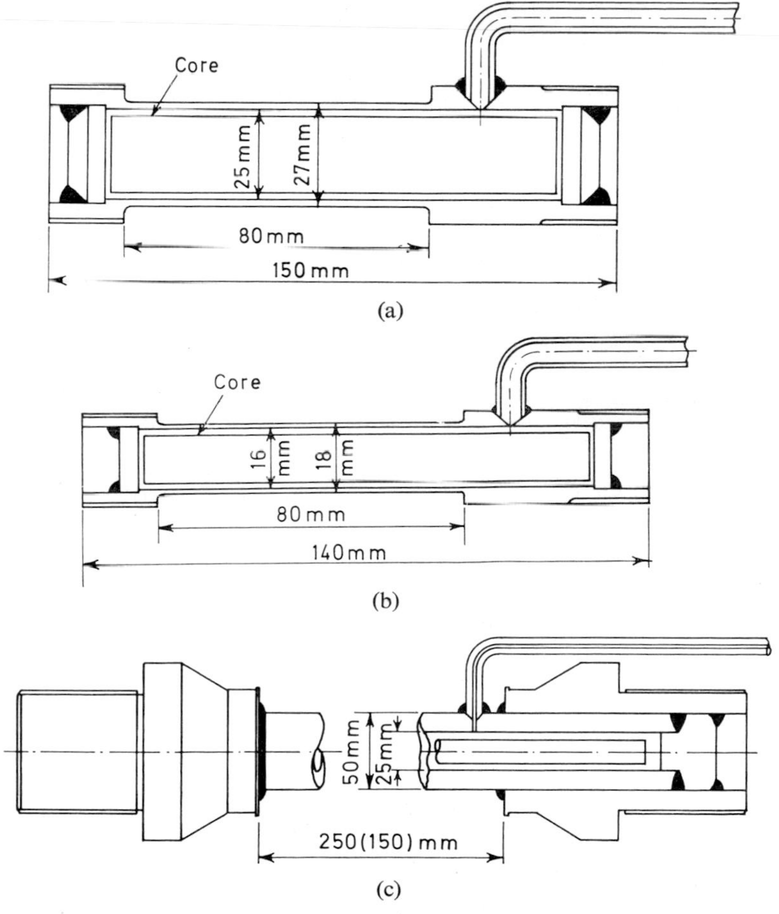

FIG. 2. Tubular specimens for combined stress creep. (a) Material A. (b) Material D. (c) Material A, arrangement for combined axial compression and internal pressure.

CREEP CURVES

Figure 3 shows simple tension creep curves (full lines) obtained with 6 mm diameter solid bar specimens cut from the tubes, as well as internal pressure creep curves (dotted lines) for the thick walled tubular specimens illustrated in Fig. 1a. For the creep curves of internally pressurised tubes, the

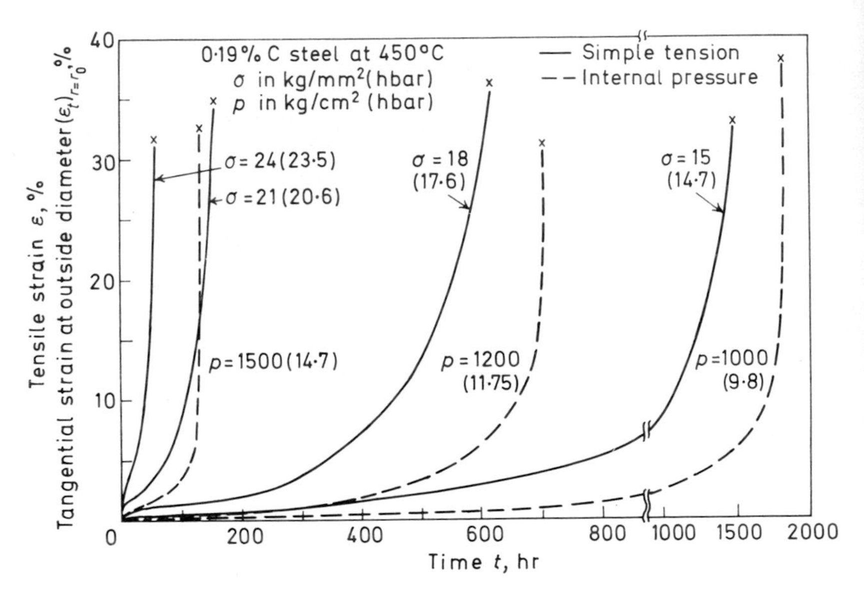

FIG. 3. Creep curves under simple tension only, and under internal pressure only.

ordinates of the diagram are the tangential strains at the outside diameter $(\varepsilon_t)_{r=r_0}$.

Figures 4a and b show creep curves for the thin walled tubular specimens in Figs. 2a and b, respectively, which are subjected to combined internal pressure and axial tension. The initial effective stress was chosen to have the same value of 18·0 kg/mm² (17·6 hbar), based on the von Mises stress criterion:

$$\sigma^* = \frac{1}{\sqrt{2}}[(\sigma_t - \sigma_z)^2 + (\sigma_z - \sigma_r)^2 + (\sigma_r - \sigma_t)^2]^{\frac{1}{2}} \tag{1}$$

in which the subscripts t, z and r mean tangential, axial and radial component, respectively. The ordinates of these diagrams are the effective strain of the von Mises type defined as follows:

$$\varepsilon^* = \frac{\sqrt{2}}{3}[(\varepsilon_t - \varepsilon_z)^2 + (\varepsilon_z - \varepsilon_r)^2 + (\varepsilon_r - \varepsilon_t)^2]^{\frac{1}{2}} \tag{2}$$

In Fig. 5, the results of creep tests carried out at the temperature of 500°C on the specimens shown in Fig. 2c are illustrated in the creep curves.

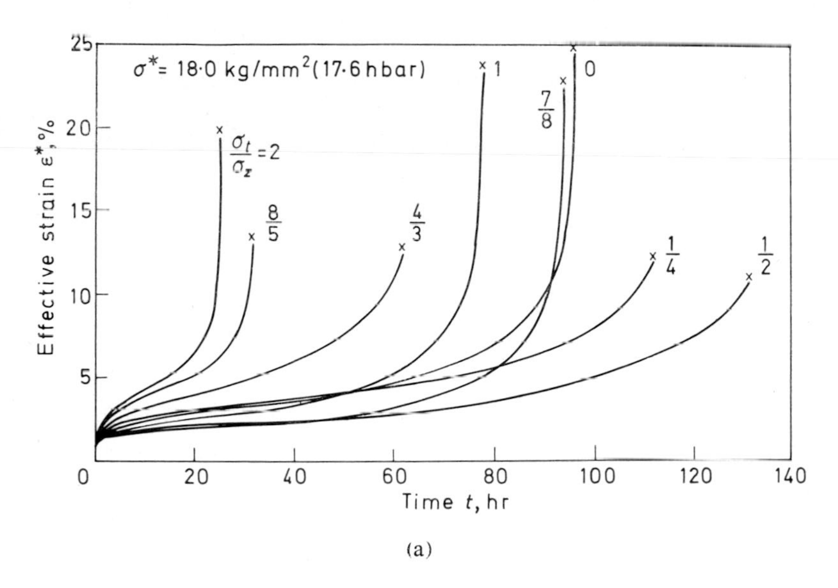

(a)

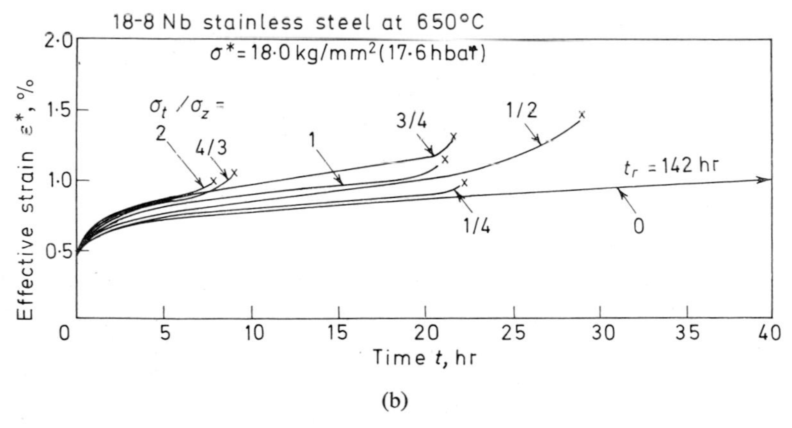

(b)

FIG. 4. Creep curves for the specimens under combined axial tension and internal pressure. (a) Specimens in Fig. 2(a). (b) Specimens in Fig. 2(b).

The stress conditions employed in the tests are as follows:

1 Tension-internal pressure: $p = 1000$ kg/cm^2 (9·8 hbar), $\sigma_a = 10{\cdot}0$ kg/mm^2 (9·8 hbar)

2 Pure internal pressure: $p = 1290$ kg/cm^2 (12·6 hbar), $\sigma_a = 0$

3 Compression-internal pressure: $p = 1000$ kg/cm^2 (9·8 hbar), $\sigma_a = -10{\cdot}0$ kg/mm^2 ($-9{\cdot}8$ hbar)

where p is an internal pressure and σ_a is an additional axial stress. These applied stresses were chosen for each case to be the same value as the von Mises effective stress throughout the initial cross-section of 15·9 kg/mm^2 (15·6 hbar).

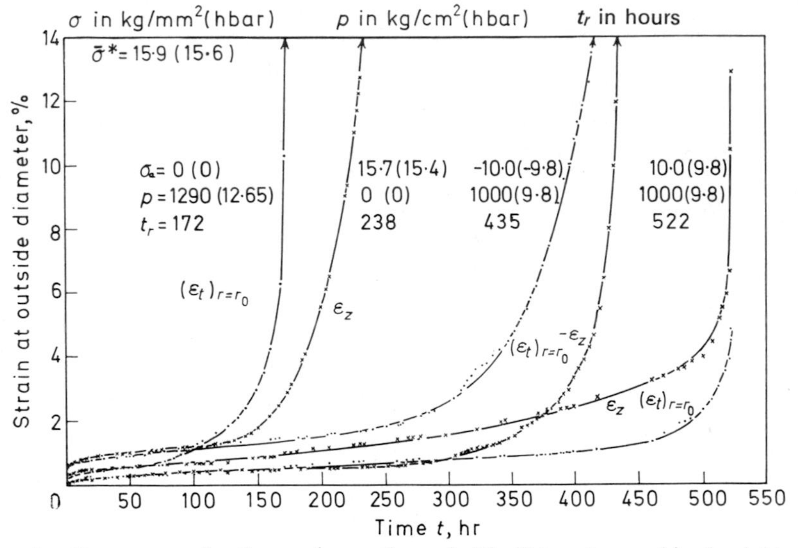

FIG. 5. Creep curves for the specimens shown in Fig 2(c) under combined axial load and internal pressure.

In these figures, there are considerable discrepancies among the creep curves, and the shapes of the creep curves under internal pressure and under combined axial load and internal pressure do not coincide with those under simple tension. When the data in Fig. 3 were replotted as true stress against creep strain and creep strain against time diagrams, Fig. 6a and b, respectively, were obtained. The parts of these lines corresponding to the transient and the steady stages of creep are parallel with each other.

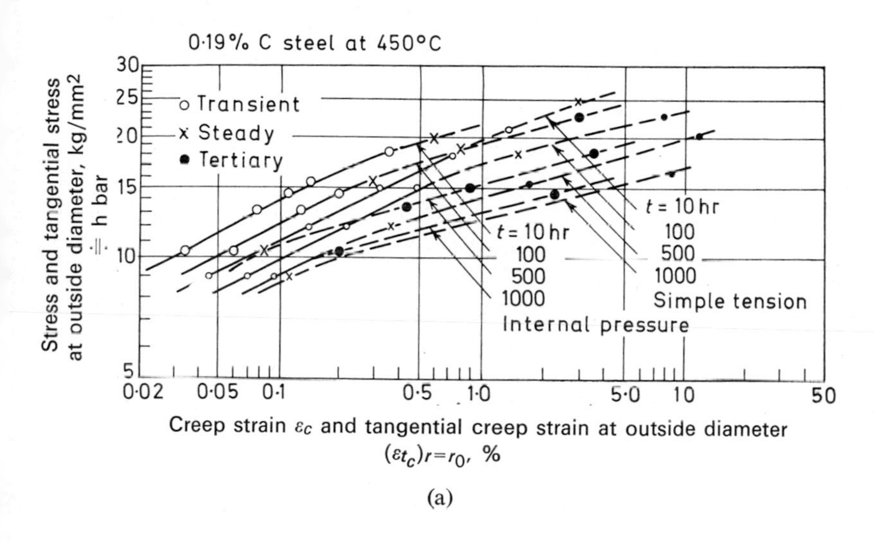

(a)

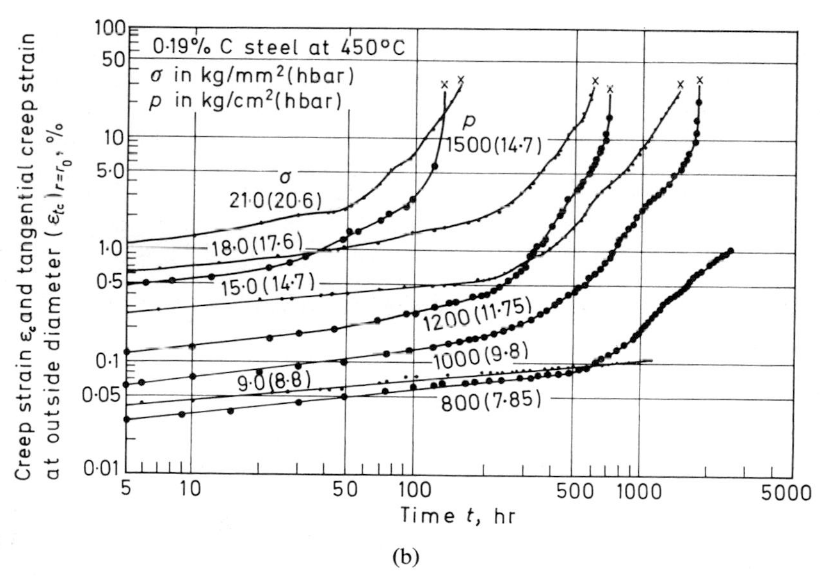

(b)

FIG. 6. Curves under simple tension only and internal pressure only for material B. (a) True stress/creep strain. (b) Creep strain/time.

This indicates that, if the true effective stress σ^*, the natural effective creep strain ε_c^* and the effective creep rate $\dot{\varepsilon}_c^*$ are related through the material constants b, α and n in the form

$$\dot{\varepsilon}_c^* = b\sigma^{*\alpha}\varepsilon_c^{*n} \tag{3}$$

the magnitudes of these material constants for the creep of internally pressurised cylinders are equal to those for simple tensile creep. The same can be found in the test results under combined stress. Figure 7 shows the parts of the creep curves in Fig. 5 by taking the von Mises

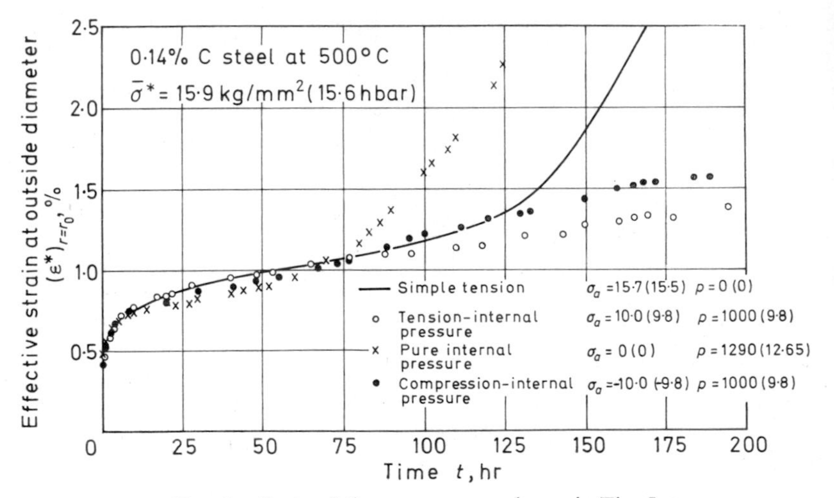

FIG. 7. Parts of the creep curves shown in Fig. 5.

effective strain at the outside diameter $(\varepsilon^*)_{r=r_0}$ as the ordinate of the diagram. It is found that they agree well among themselves up to the beginning of the tertiary creep. The results suggest that behaviour or change of microstructure of a material in the creep condition under multiaxial stress is almost the same as that of the material in creep under uniaxial stress. In other words, in general, creep curves under simple tension as well as those under combined stress will be similar in form at the same testing temperature under the same magnitude of the effective stress σ^*.

However, Figs. 3, 4 and 5 show that, especially at the tertiary stage of creep, the creep curves for internal pressure and for combined axial load and internal pressure do not coincide with those for simple tension. From Fig. 4 the higher the stress ratio σ_t/σ_z, the larger the strain at a given time

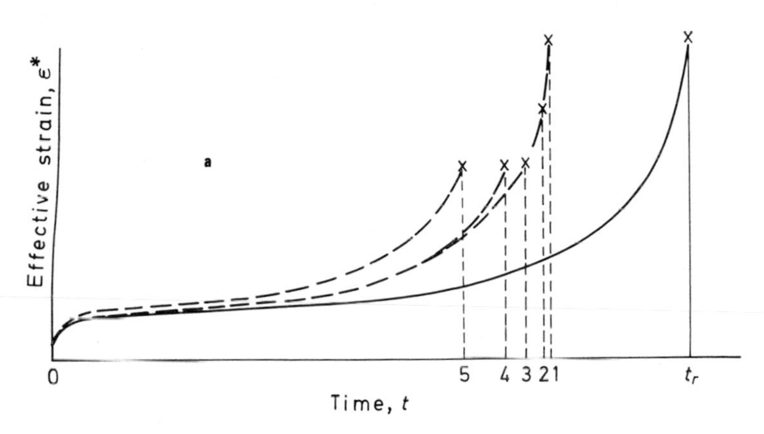

—— Simple tension for bar specimen under constant axial load

- - - Tubular specimen under constant internal pressure

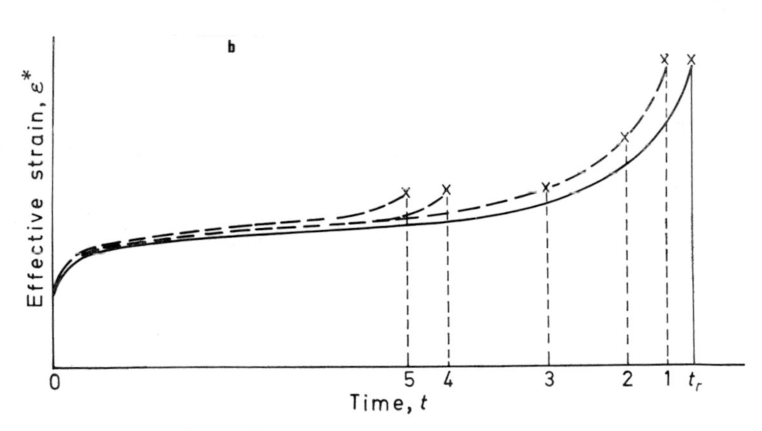

FIG. 8. Schematic creep curves on two kinds of materials, showing the influence of several factors on the shape and the time to rupture of pressurised tubes. (a) Large rupture elongation: a few or no visible cracks at the third stage of creep. (b) Small rupture elongation: propagation of grain boundary cracks at the early stage of creep.

and the shorter the time to rupture. Several possibilities are conceivable in explanation of these discrepancies [35]. The main factors which cause such disagreement in the creep curves are presumed to be as follows:

1 Different ratios of combined stress resulting in different increases in true stress.
2 Influence of the hydrostatic component of stress.
3 Size effect of specimens.
4 Influence of the maximum tensile stress on the crack initiation and propagation.
5 Influence of the anisotropy of materials.

Figure 8 shows schematically how the shape of the creep curve and the time to rupture of internally pressurised tubes are affected by the factors mentioned above for two kinds of materials: (a) with large rupture elongation and (b) with low ductility exhibiting grain boundary fracture. Detailed discussions on these factors will be made in turn.

INCREASE IN TRUE STRESS DURING CREEP

Let us first consider the increase in true stress in the thin walled tube shown in Fig. 2a and b subjected to combined constant tensile load and internal pressure. When the radial stress σ_r is regarded as being zero, the true tangential stress σ_t and axial stress σ_z can be described by the initial stresses σ_{t0}, σ_{z0} and the natural strain components ε_t, ε_z, ε_r as follows:

$$\sigma_t = \sigma_{t0} \exp(\varepsilon_t - \varepsilon_r) \qquad \sigma_z = \sigma_a \exp \varepsilon_z + \sigma_t/2 \tag{4}$$

An example of calculated results for the tubular specimen in Fig. 2a of the material A is shown in Fig. 9 [33]. The diagram indicates the variation of stress state during creep, that is, a part of the ellipse at $t = 0$ hour represents the initial state of the von Mises effective stress of 18·0 kg/mm^2 (17·6 hbar), and it changes in the manner shown by the full lines corresponding to 20, 40, 50, 60 and 80 hr. It is noted that the increase in true stress is greatest in the case of pure internal pressure ($\sigma_t/\sigma_z = 2$) and is smallest for the stress ratio $\sigma_t/\sigma_z = 1/2$. The dotted lines in the figure indicate the traces of the change in each stress ratio, which is found to become larger or smaller than the initial value with the increase in creep deformation of the tube except for the cases of pure internal pressure, simple tension and combined stress of $\sigma_t/\sigma_z = 1/2$. Such a change in stress state during creep has a close relation to the discrepancy in the creep curves shown in Fig. 4a. The difference in shape among the creep curves shown in Figs. 3 and 5 is also partly attributable to the different rate of increase in true stress.

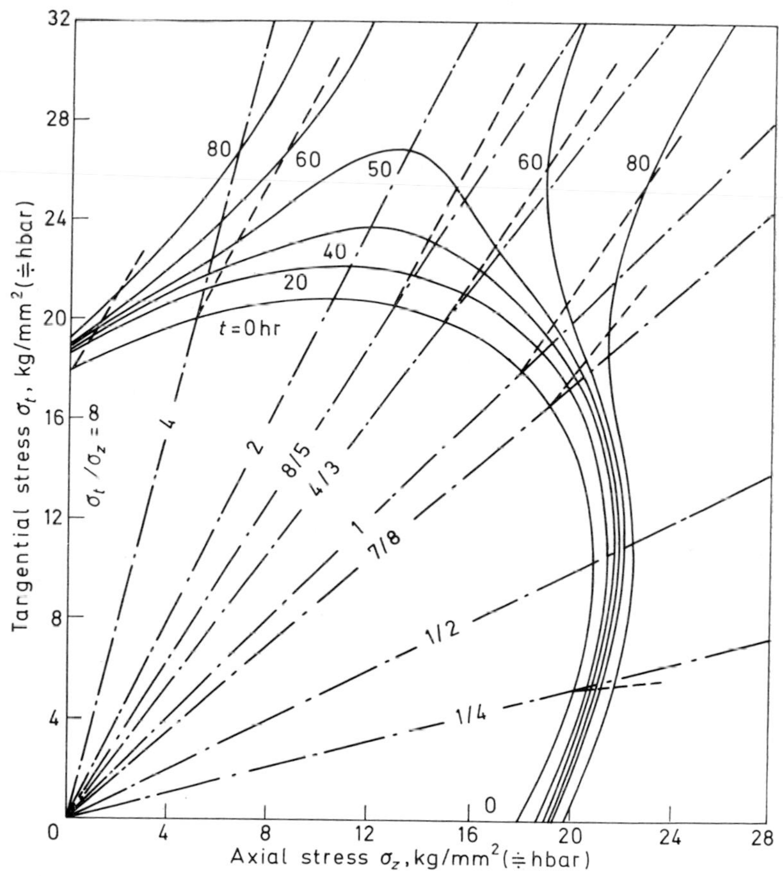

FIG. 9. Time dependent variation of stress state presented as a part of the ellipse of the von Mises effective stress.

To proceed to the analysis of the stress and strain in a thick walled tube under combined axial load P and internal pressure p, let us now assume that the tube material creeps according to a power function

$$\dot{\varepsilon}^* = b\sigma^{*\alpha} \tag{5}$$

Then, the equilibrium equation of stress and the equation of axial force balance in a capped-end tube [34] are found to be

$$\int_{r_i}^{r_o} \frac{K\sigma^*}{r} \frac{1}{1 + (C/r^2)} \,\mathrm{d}r = p \tag{6}$$

$$\frac{2\pi}{\exp \varepsilon_z} \int_{r_i}^{r_o} \left(1 - \frac{3}{4} K^2\right)^{\frac{1}{2}} \sigma^* r \,\mathrm{d}r = P \tag{7}$$

where

$$\begin{aligned} K &= \frac{2}{\sqrt{3}} \left[1 - \left(\frac{\dot{\varepsilon}_z}{\dot{\varepsilon}^*}\right)^2\right]^{\frac{1}{2}} \\ C &= r_o{}^2[\exp \sqrt{3}(\varepsilon_o{}^{*2} - \varepsilon_z{}^2)^{\frac{1}{2}} - 1] \\ \dot{\varepsilon}^* &= \frac{1}{\sqrt{3}} \left[\left(\frac{\dot{C}}{r^2 - C}\right)^2 + 3\dot{\varepsilon}_z{}^2\right] \end{aligned} \tag{8}$$

and the subscripts i and o indicate the inner and the outer surfaces of the tube, respectively. Equations (6) and (7) can then be integrated with respect to r, by using eqns (5) and (8). The principal stresses in the tube result in

$$\begin{aligned} \sigma_r &= \int_{r}^{r_i} \frac{K\sigma^*}{r} \frac{1}{1 + (C/r^2)} \,\mathrm{d}r - p \\ \sigma_t &= K\sigma^* + \sigma_r \\ \sigma_z &= \left[1 - \frac{3}{4} K^2\right]^{\frac{1}{2}} \sigma^* + \frac{1}{2}(\sigma_t + \sigma_r) \end{aligned} \tag{9}$$

Numerical calculations are necessary to obtain the stress distributions. In the case of a tube under internal pressure ($P = 0$), however, the stresses are given in the analytical form of the following equations [36]:

$$\begin{aligned} \sigma_t &= p \frac{[(2/\alpha) - 1](\dot{\varepsilon}_t/\dot{\varepsilon}_{t_o})^{1/\alpha} + 1}{(\dot{\varepsilon}_{t_i}/\dot{\varepsilon}_{t_o})^{1/\alpha} - 1} \\ \sigma_z &= p \frac{[(1/\alpha) - 1](\dot{\varepsilon}_t/\dot{\varepsilon}_{t_o})^{1/\alpha} + 1}{(\dot{\varepsilon}_{t_i}/\dot{\varepsilon}_{t_o})^{1/\alpha} - 1} \\ \sigma_r &= -p \frac{(\dot{\varepsilon}_t/\dot{\varepsilon}_{t_o})^{1/\alpha} - 1}{(\dot{\varepsilon}_{t_i}/\dot{\varepsilon}_{t_o})^{1/\alpha} - 1} \end{aligned} \tag{10}$$

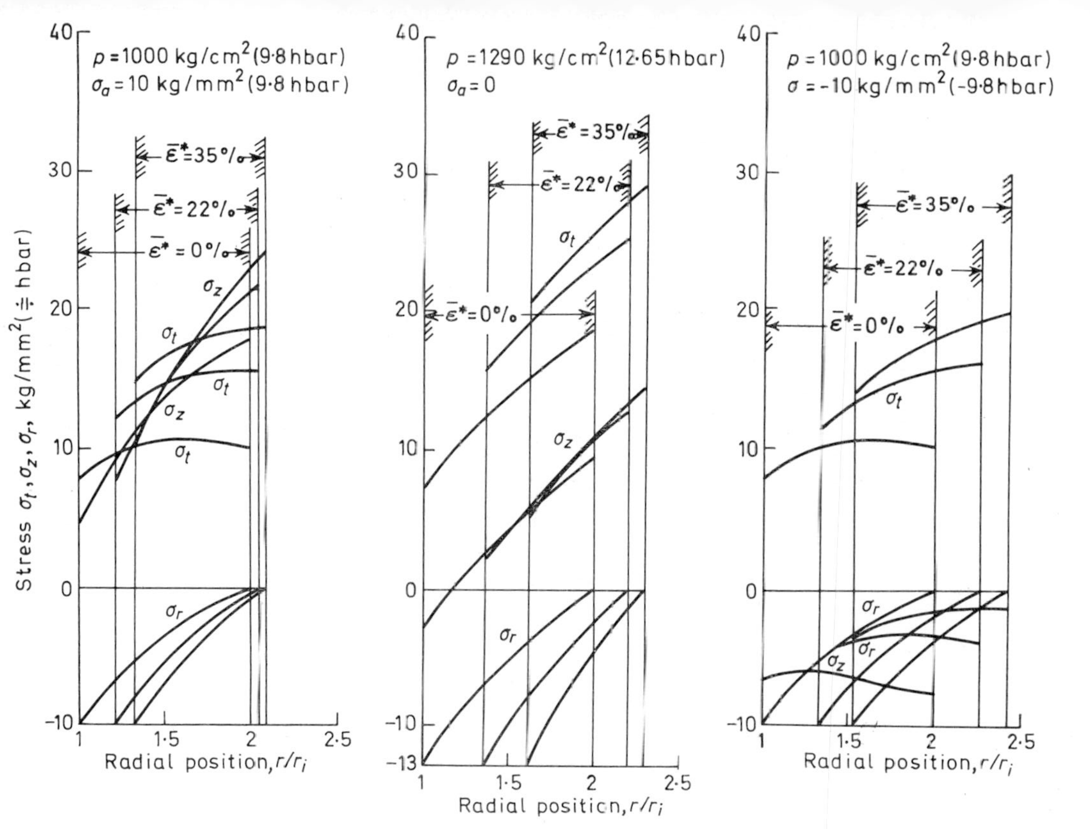

FIG. 10. Analytical stress redistributions with the increase in strains in the axially loaded pressurised tubes.

where

$$\frac{\dot{\varepsilon}_t}{\dot{\varepsilon}_{t_o}} = \frac{(r_o/r)^2 \exp 2\varepsilon_{t_o}}{1 + (r_o/r)^2(\exp 2\varepsilon_{t_o} - 1)}$$

Figure 10 shows an example of the analytical results on the successive redistributions of the stress with the increase in strain for three kinds of combined stress systems in the thick walled cylinders shown in Fig. 2c. Of these three, the greatest increase in diameter as well as the decrease in wall thickness result in the most remarkable increase in true stresses in the tube under pure internal pressure.

In this study, measurements were made of the residual stresses on the outer surface of the cylinders after creep test by the X-ray method [28, 37] as well as the residual stress distributions by means of the Sachs method [28, 38]. From the results, experimental stress distributions in the creep condition are derived by adding the elastic stress to the residual stress, and are illustrated in Fig. 11a (pure internal pressure) and b (combined axial tension and internal pressure) together with the calculated stress distributions. It is found that the theoretical curves show fairly good agreement with the experimental. It is also evident, from Fig. 11a, that the tangential stress at 1780 hr is twice as large as that at 1 hr. The corresponding increase in axial stress in simple tensile creep is about 1·35

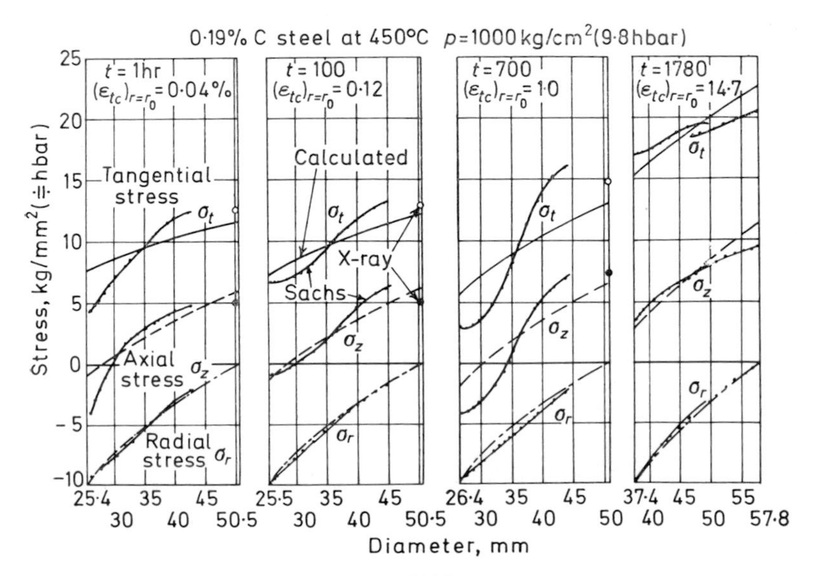

FIG. 11(a)

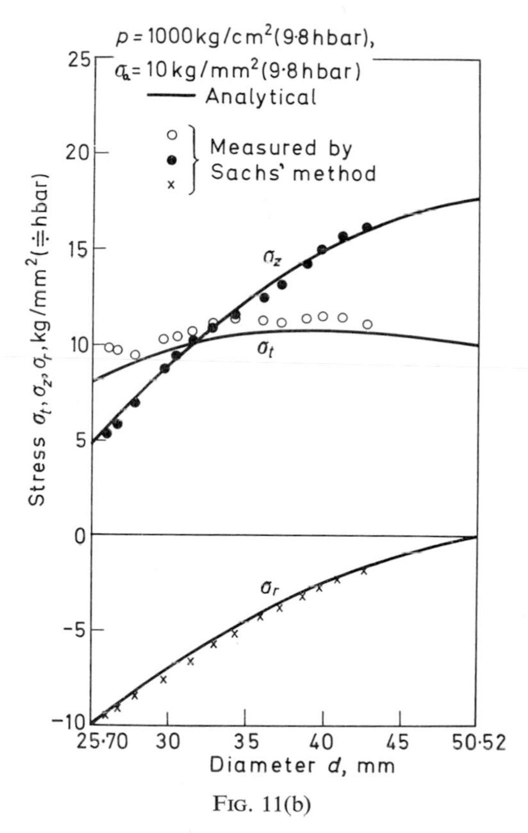

FIG. 11(b)

FIG. 11. Comparison of experimentally determined stress distributions with theoretical results for a thick walled tube. (a) Internal pressure only. (b) Combined axial tension and internal pressure.

times for the same interval of the testing time with the increase in the effective creep strain of 35 per cent.

Based on the large strain theory, a 'creep failure time' can be obtained by letting a creep deformation approach infinity [39, 40]. It is natural that the application of the concept to the prediction of rupture life is valid only when the tertiary creep process is mainly due to reduction of specimens and the rupture elongation of a material is large enough for the differences between the time to rupture and the time to reach infinite deformation to be insignificant. Figure 12 [33] shows the time to fracture, under combined stress, of the specimens in Fig. 2a, as a function of the original stress ratio. The full and dotted lines indicate the analytical rupture lives derived on the basis of the large strain theory with the von Mises and the maximum principal stress criterion, respectively, and the circles show the test results under the von Mises effective stress of 18·0 kg/mm^2 (17·6 hbar). The values based on the von Mises criterion agree with the test results better than those based on the maximum principal stress. This substantiates the concept of the large strain theory, and it is also appropriate to adopt the von Mises criterion for the prediction of rupture life

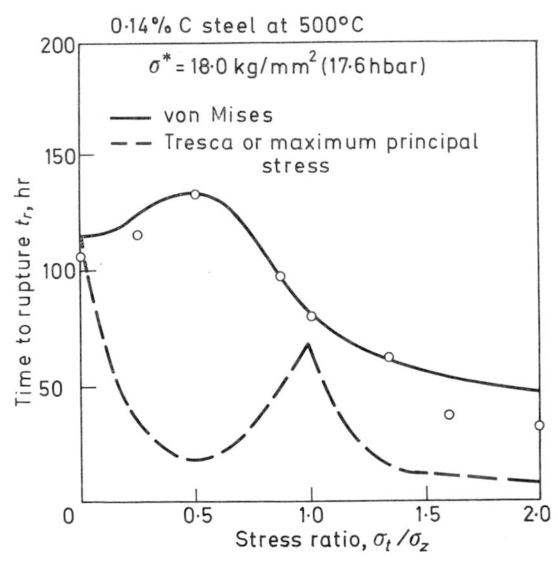

FIG. 12. Test results of combined stress creep rupture and analytical curves for the prediction of rupture life obtained on the basis of the large strain theory.

as well as of creep deformation. This implies that the material tested deforms by means of slip in the individual crystal grains according to the rule of critical shear stress [41–43], and that there is little chance of void or crack formation up to near-fracture, and also that tertiary creep appears owing to the increase in true stress caused by the reduction of cross-sectional area of the specimen.

When eqn (5) is taken as the true stress–natural creep rate relation in a tube material [18], the time t can be obtained as a function of strain for a constant internal pressure [35]

$$t = K_1{}^{\alpha} \frac{(\alpha/2)^{\alpha}}{b} \left(\frac{1}{p}\right)^{\alpha} I \tag{11}$$

where

$$I = \int_0^{\varepsilon_0^*} \left\{ \left[\frac{(r_o/r_i)^2 \exp 2K_2\varepsilon_0^*}{1 + (r_o/r_i)^2(\exp 2K_2\varepsilon_0^* - 1)} \right]^{1/\alpha} - 1 \right\}^{\alpha} \mathrm{d}\varepsilon_0^*$$

and

$$K_1 = 2/\sqrt{3},\ K_2 = \sqrt{3}/2;\ \text{(von Mises)}$$
$$K_1 = 1, \qquad K_2 = 3/4; \quad \text{(Tresca)}$$

Thus, the following relations which express the time to rupture of the cylinder can be obtained on the basis of the large strain theory [36, 44]:

$$t_r = K_1{}^{\alpha} \frac{(\alpha/2)^{\alpha}}{b} \left(\frac{1}{p}\right)^{\alpha} I_{\infty} \tag{12}$$

$$I_{\infty} = \int_0^{\infty} \left\{ \left[\frac{(r_o/r_i)^2 \exp 2K_2\varepsilon_0^*}{1 + (r_o/r_i)^2(\exp 2K_2\varepsilon_0^* - 1)} \right]^{1/\alpha} - 1 \right\}^{\alpha} \mathrm{d}\varepsilon_0^*$$

In the case of the creep under simple tension, relations similar to those in eqns (11) and (12) are found to be

$$t = \frac{1}{\alpha b} \left(\frac{1}{\sigma_0}\right)^{\alpha} [1 - \exp(-\alpha\varepsilon)] \tag{13}$$

$$t_r = \frac{1}{\alpha b} \left(\frac{1}{\sigma_0}\right)^{\alpha} \tag{14}$$

in which σ_0 is the initial tensile stress.

When the time to rupture of a tube under an internal pressure of p is equal to that of a bar in simple tension under an initial tensile stress of σ_0, the relation between σ_0 and p is obtained from eqns (12) and (14) as follows:

$$\sigma_0 = p\left(\frac{2}{K_1}\right) \alpha^{-[1+(1/\alpha)]} I_{\infty}{}^{-1/\alpha} \tag{15}$$

A number of formulae for the calculation of rupture stress in pressure vessels and boiler tubes have so far been proposed [45, 46]. Let us now take the following five kinds of well known design formulae and discuss their correlation with the test results.

$$\sigma = p\left(\frac{1}{2}\frac{D}{t} - y\right) \tag{16}$$

where $y = 0$ for outside diameter formula
$y = 0{\cdot}4$ for modified Lamé formula
$y = 0{\cdot}5$ for mean diameter formula
$y = 0{\cdot}7$ for creep common formula
$y = 1{\cdot}0$ for thin walled formula.

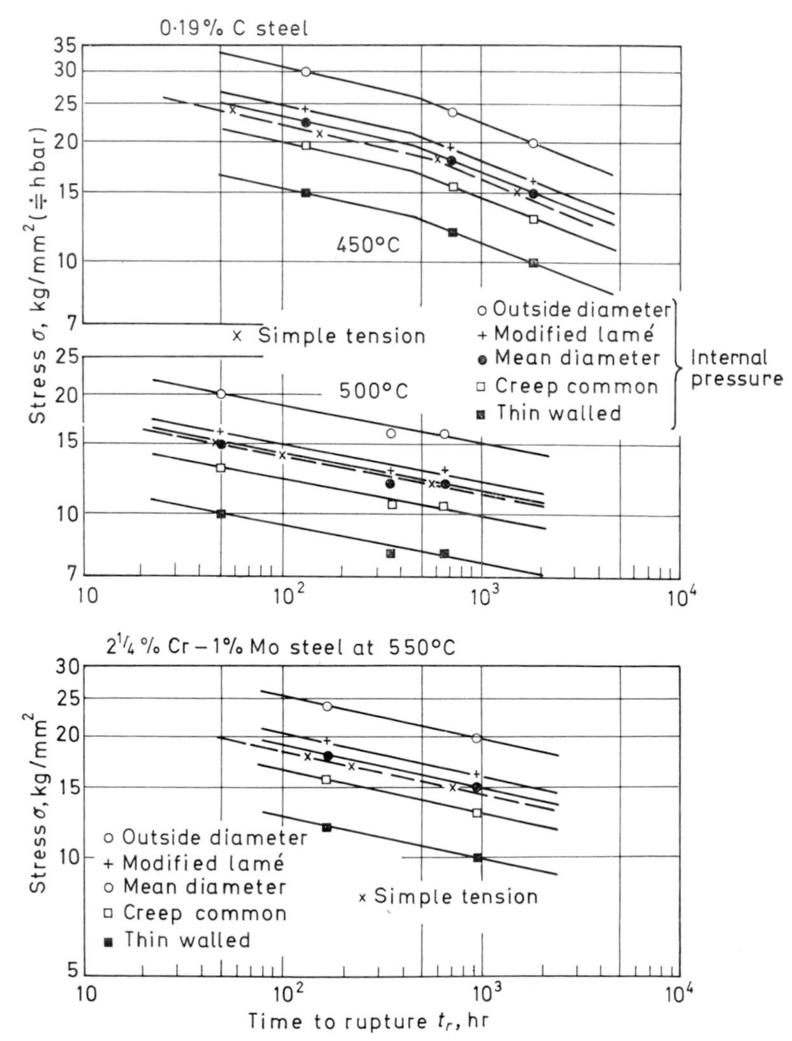

FIG. 13. Stress–rupture diagrams for simple tension creep and internal pressure creep.

Figure 13 [35, 36] shows the stress rupture diagrams of the materials B and C under simple tension only and under internal pressure only. It is found that the mean diameter formula has the best correlation with the results of uniaxial creep testing. In Fig. 14 most of the existing data on carbon steels, Cr–Mo alloy steels and 18–8 austenitic stainless steels

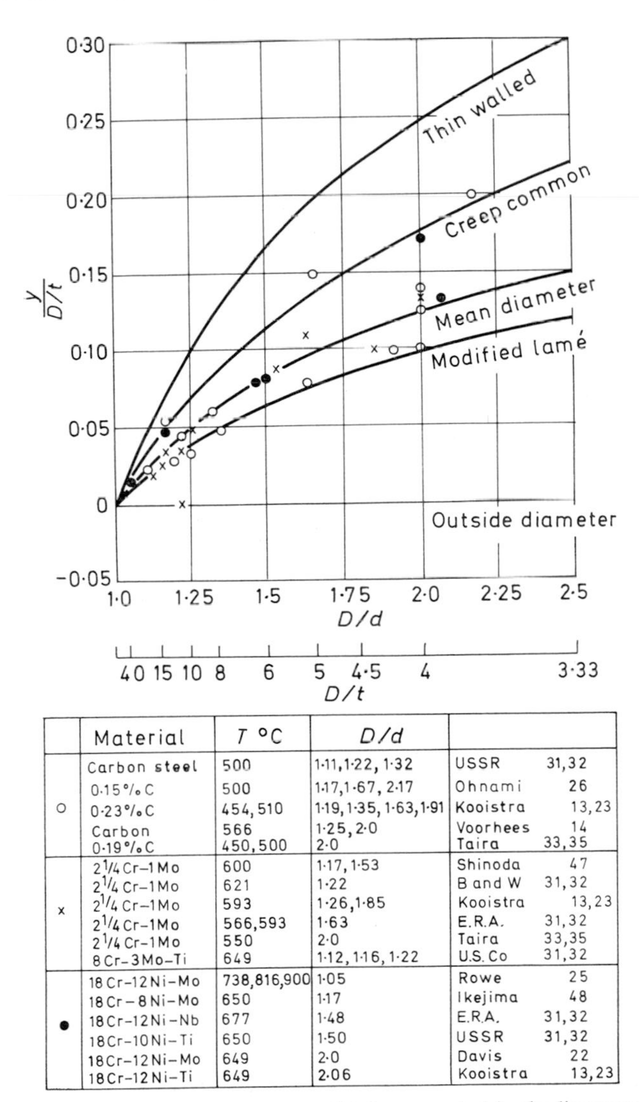

	Material	T °C	D/d		
○	Carbon steel	500	1·11, 1·22, 1·32	USSR	31, 32
○	0·15% C	500	1·17, 1·67, 2·17	Ohnami	26
○	0·23% C	454, 510	1·19, 1·35, 1·63, 1·91	Kooistra	13, 23
○	Carbon	566	1·25, 2·0	Voorhees	14
○	0·19% C	450, 500	2·0	Taira	33, 35
×	2¼ Cr–1 Mo	600	1·17, 1·53	Shinoda	47
×	2¼ Cr–1 Mo	621	1·22	B and W	31, 32
×	2¼ Cr–1 Mo	593	1·26, 1·85	Kooistra	13, 23
×	2¼ Cr–1 Mo	566, 593	1·63	E.R.A.	31, 32
×	2¼ Cr–1 Mo	550	2·0	Taira	33, 35
×	8 Cr–3 Mo–Ti	649	1·12, 1·16, 1·22	U.S. Co	31, 32
●	18 Cr–12 Ni–Mo	738, 816, 900	1·05	Rowe	25
●	18 Cr–8 Ni–Mo	650	1·17	Ikejima	48
●	18 Cr–12 Ni–Nb	677	1·48	E.R.A.	31, 32
●	18 Cr–10 Ni–Ti	650	1·50	USSR	31, 32
●	18 Cr–12 Ni–Mo	649	2·0	Davis	22
●	18 Cr–12 Ni–Ti	649	2·06	Kooistra	13, 23

FIG. 14. Data on creep rupture of tubes reported in the literature.

reported in the literature are plotted in the diagram of $y/(D/t)$ against D/t. A number of data lie along the line of the mean diameter formula, but most are scattered between the creep common and the modified Lamé formula.

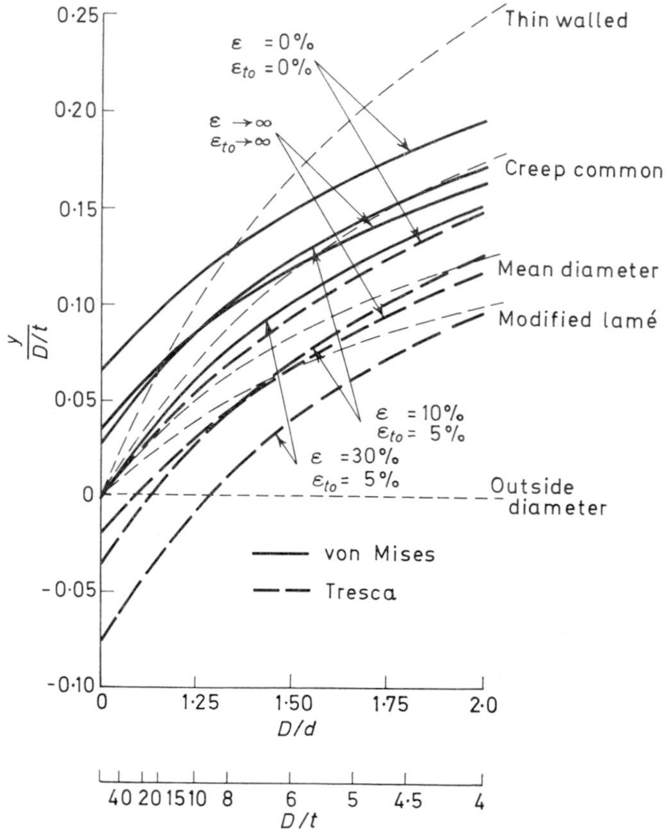

FIG. 15. Comparison of the design formulae of pressure vessels with the von Mises and the Tresca criteria in infinitesimal strain, finite strain and large strain theories.

In order to determine the physical meaning of the design formulae and to get the rupture criterion of pressurised tubes, the correlation between the design formulae and the stress criteria in the large strain theory was investigated. Comparison of eqn (15) with eqn (16) leads to

$$\frac{y}{D/t} = \frac{1}{2} - \left(\frac{2}{K_1}\right) \alpha^{-[1+(1/\alpha)]} \frac{I_\infty^{\;-1/\alpha}}{D/t} \tag{17}$$

The relations between D/t or D/d and $y/(D/t)$ obtained from eqn (17) are shown in Fig. 15, being indicated by infinite strains of $\varepsilon \rightarrow \infty$ and $\varepsilon_{t0} \rightarrow \infty$. Similar relations derived from the infinitesimal strain theory are also illustrated in Fig. 15, represented by $\varepsilon = 0$ per cent and $\varepsilon_{t0} = 0$ per cent. The von Mises and the Tresca criteria for the large strain theory give smaller values of y than those for the infinitesimal strain theory, that is, the former is on the safe side as compared with the latter. It is shown that the stress value in the von Mises criterion in the large strain theory is nearly equal to that in the creep common formula. Hence the von Mises criterion results in the unsafe prediction of the rupture life of tubes. The departure from the von Mises criterion can be attributed to the other factors mentioned later.

INFLUENCE OF THE HYDROSTATIC COMPONENT OF STRESS

In the theory of plasticity, it is assumed that the hydrostatic pressure does not affect the plastic deformation of solid materials. The hypothesis is based on the experimental results of Bridgman [49] on various kinds of metallic and non-metallic materials. However, strict examination indicates the existence of the influence of the hydrostatic component of stress on the plastic flow of materials [50–53]. In regard to creep, a few experiments have been carried out on some pure metals under hydrostatic pressure [54–57], and the creep rate was found to decrease with increase in hydrostatic pressure. It seems, however, that the extent of the influence depends on the material and the testing conditions.

If the coefficient of the hydrostatic component of stress K is defined [22] by the expression

$$K = 3\sigma_m/\sigma^* \tag{18}$$

in which σ_m is the hydrostatic component of stress and σ^* is the von Mises effective stress, then for uniaxial tension $K = 1$, whereas, at the outer surface in cylinders under internal pressure, $K = 1{\cdot}73$ and is constant regardless of wall thickness. At the bore it decreases to a negative value with increase in wall thickness. This is shown in Fig. 16, as a diagram of K against D/d. Therefore, the hydrostatic stress becomes less as the wall thickness of a tube increases.

In order to determine the effect of the hydrostatic pressure on the creep of tubes, four kinds of thin and thick walled tubular specimens of material

C were prepared as shown in Fig. 1b. Figure 17 indicates the results of the creep tests for these specimens. The relation between von Mises effective stress and minimum effective creep rate at the outside surface of the cylinder can be represented by a straight line parallel to that of the simple tension creep data irrespective of the wall thickness—in other words, regardless of

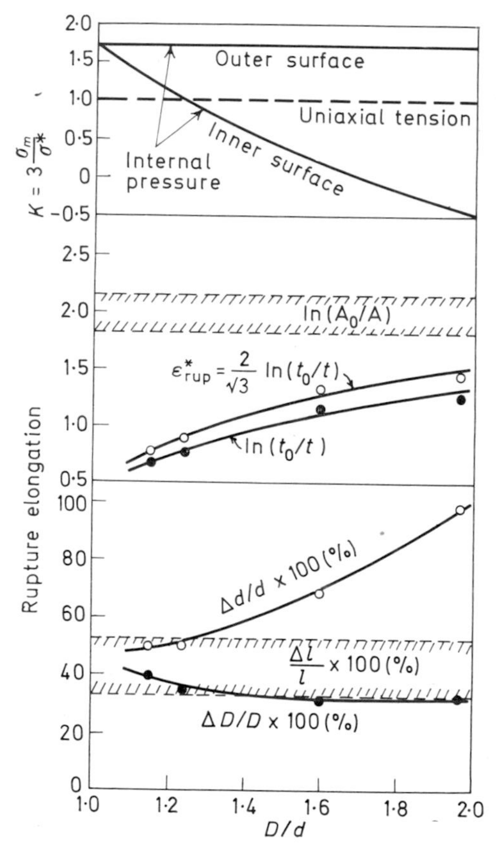

FIG. 16. Influence of wall thickness of tubes on the rupture elongation and relation between the wall thickness and the magnitude of hydrostatic component of stress.

the magnitude of hydrostatic pressure. This leads to the conclusion that one can neglect the influence of the hydrostatic pressure on the creep deformation of the material.

On the other hand, the creep rupture properties may be affected by the hydrostatic pressure. Figure 16 also shows the rupture elongation of the four kinds of cylinders. The percentage rupture elongation at the outside surface is nearly the same in all tubes, while that at the inner surface of the thick walled cylinder ($D/d = 1{\cdot}961$) is twice as large as that of the thin walled cylinder ($D/d = 1{\cdot}156$ or $1{\cdot}242$). Natural logarithms of the ratio of initial wall thickness t_0 to minimum wall thickness t at rupture indicate the absolute value of true rupture strain in the radial direction. Therefore, $2/\sqrt{3}$ of this value gives the effective true rupture strain ε^*_{rup} of the cylinder. The rupture elongation and the true rupture strain of bar

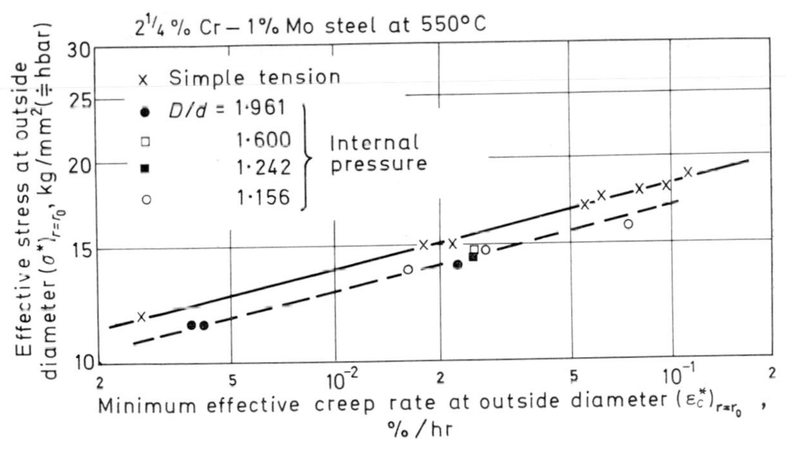

FIG. 17. Influence of wall thickness of tubes on the minimum effective creep rates.

specimens under simple tension are shown by the hatched bands in the figure. It is recognised that the thicker the tube wall, the larger the rupture elongation at the bore and the effective rupture strain. These experimental facts indicate the influence of hydrostatic stress on the ductility of the material.

Figure 18 shows typical fractures of thick walled cylinders of material B at temperatures of 450° and 500°C. Tangential cracks are visible on the outer surface at the end of the tertiary creep, and the destructive cracks start at the outer surface and progress inwards forming transcrystalline cracking. On the inner surface there are also some tangential cracks, but these are not so deep as the external cracks. On the other hand, in the fractured thin walled tubular specimens, such cracks cannot be seen on either the outer or the inner surface, and there is a tendency for the fracture to be more pronounced at the outside. These facts may also be rationalised by the effect of hydrostatic component of stress [22].

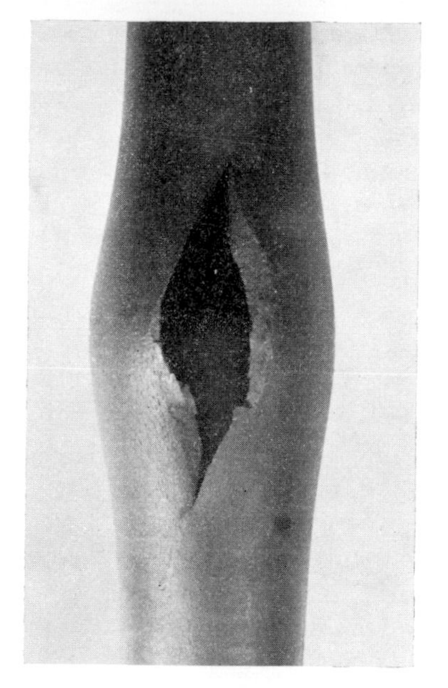

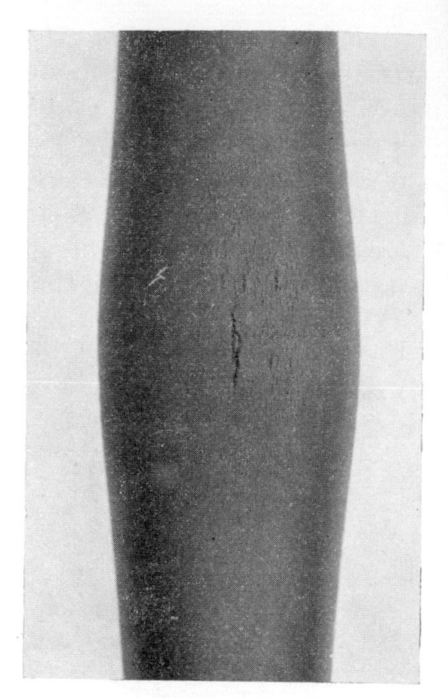

FIG. 18. Typical fractures in thick walled tubes of a 0·19 per cent C steel. (a) 450°C, 1200 kg/cm^2 (11·75 hbar). (b) 500°C, 800 kg/cm^2 (7·85 hbar).

SIZE EFFECT OF SPECIMENS

The problem on the size effect of specimens will be divided into three cases: first, the size effect in uniaxial tension specimens; next, that under combined stress; and finally, the correspondence between these.

In materials A, B and C, the size effect of specimens could not be detected in uniaxial tensile creep. Examples are shown in Fig. 19a and b for the material A, which indicates little difference in creep strength between the two kinds of tubular specimens shown in Fig. 2a and c and a solid bar specimen of 6 mm diameter.

On the other hand, the material D yielded a remarkable difference between the tensile creep curves for a solid specimen of 10 mm diameter and those for a thin walled tubular specimen, as shown in Fig. 2b. An example is shown in Fig. 20. The creep curves for the tubular specimen (dotted lines) exhibit a gradual increase in creep rate with time, although the creep in the bar specimen (solid lines) still remains in the secondary stage.

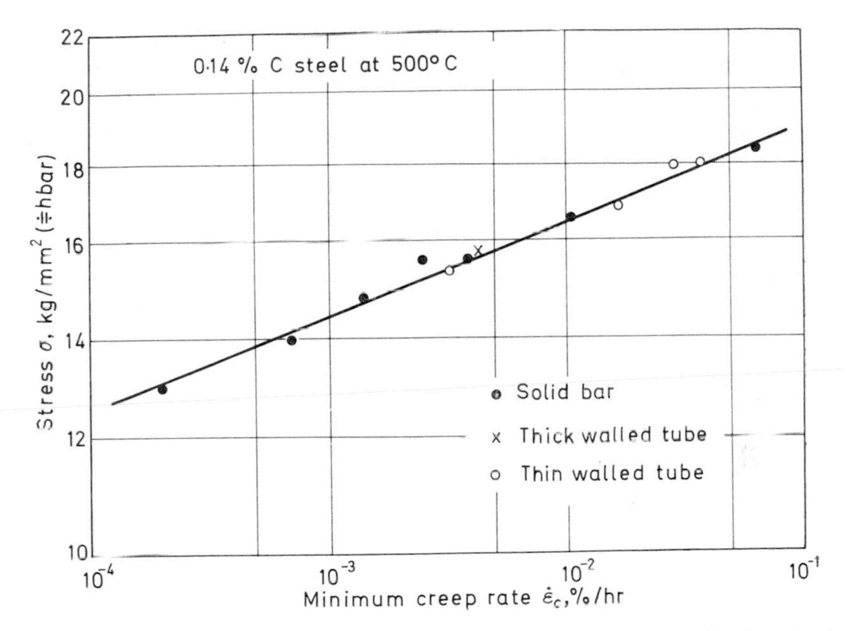

FIG. 19(a). Stress/minimum creep rate diagram in simple tension creep for three kinds of specimen.

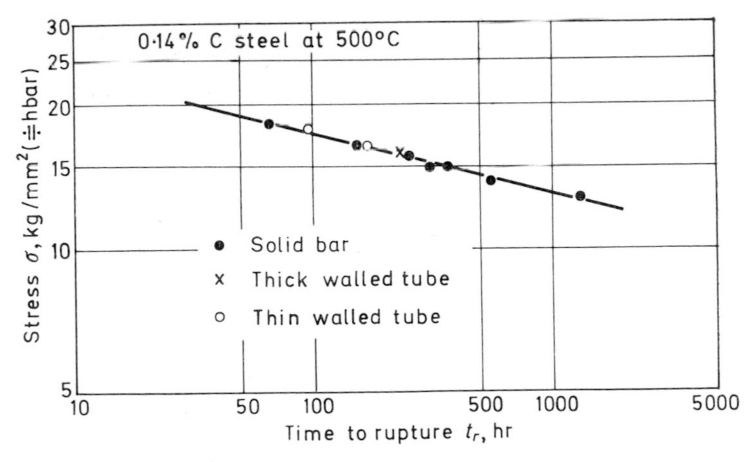

FIG. 19(b). Stress/rupture diagram of a simple tension creep for three kinds of specimen.

Such contradictory results between the materials A, B and C and the material D are attributed to the characteristics of the crack propagation and growth. The former steels are ductile materials, in which little or no cracking can be found on the surface of the fractured specimen. The latter 18–8 austenitic stainless steel, shows brittle fracture in creep, and many cracks can be detected, not only in the fractured specimens but also on the surfaces of specimens during creep. Figure 21 shows micrographs of the outer surface of tubular specimens of the material D subjected to the tensile

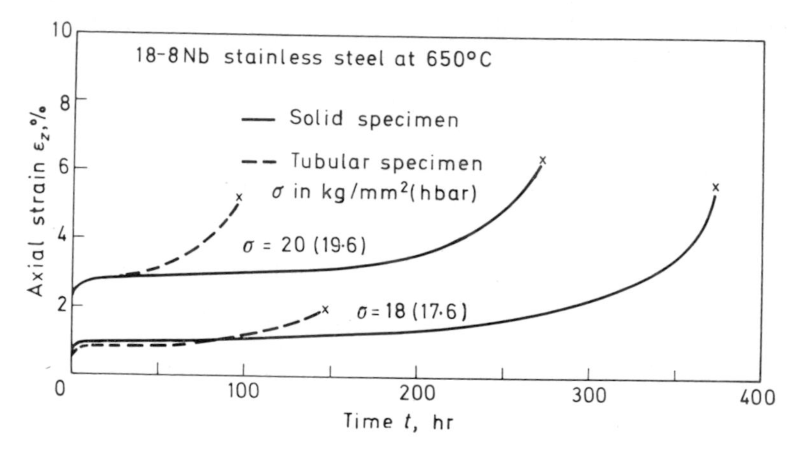

FIG. 20. Creep curves in simple tension for solid bar specimens and thin walled tubular specimens.

stress 18·0 kg/mm^2 (17·6 hbar), illustrating gradual propagation of the grain boundary cracks during creep. Creep deformation makes them increase in length, depth and width, so that the true stress in a specimen will increase with the decrease in its effective cross-sectional area.

In materials A, B and C, the effect of the wall thickness of tubes can be found only in the difference of their rupture elongations as shown in Fig. 16. One of the factors causing the increase in rupture elongation with increase in wall thickness is the difference of magnitude of the hydrostatic component of stress. Another factor is the difference of ductility between thin walled and thick walled tubes [24]. If the time to rupture of a tube is defined as the time to burst, there will be some discrepancies in the period of propagation of destructive cracks throughout the tube wall. In general, the propagation period in a thick walled cylinder is longer than that in a thin walled cylinder. Thus, an increase in rupture elongation with increase in wall thickness results, and therefore, in material D with low ductility,

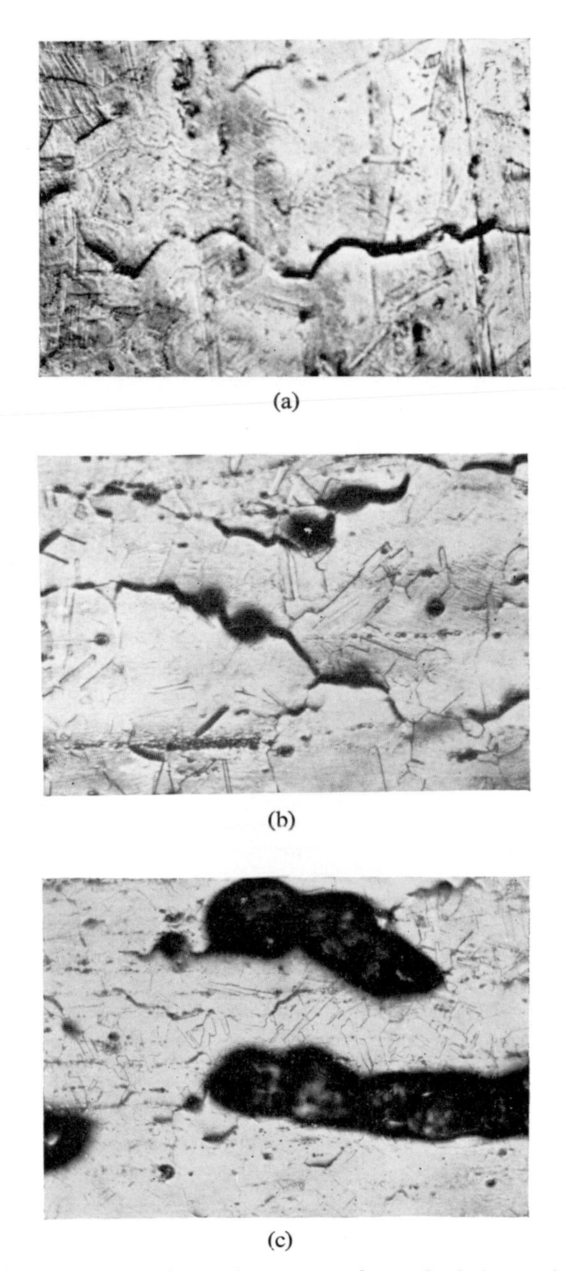

(a)

(b)

(c)

Fig. 21. Grain boundary cracks on the outer surfaces of tubular specimens in simple tension. 18–8 Nb steel, 650°C, 18·0 kg/mm^2 (17·6 hbar). (a) 6 hr, (b) 50 hr, (c) 100 hr.

the rupture elongation as well as the rupture life will be considerably affected by the size and wall thickness of the tubes.

Let us consider a small element including a crack of a given size on the outer surface of a tube, as is shown in Fig. 22. It is well understood that the extent of the increase in tangential stress caused by the crack in the axial direction (a) is larger than the increase in axial stress due to the crack of the same size in the tangential direction (b), because an effective wall thickness to carry the tangential force decreases more than an effective

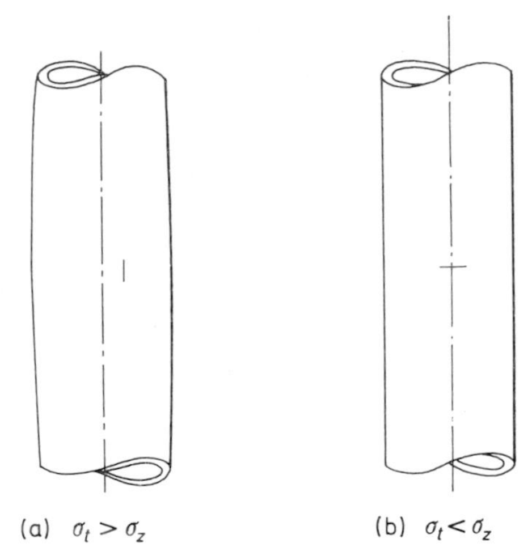

FIG. 22. A crack of a given size on the outer surface of a tube subjected to combined stress.

cross-sectional area to carry the axial force. Moreover, the tangential stress concentration at the tip of an axially distributed crack is higher than the axial stress concentration at the tip of a tangentially distributed crack. Thus, the local tangential stress near the tip of a crack becomes large enough to increase the rate of its growth.

Figure 23 shows the results of creep rupture tests carried out on thin walled tubular specimens shown in Fig. 2b of material D under combined stresses at a temperature of 650°C. The diagram is of a non-dimensional form, obtained by plotting the ratios of axial stress σ_z and tangential stress σ_t to the simple tensile stress σ_f causing fracture in the same time. The experimental results indicate that the maximum tensile stresses shown by the solid lines give greater strength for rupture. It is considered that the discrepancy is attributable to the factors mentioned above: these are hydrostatic stress, reduction of the effective cross-sectional area and notch effect of the crack. It is clear that, in this material, such factors greatly influence the creep rupture strength, because the crack initiates in the grain boundary so early during creep that the period of its propagation is nearly equal to the rupture life.

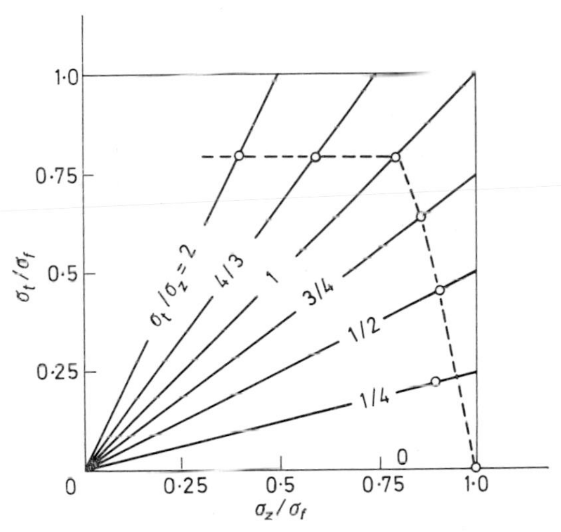

FIG. 23. Test results on combined stress creep rupture of an 18–8 Nb steel at 650°C.

INFLUENCE OF THE MAXIMUM TENSILE STRESS ON THE CRACK PROPAGATION

Figure 24 indicates typical cracks propagating inwardly from the outer surface (a) and (b) and from the inner surface (c) and (d) observed in the fractured thick walled tubes in Fig. 1a of material B. The fractures start at the outside in all cases, although the deformation in the tangential direction at the outside is smaller than that at the bore. This was explained in the foregoing discussions by the effect of a smaller hydrostatic stress at the outside. Other factors that are likely to cause the cracking to be more pronounced at the outer surface are as follows.

Firstly, the fluid under high internal pressure percolates into the cracks at the bore, making the tip of the crack round as shown in Fig. 24c and d. This will relieve the stress concentration at the cracks and prevent them from propagating into the wall.

Secondly, tensile stresses at the outside are higher than those at the bore, and the maximum tensile stress is the tangential stress at the outer surface, as shown in the stress distribution curves of Figs. 10 and 11a.

In the thick walled tubes (Fig. 2c) of material A subjected to combined axial load and internal pressure, the cracking also starts, as transgranular cracks, from the outer surface. This may be explained by the maximum tensile stress at the outside, which seems to overshadow the higher effective stress and the greater strain at the bore as the cause of fractures [22].

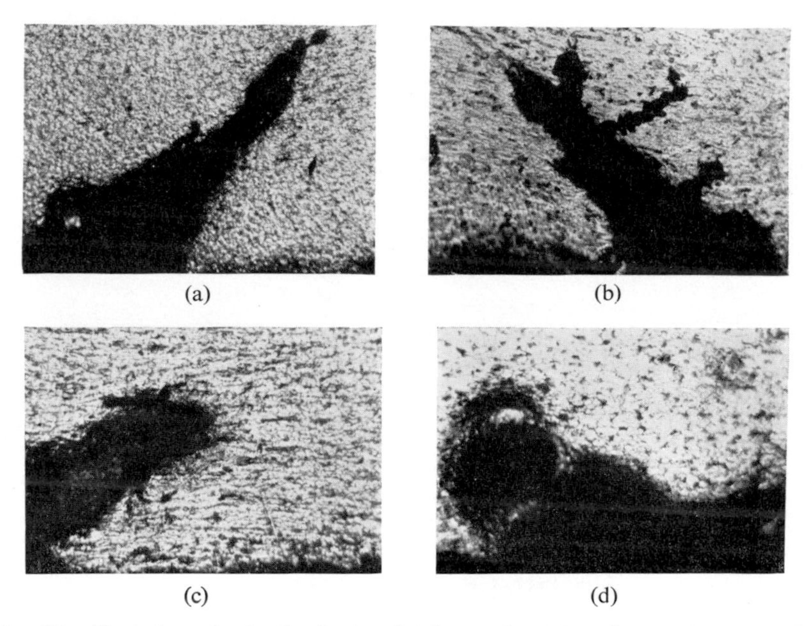

FIG. 24. Typical cracks in the fractured tubes under internal pressure propagated from the outer surface (a (×90) and b (×60)) and from the inner surface (c (×90) and d (×90)).

Another interesting point suggesting that the maximum tensile stress controls the crack growth can be found in the fractured specimen under combined tension and internal pressure. An example of the typical fracture is shown in Fig. 25. In this case, as was shown in Fig. 10, the curve of the tangential stress distribution crosses that of the axial stress distribution at the mid-section of the specimen wall. The fracture was found to start from the outside in the tangential direction through cracks, to progress inwards and to stop near the bore. The inside surface contains no destructive cracks, although the internal pressure was released at fracture.

In the ductile materials A, B and C, the crack propagation period is a part of the period at the end of the tertiary creep before fracture, where the strain is so large and the rate of straining is so high that the interval between the time to crack and the time to rupture is negligibly short. Therefore, only an insignificant difference may be found between the actual rupture time and the theoretical rupture time predicted by the large strain theory on the basis of the von Mises criterion, even if the propagation of the crack is caused by the maximum tensile stress.

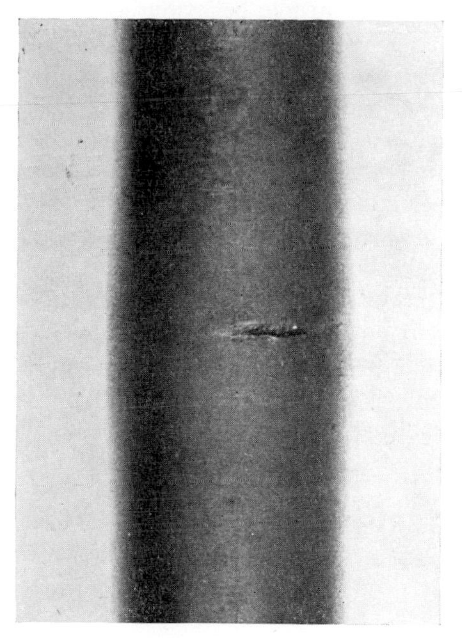

FIG. 25. A thick walled cylinder fractured under combined axial tension and internal pressure at 500°C.

On the other hand, in the thin walled tubular specimens, Fig. 2b, made of material D, grain boundary cracks like those shown in Fig. 21 can be detected even at the secondary stage under combined tension and internal pressure. The cracks are seen to progress perpendicularly to the axis of maximum tensile stress in all cases of stress ratios. If void formation or relative slip of adjacent grains along the grain boundary leads to the initiation and growth of cracks at the early stage of creep, as was observed in the material D and is often seen in other materials under relatively high temperatures and low stresses [58–60], and such a situation is activated by the maximum tensile stress, it may well be expected that the rupture life can be predicted from the maximum tensile stress, while the von Mises effective stress is correct as the stress criterion for creep deformation.

ANISOTROPY OF THE MATERIAL

Many ordinary polycrystalline metallic materials with which engineers deal have approximately isotropic behaviour. This is because strains are measured over volumes including a great many crystal grains, which are often random in orientation. It is true, however, that rolled or drawn materials may develop a preferred orientation of grains, and that anisotropy may be introduced during creep deformation from a virgin state of isotropy. Within the limit of the present data on the materials B and C, the von Mises criterion is not appropriate for the stress criterion of creep rupture as well as of creep deformation as shown in Fig. 17. This is considered to be due to the anisotropy of the materials [28], which weakens the creep strength in the tangential direction of the tube and results in the shorter rupture life. In any case, so far as existing data on multiaxial stress creep are concerned, the discrepancy from the von Mises criterion is so small [11] that it might be well to mention that this influence on the time to rupture is less than the effect of other factors stated above.

CRITERION FOR CREEP FRACTURE OF PRESSURISED TUBES AND ITS RELATION TO DESIGN FORMULAE

In the materials A, B and C tested in this study, transgranular creep deformation was dominant and the time to crack initiation was nearly equal to the time to rupture [35]. Creep curves for such ductile materials are like those shown in Fig. 8a. As was already mentioned, the most remarkable factor that causes the discrepancy in rupture life between simple tension (p. 300) and internal pressure creep is factor (1): different increases in true stress between the two systems. This yields a shorter rupture life of tubes as compared with that of simple tension bar specimens under the same magnitude of the initial von Mises effective stress. Other factors will also make the rupture life short, but these are not so pronounced as factor (1).

Taking these factors as effective, we are able to establish the criterion for creep fracture of tubes of ductile materials. This is shown in the

$p/\sigma - D/t$ diagram of Fig. 26, indicated by the vertical hatched band (a). The upper limit of the band is the von Mises criterion for the large strain theory, and the lower limit is the Tresca criterion. The so-called mean diameter formula lies within the band.

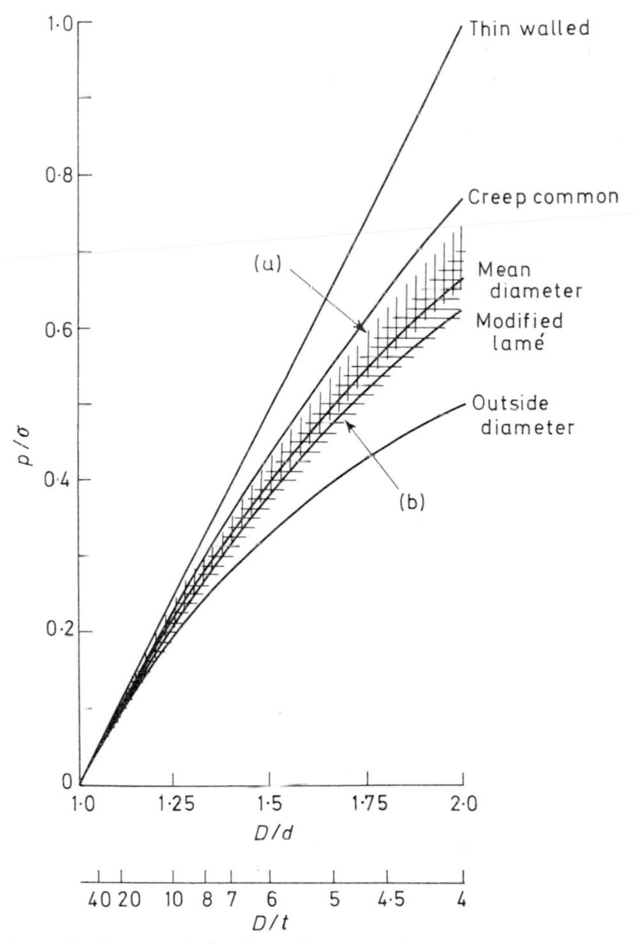

FIG. 26. $p/\sigma - D/t$ diagram indicating the stress criterion for creep fracture of tubes under internal pressure.

In materials with little ductility, on the other hand, the effect of the factors from (1) to (5) on the rupture life is different from that in ductile materials, as shown in Fig. 8b. In this case, as was found in the test results on material D, the decrease in the rupture life of tubes is closely related to the reduction of the effective wall thickness due to gradually progressing cracks. Propagation of these cracks will be controlled by the maximum tensile stress, and the strains in the tube concentrate at the tip of the crack, yielding the small rupture elongation and the short rupture life as compared with those of bar specimens under uniaxial tension [24, 25].

With these points in mind, an attempt was made to investigate the rupture criterion for brittle materials on the basis of the finite strain theory. The basic concept is that both simple tensile bar specimens and pressurised tubular specimens will fracture when the creep deformations reach some finite values. The rupture stress can then be obtained from eqns (11) and (13) as follows:

$$\sigma_0 = p\left(\frac{2}{K_1}\right)\alpha^{-[1+(1/\alpha)]}\left[\frac{1-\exp(-\alpha\varepsilon)}{I}\right]^{1/\alpha}$$

$$I = \int_0^{\varepsilon_{t_o}}\left\{\left[\frac{(r_o/r_i)^2\exp\sqrt{3}\varepsilon_{t_o}}{1+(r_o/r_i)^2(\exp\sqrt{3}\varepsilon_{t_o}-1)}\right]^{1/\alpha}-1\right\}^{\alpha}\mathrm{d}\varepsilon_{t_o} \qquad (19)$$

As there is a fairly large discrepancy in the rupture elongations between the two, $\varepsilon = 30$ per cent and $\varepsilon_{t_o} = 5$ per cent are taken as the rupture elongations for an example of the typical case. The results are indicated in Fig. 26 by the lower hatched band (b). It is found from the figure that not only the mean diameter formula but also the modified Lamé formula lie within this band.

CONCLUSIONS

Creep and creep rupture tests were carried out with thin and thick walled tubular specimens of four types of steel subjected to combined stress at elevated temperatures. The experimental results on creep deformation proved the validity of the von Mises effective stress and its associated flow rule. It is, however, necessary to check the existence of the anisotropy of the materials not only in their virgin state but also during creep.

Although the von Mises criterion was applicable to the prediction of rupture life of the ductile materials A, B and C, the maximum tensile stress criterion was more correct for the brittle material D. The discrepancy is considered to be attributable to the difference between the strength in the grain boundary and the matrix. When cracks initiate at an early stage of creep and progress gradually along the grain boundary subjected to the maximum tensile stress, leading to brittle fracture, the maximum tensile stress is a dominant factor in the rupture strength of the material. On the other hand, for the ductile material in which creep deformation depends on slip in grains until cracks initiate at the end of the tertiary stage, the von Mises criterion defining the rule of deformation in polycrystalline materials is a main factor in the fracture strength.

In general, the time to rupture of tubes under internal pressure is shorter than that of bar specimens in simple tension for the same magnitude of the initial von Mises effective stress. For the ductile material with comparatively large rupture elongation, this results from the greater increase in true stress in the tube than in the tensile bar specimen. It was found that, in this case, the large strain theory was applicable to the prediction of the rupture life. In the brittle material, the decrease in the rupture strength of tubes is mainly attributed to the maximum tensile stress and the size effect of the specimen. Although the maximum tensile stress is one of the more reliable criteria, it gives an unsafe prediction in some cases. This may be attributed to the size effect inherent in the tube, which is classified into three factors; the reduction of the effective wall thickness due to the formation of axially distributed cracks, the notch weakness effect due to the crack and the high hydrostatic component of stress.

Taking these factors into account, the stress criterion for the prediction of rupture life of tubes has been considered and the investigation led to the conclusion that the mean diameter formula is reasonably valid for the design of pressure vessels and boiler tubes, irrespective of the materials and working conditions.

REFERENCES

1. BAILEY, R. W. (1935). 'The utilisation of creep test data in engineering design.' *Proc. Instn Mech. Engrs*, **131,** 131–349.
2. SODERBERG, C. R. (1936). 'Interpretation of creep tests for machine design.' *Trans. Amer. Soc. Mech. Engrs*, **58**(8), 733–743.
3. MARIN, J. (1937). 'Design of members subjected to creep at high temperatures.' *J. Appl. Mech.*, **4**(1), A21–A24.
4. NADAI, A. (1937). 'Creep of solids at elevated temperatures.' *J. Appl. Phys.*, **8**(6), 418–432.
5. NAKAHARA, M. (1939). 'New theories of creep of metals.' *Soc. Mech. Engrs Japan*, **5**(18), 12–20.

6a. NORTON, F. H. (1939). 'Creep in tubular pressure vessels.' *Trans. Am. Soc. Mech. Engrs*, **61**(3), 239–244.

6b. NORTON, F. H. (1941). 'Progress report on tubular creep tests.' *Trans. Amer. Soc. Mech. Engrs*, **63,** 735–736.

7. NORTON, F. H. and SODERBERG, C. R. (1942). 'Report on tubular creep tests.' *Trans. Amer. Soc. Mech. Engrs*, **64**(8), 769–777.

8. Coffin, L. F., Shepler, P. R. and Cherniak, G. S. (1949). 'Primary creep in design of internal-pressure vessels.' *J. Appl. Mech.*, **16**(3), 229–241.
9. Suezewa, Y. (1951). Bull. Tokyo Inst. Tech. (Ser. B), 53.
10a. Bailey, R. W. (1951). 'Steam piping for high pressures and high temperatures.' *Proc. Instn Mech. Engrs*, **164**, 324–424.
10b. Bailey, R. W. (1951). 'Creep relationships and their application to pipes, tubes and cylindrical parts under internal pressure.' *Proc. Instn Mech. Engrs*, **164**, 425–431.
11. Johnson, A. E. (1951). 'Creep under complex stress systems at elevated temperatures.' *Proc. Instn Mech. Engrs*, **164**, 432.
12. Johnson, A. E. and Frost, N. E. (1951). 'Fracture under combined stress creep conditions of a 0·5 per cent Mo steel.' *Engineer*, **191**, 434–437.
13. Kooistra, L. F., Blaser, R. J. and Tucker, J. T. (1952). 'High-temperature stress-rupture testing of tubular specimens.' *Trans. Amer. Soc. Mech. Engrs*, **74**(5), 783–792.
14. Voorhees, H. R., Sliepcevich, C. M. and Freeman, J. W. (1956). 'Thick-walled pressure vessels.' *Ind. & Engng Chem.*, **48**(5), 872–881.
15. Wahl, A. M. (1956). 'Analysis of creep in rotating disks based on Tresca criterion and associated flow rule.' *J. Appl. Mech.*, **23**, 231–238.
16. Johnson, A. E., Henderson, J. and Mathur, V. D. (1956). 'Combined stress creep fracture of a commercial copper at 250 deg. Cent.' *Engineer*, **202**, 261–265; 299–301.
17. Kennedy, C. R., Harms, W. O. and Douglas, D. A. (1959). 'Multiaxial creep studies on Inconel at 1500 F.' *Trans. Amer. Soc. Mech. Engrs*, **81**(4), 599.
18. Rimrott, F. P. J. (1959). 'Creep of the thick-walled tubes under internal pressure considering large strains.' *J. Appl. Mech.*, **26**(2), 271–275.
19. Finnie, I. and Heller, W. R. (1959). *Creep of Engineering Materials.* New York: McGraw-Hill.
20. Johnson, A. E. (1960). 'Complex-stress creep of metals.' *Metall. Rev.*, **5**(20), 447–506.
21. Johnson, A. E. and Henderson, J. (1960). 'Complex stress creep fracture of aluminium alloy.' *Aircr. Engng*, **32**(376), 161–170.
22. Davis, E. A. (1960). 'Creep rupture tests for design of high-pressure steam equipment.' *Trans. Amer. Soc. Mech. Engrs*, **82**(2), 453–461.
23. Tucker, T. J., Coulter, E. E. and Kooistra, L. F. (1960). 'Effect of wall thickness on stress-rupture life of tubular specimens.' *Trans. Amer. Soc. Mech. Engrs*, **82**(2), 465–476.
24. Sawert, W. and Voorhees, H. R. (1961). 'Influence of ductility on creep rupture under multiaxial stresses.' *ASME Paper* No. 61-WA-233. New York: American Society of Mechanical Engineers.
25. Rowe, G. H., Stewart, J. R. and Burgess, K. N. (1963). 'Capped end, thin-wall tube creep-rupture behaviour for type 316 stainless steel.' *Trans. Amer. Soc. Mech. Engrs*, **85**(1), 71–86.
26. Ohnami, M. and Awaya, Y. (1963). 'Creep and creep rupture of cylindrical tube under combined axial tension and internal pressure. *Proc. Japan Cong. Test. Mater.*, **11**, 76–81.

27. INSTITUTION OF MECHANICAL ENGINEERS (1963–64). Joint International Conference on Creep, New York and London, 1963. *Proc. Instn Mech. Engrs*, **178**(Pt 3A).
28. TAIRA, S., KOTERAZAWA, R. and OHTANI, R. (1965). 'Creep of thick-walled cylinders under internal pressure at elevated temperature.' *Proc. Japan Cong. Test. Mater.*, **8,** 53–60.
29. ODQVIST, F. K. G. (1966). *Mathematical Theory of Creep and Creep Rupture.* Oxford: Clarendon Press.
30. RABOTNOV, N. Y. (1966). *Polzuchest' elementor Konstruktsiy* (Creep of Structural Elements). Moscow: Nauka.
31. CARLSON, W. B. and DUVAL, D. (1962). 'Rupture data and pipe design formulae' *Engineering*, **193,** 829–831.
32. CHITTY, A. and DUVAL, D. (1963). 'Creep-rupture properties of tubes for high temperature steam power plant.' *Proc. Instn Mech. Engrs*, **178** (Pt 3A), 4-1–4-11.
33. TAIRA, S., OHTANI, R. and ISHISAKA, A. (1968). 'Creep and creep fracture of a low carbon steel under combined tension and internal pressure.' *Proc. Japan Cong. Test. Mater.*, **11,** 76–81.
34. TAIRA, S., OHTANI, R. and ISHISAKA, A. (1967). 44th Annual Meeting of JSME. No. 167, p. 89.
35. TAIRA, S. and OHTANI, R. (1968). 'Creep rupture of internally pressurised cylinders at elevated temperatures.' *Bull. JSME*, **11**(46), 593–604.
36. TAIRA, S. and OHTANI, R. (1967). 'Creep rupture of pressurised (steel) cylinders at elevated temperature.' *J. JSME*, **70**(587), 1737–1744.
37a. TAIRA, S. and YOSHIOKA, Y. (1964). 'X-ray investigation on residual stress of metallic materials (on residual stress of stretched carbon steel).' *Proc. Japan Cong. Test. Mater.*, **7,** 31–37.
37b. TAIRA, S. and YOSHIOKA, Y. (1966). *Proc. 9th Japan Cong. Test. Mater.*, 48.
38. SACHS, G. (1927). 'Detection of internal stress in brass rods and tubes' (in German). *Z. Metallkunde*, **19,** 352–357.
39. HOFF, N. J. (1953). 'Necking and rupture of rods subjected to constant tensile loads.' *J. Appl. Mech.*, **20**(1), 105–108.
40. HOFF, N. J. (1959). 'Mechanics applied to creep testing.' *Proc. Soc. Exp. Stress Analysis*, **17**(2), 1–32.
41. COX, H. L. and SOPWITH, D. G. (1937). 'The effect of orientation on stress in single crystals and of random orientation on strength of polycrystalline aggregates.' *Proc. Phys. Soc.*, **49**(2), 134–151.
42. TAIRA, S., ABE, T. and NAGAO, M. (1967). (a) 'Crystallographic study of yield condition of polycrystalline metals.' 1st Report bcc metal (in Japanese). *Trans. JSME*, **33**(246), 199–205. (b) English Translation 1967. *Bull. JSME*, **10**(91), 711–717.
43. TAIRA, S. and ABE, T. (1967). (a) 'Crystallographic study of yield condition of polycrystalline metals.' 2nd Report hcp metal (in Japanese). *Trans. JSME*, **33**(254), 1535–1541. (b) English Translation 1968. *Bull. JSME,* **11**(45), 414–425.
44. RIMROTT, F. P. J., MILLS, F. J. and MARIN, J. (1960). 'Prediction of creep failure time for pressure vessels.' *J. Appl. Phys.*, **27**(2), 303–307.

45. Buxton, W. J. and Burrows, W. R. (1951). 'Formula for pipe thickness.' *Trans. Amer. Soc. Mech. Engrs*, **73**(2), 575–581.
46. Burrows, W. R., Michel, R. and Rankin, A. W. (1954). 'A wall-thickness formula for high-pressure, high-temperature piping.' *Trans. Amer. Soc. Mech. Engrs*, **76**, 427–444.
47. Shinoda, N., Tamada, I. and Inove, Y. (1965). 'Creep rupture of tubular specimens under internal pressure at high temperature' (in Japanese). *Zairyo Shiben* (*Jap. Soc. Test. Mater.*), **14**(137), 78–82.
48. Ikejima, T. and Maruoka, H. (1962). 'On a creep rupture test applied to the tubular specimen.' *J. JSTM*, **11**(102), 165–168.
49. Bridgman, P. W. (1964). *Studies in Large Plastic Flow and Fracture*. Cambridge, Mass.: Harvard University Press.
50a. Pugh, H. Ll. D. and Green, D. (1954). 'Progress report on the behaviour of materials under high hydrostatic pressure.' *NEL Plasticity Report* No. 103. East Kilbride, Glasgow: National Engineering Laboratory.
50b. Pugh, H. Ll. D. and Green, D. (1956). 'The behaviour of metals under high hydrostatic pressure. II Tensile and torsion tests.' *NEL Plasticity Report* No. 128. East Kilbride, Glasgow: National Engineering Laboratory.
51. Nishihara, M., Tanaka, K. and Muramatsu, T. (1964). 'Effect of hydrostatic pressure on mechanical behaviour of materials, I.' *Proc. Japan Cong. Test. Mater.*, **7**, 154–159.
52. Nishihara, M., Tanaka, K. and Hamada, H. (1965). 'Effect of hydrostatic pressure on mechanical behaviour of materials, II.' *Proc. Japan Cong. Test. Mater.*, **8**, 73–77.
53. Hu, L. W. and Pae, K. D. (1963). 'Inclusion of hydrostatic stress component in formulation of yield condition.' *J. Franklin Inst.*, **275**(6), 491–502.
54. Hull, D. and Rimmer, D. E. (1959). 'Growth of grain-boundary voids under stress.' *Phil. Mag.*, **4**(42), 673–687.
55. Devries, K. L., Baker, G. S. and Bibbs, P. (1963). 'Pressure dependence of creep of lead.' *J. Appl. Phys.*, **34**(8), 2254–2257.
56. Kohler, C. R. and Ruoff, A. L. (1965). 'Pressure and temperature dependence of creep in potassium.' *J. Appl. Phys.*, **36**(8), 2444–2445.
57. Ohnami, M. (1967). 'Plasticity laws in creep of polycrystalline metallic materials at elevated temperatures: effects of hydrostatic stress and of strain history on creep deformation' (in Japanese). *Zairyo*, **16**(162), 161–168.
58. Garofalo, F. (1965). *Fundamentals of Creep and Creep-rupture in Metals*. New York: Macmillan.
59. Chang, H. C. and Grant, N. J. (1956). 'Mechanism of intercrystalline fracture.' *Trans. Amer. Inst. Mech. Engrs*, **206**, 544–551.
60. McLean, D. and Farmer, M. H. (1956). 'Relation during creep between grain-boundary sliding, sub-crystal size, and extension.' *J. Inst. Metals*, **85**(2), 41–50.

Chapter 16

CREEP AND CREEP-RUPTURE OF COPPER TUBES UNDER MULTIAXIAL STRESS

I. FINNIE AND M. M. ABO EL ATA

SUMMARY

Blocks of electrolytic tough pitch copper were prepared in such a way as to minimise anisotropy. Constant stress tests on specimens cut from the blocks in three orthogonal directions showed highly isotropic creep behaviour and virtually identical behaviour in tension and compression until the beginning of tertiary creep. The results of tests on tubes loaded by internal pressure and by torsion are compared with tensile data. Plotting maximum shear strain as a function of time for a given maximum shear stress leads to considerably better correlation of the tests than a similar plot based on effective stress and effective strain (octahedral shear stress and octahedral shear strain). Despite this result, in tubes loaded by combinations of internal pressure and axial load, the ratios of the principal strains were found to be approximated quite well by the Levy–von Mises equations.

Creep–rupture studies indicate that no simple criterion is adequate for predicting life under multiaxial stress. It appears that rupture life depends on both effective stress and maximum tensile stress. There are indications that the former may determine the nucleation of cracks and the latter their subsequent growth.

NOTATION

A	Coefficient in eqn (2).
B	Constant in eqn (8).
b_1, b_2	Constants in eqn (9).

D	Direction function.
J_2', J_3'	Invariants of stress deviator tensor.
K	Constant in eqn (7).
M	Magnitude function.
m	Exponent in Bailey's equation (eqn 5).
n	Exponent of stress.
p	Hydrostatic stress.
t	Time.
x, y, z	Direction of axes.
$\dot{\gamma}$	Shear strain rate.
$\dot{\varepsilon}$	Creep strain rate.
$\dot{\bar{\varepsilon}}$	Effective creep strain rate.
σ	Tensile stress.
$\bar{\sigma}$	Effective stress.
σ^*	Effective stress defined by eqn (6).
τ	Shear stress.

Subscripts

max	Maximum value.
θ, a, r	Hoop, axial and radial directions.
1, 2, 3	Principal stress directions.

INTRODUCTION

Although the phenomenon of creep has been known for over a century, it was only in the 1930s that the need to design parts for service at elevated temperature led to detailed analytical and experimental studies of creep under multiaxial stress. On the analytical front, progress initially was rapid. By combining the procedures developed earlier by von Mises for time independent plasticity with a constitutive equation relating creep rate to stress, Odqvist [1] in 1934 developed the expressions for creep rate prediction which are in general use today. Somewhat different approaches were later taken by Soderberg [2], Bailey [3] and others. The test of any method which attempts to predict the creep rates $\dot{\varepsilon}_1$, $\dot{\varepsilon}_2$, $\dot{\varepsilon}_3$ from a knowledge of the principal stresses σ_1, σ_2, σ_3 and tensile creep data lies, of course, in comparison with experiment. However, the experimental difficulties in creep testing are great, the tests are inherently time consuming and progress in this direction would have come about very slowly were it not for the work of Dr Johnson and his colleagues. Fortunately, with great insight he carried out, over a period of more than 30 years, a very extensive and careful series of creep and creep–rupture studies on a variety of metals. This work is unique in its scope and into the foreseeable future it will serve as a reference, against which new ideas in creep and creep–rupture may be compared. Dr Johnson, Henderson and Khan summarised this work up to 1962 [4].

We recall that in studying primary creep under constant multiaxial stresses, Dr Johnson found that the stress and time dependence of creep strain were separable so that the creep strain ε could be expressed in the form $\varepsilon = F(\sigma)\ G(t)$. Of necessity, he adopted two other expedients to simplify the reduction of experimental data. First, he subtracted all of the initial strain before constructing creep curves, and secondly he used the creep rates at a fixed time of 150 hr in comparing the effect of different stress states. On this basis he found that the 'octahedral' or 'effective' strain rate† $\dot{\bar{\varepsilon}}$ was a function of the 'effective' stress‡ $\bar{\sigma}$ when thin walled tubes were tested under various combinations of tension and torsion.

For an aluminium alloy at 150° and 200°C, a magnesium alloy at 20° and 50°C, a 0·17 per cent C steel at 350°, 450° and 550°C, a nickel–chromium alloy at 550° and 650°C, and commercially pure copper at 250°C, the plots of logarithm $\dot{\bar{\varepsilon}}$ as a function of logarithm $\bar{\sigma}$ were strikingly similar. At low creep rates, say $\dot{\bar{\varepsilon}} \lesssim 10^{-6}\ \text{hr}^{-1}$, all materials except copper, which was not tested in this region, show a stress dependence of the form $\dot{\bar{\varepsilon}} \sim \bar{\sigma}^n$ with n values lying between 1 and 3. By contrast, for higher stresses leading to creep rates, say $\dot{\bar{\varepsilon}} \gtrsim 10^{-5}\ \text{hr}^{-1}$, the creep rate is considerably more sensitive to stress. The data can again be approximated by the expression $\dot{\bar{\varepsilon}} \sim \bar{\sigma}^n$ but the values of the exponent n now generally lie between 6 and 8. For example, for commercially pure copper in this region, Dr Johnson found $n = 7{\cdot}05$.

For all materials except the 0·17 per cent C steel at 450° and 550°C in the low creep rate range and for copper in the high creep rate range, Dr Johnson found that the creep rates at 150 hr could be represented adequately by a generalised form of the equations proposed by Odqvist for steady creep. That is, by

$$\dot{\varepsilon}_1 = \frac{\bar{\sigma}^{(n-1)}}{2}[(\sigma_1 - \sigma_2) - (\sigma_3 - \sigma_1)]\frac{dG(t)}{dt} \tag{1}$$

† $\dot{\bar{\varepsilon}} = (\sqrt{2}/3)[(\dot{\varepsilon}_1 - \dot{\varepsilon}_2)^2 + (\dot{\varepsilon}_2 - \dot{\varepsilon}_3)^2 + (\dot{\varepsilon}_3 - \varepsilon_1)^2]^{\frac{1}{2}}$.

‡ $\bar{\sigma} = (1/\sqrt{2})[(\sigma_1 - \sigma_2)^2 + (\sigma_2 - \sigma_3)^2 + (\sigma_3 - \sigma_1)^2]^{\frac{1}{2}}$.

and two similar equations for ε_2, ε_3 which become $\varepsilon = \sigma^n G(t)$ for constant stress uniaxial tension tests. In the high creep rate range, Dr Johnson found that these equations applied only to copper and the nickel–chromium alloy at 650°C with anisotropy developing in all other cases. In situations for which he deduced that anisotropy was occurring, the creep rate equations used by Dr Johnson were of the general form

$$\dot{\varepsilon}_1 = \frac{\bar{\sigma}^{(n-1)}}{2}[A_{12}(\sigma_1 - \sigma_2) - A_{13}(\sigma_3 - \sigma_1)]\frac{dG(t)}{dt} \tag{2}$$

where $\bar{\sigma}$ has an anisotropic value and the coefficients A_{12}, A_{23}, A_{13} are constant for a given stress system. To satisfy the condition that there are no volume changes during creep, $A_{12} = A_{21}$, etc.

With this very limited review of but one facet of Dr Johnson's comprehensive work on creep as background, we turn now to consider some other observations which have been made about creep under constant multiaxial stresses.

By contrast to Dr Johnson's experiments which were carried out in tension–torsion, that is the second quadrant of the σ_1, σ_2 plane, in work carried out by Wahl and his colleagues [5] attempts were made to predict the creep strain in rotating discs (biaxial tension) from uniaxial creep data. They found that the Odqvist expressions, eqn (1), consistently underestimated the creep strains in the discs and examined several other methods of strain prediction. These were as follows:

1. The maximum shear strain rate was assumed to be a function only of time and a power of the maximum shear stress, which leads to

$$\dot{\gamma}_{max} = 3(2)^{(n-1)}(\tau_{max})^n \frac{dG(t)}{dt} \tag{3}$$

We have introduced the numerical factor so that this expression reduced to $\varepsilon = \sigma^n G(t)$ for uniaxial tension. Assuming in addition that strain only occurs in the maximum shear stress direction leads, for $\sigma_1 > \sigma_2 > \sigma_3$, to

$$\dot{\varepsilon}_2 = 0 \qquad \dot{\varepsilon}_1 = -\dot{\varepsilon}_3 = \dot{\gamma}_{max}/2$$

This approach is equivalent to the classical Tresca equations of time independent plasticity.

2. Making the first of the preceding assumptions, and then invoking the Levy–von Mises equations to calculate the individual strain components, leads to

$$\dot{\varepsilon}_1 = \frac{(2\tau_{max})^{(n-1)}}{2}[(\sigma_1 - \sigma_2) - (\sigma_3 - \sigma_1)]\frac{dG(t)}{dt} \tag{4}$$

and two similar equations, in which $2\tau_{max} = (\sigma_1 - \sigma_3)$. This approach was used in some early plasticity studies and by Coffin and colleagues [6] in predicting the deformation of thick walled cylinders due to internal pressure.

3. The equations proposed by Bailey [3]

$$\dot{\varepsilon}_1 = \bar{\sigma}^{2m}[(\sigma_1 - \sigma_2)^{(n-2m)} - (\sigma_3 - \sigma_1)^{(n-2m)}]\frac{dG(t)}{dt} \tag{5}$$

where m is an additional constant to be obtained experimentally.

4. A modification of the Bailey equation, based on a suggestion by Davis [7], so that they would be consistent with the concept of a flow potential [8]. This involves redefining the effective stress as

$$\sigma^* = \left[\frac{(\sigma_1 - \sigma_2)^{(n-2m+1)} + (\sigma_2 - \sigma_3)^{(n-2m+1)} + (\sigma_3 - \sigma_1)^{(n-2m+1)}}{2}\right]^{[1/(n-2m+1)]} \tag{6}$$

and leads to an expression for $\dot{\varepsilon}_1$ similar to eqn (5) but with $\bar{\sigma}^{2m}$ replaced by $(n + 1)\sigma^{*2m}$.

In comparing predictions with experimental data, Wahl found little difference between methods (3) and (4). However, both methods gave better predictions than the Odqvist equations or methods (1) and (2), which is perhaps not surprising in view of the additional disposable constant m which the Bailey or modified Bailey equations contain. As far as the simpler equations were concerned, Wahl found that method (2) gave considerably better predictions for the creep strains in discs than either method (1) or the Odqvist equations. This finding has been criticised on the grounds that method (2) is inconsistent with the concept of a flow potential. However, as pointed out by Rabotnov [9], this concept does not follow directly from thermodynamics but requires additional hypotheses. Berman and Pai [10] have pointed out that the principal creep rates may be written in the form

$$\dot{\varepsilon}_1 = KM^n \, \partial D/\partial\sigma_1 \tag{7}$$

where, contrary to the usual concept of a flow potential, the magnitude function M and the direction function D may be of different forms. They suggested that, for an isotropic material, M could be taken as the maximum shear stress and, guided by the results of Lensky for time independent plasticity, proposed that D be taken as the effective stress $\bar{\sigma}$.

These assumptions lead to equations of the form

$$\dot{\varepsilon}_1 = B(\sigma_1 - \sigma_3)^n \frac{[\sigma_1 - \frac{1}{2}(\sigma_2 + \sigma_3)]}{\bar{\sigma}} \tag{8}$$

for secondary creep and could be extended to the case of primary creep under constant stresses by replacing the constant B by a function of time. More general equations were also presented by these authors for certain types of anisotropy; in particular, the case in which tensile behaviour differs from compression behaviour was considered.

In returning to consider Wahl's results, as with so many of the reported multiaxial creep tests in the literature, it is difficult to rule out anisotropy. The tests were made on forged 12 per cent Cr steel discs and tensile specimens cut perpendicular to the plane of the disc showed average creep rates 20 to 40 per cent greater than specimens taken in the plane of the disc. The latter specimens themselves showed considerable scatter, with strain values at a given stress and time varying by a factor of about 2·5. Still, the evidence cited by Wahl from his own tests and those of Gemma and Kennedy does indicate that the Odqvist equations underestimate creep strains in the tension–tension zone of the σ_1, σ_2 plane. In an attempt to resolve this apparent contradiction with Dr Johnson's findings, one of the present authors [11] tested aluminium and lead tubes under torsion, internal pressure and axial loading. It was found that better correlation of the tests was obtained by plotting maximum shear strain as a function of time for a given maximum shear stress than when a similar correlation was attempted based on effective stress. A similar conclusion was drawn by Rabotnov [9] in discussing tests in combined tension–torsion made on an austenitic steel at 600°C. Also, Rabotnov pointed out that in Dr Johnson's results for the 0·17 per cent C steel at 450°C the tests at the four highest stress levels agree well on the basis of a maximum shear strain correlation while the other results more nearly follow the effective strain correlation. In the tests on aluminium tubes at 250°C and lead tubes at 60°C it was found that the internal pressure specimens had considerably higher creep rates than would be expected from torsion creep data. This discrepancy cannot be resolved by invoking any of the equations we have discussed for isotropic behaviour and since the materials tested appeared to be reasonably isotropic, the suggestion was made that, at temperatures above half the melting point in degrees absolute, the hydrostatic stress had an effect on creep [11]. However, subsequent tests by Johansen [12] on aluminium tubes at 200° and 150°C showed behaviour similar to the previous tests at 250°C. Thus, the discrepancy between the internal pressure and torsion tests may be due to anisotropy.

From this brief review of some of the relevant literature it is clear that there is uncertainty in choosing an equation for creep rate prediction, particularly in the high stress region. At the present time, in a practical sense, this may not be a serious problem because in many members that operate under creep conditions the stresses are not known precisely and since creep rates are very sensitive to stress, under these conditions no method of creep strain prediction will be exact. On the other hand, techniques for computing stresses in structures which are subject to elastic, plastic and creep deformation have made enormous progress in recent years. For this reason, the uncertainities in stress and strain prediction will lie, in the future, to an increasing extent in the equations describing material behaviour rather than in the approximations required for a tractable analysis. For this reason, it appears essential, despite the experimental difficulties, that experiments continue to be performed on multiaxial creep under both constant and changing stress systems.

We turn now to the question of predicting creep–rupture life under multiaxial stress, from data obtained in uniaxial tension. Again, in surveying the literature, one finds that the most complete experimental data are those of Dr Johnson and his colleagues. We recall that they found that the materials tested fell into two categories.

1. Life depends on the effective stress, $\bar{\sigma}$. In these materials, creep occurs without cracking. The cracks that form just prior to failure are confined to the region in which fracture takes place.
2. Life depends on the maximum tensile stress, σ_{max}. In these materials, failure is preceded by gradual and general crack propagation.

Since it was found that a material would not change its mode of failure over the normal working temperature range, Dr Johnson suggested that the rupture life criterion appropriate to a given material could be selected by metallographic examination of a tensile creep–rupture specimen.

These findings represented a great step forward in the engineering treatment of creep–rupture in that they combined multiaxial tests with metallographic examination of the fracture process. Clearly, it is important to continue this type of investigation and determine whether all materials fall into the two categories discussed by Dr Johnson or if these merely represent extreme types of material behaviour with some intermediate type of creep–rupture criterion being the more general case. In this connection, we will consider, very briefly, some of the metallurgical studies of creep–rupture, the phenomenological studies and some observations based on notched bar tests.

In metallurgical studies, most of the mechanisms suggested for the nucleation of cracks are based on the concept that fracture initiation depends on plastic deformation. Intergranular cracks are generally thought to be initiated by the nucleation of voids at locations of high stress concentration such as three grain junctions (triple points), pre-existing ledges in the grain boundaries or ledges produced by slip bands penetrating the boundaries. For this kind of nucleation to take place, grain boundary sliding is a necessary prerequisite. Grain boundary sliding and the formation of ledges at the grain boundaries by slip bands are shear processes and should be governed by shear stresses. For polycrystalline materials under a multiaxial state of stress the effective stress is a simple measure of shear stresses. Hence, it might be assumed as a first approximation that the time required for crack initiation is a function of the effective stress, $\bar{\sigma}$.

The mechanism usually invoked for the growth of voids, once they have formed, is the condensation of vacancies. During creep an excess concentration of vacancies can be maintained as a result of deformation or by a Nabarro–Herring mechanism. Balluffi and Seigle [13] used the latter mechanism to estimate the stress necessary to grow a spherical void by receiving vacancies from nearby vacancy sources. They showed that stable voids should be found at boundaries normal to the tensile stress axis, which is in agreement with experimental observations. Hull and Rimmer [14] extended the analysis of Balluffi and Seigle and derived an expression for the creep rupture life of a specimen under combined tensile stress σ and hydrostatic pressure p. By assuming pre-existing spherical voids at the grain boundaries, they predicted that the time to rupture would be approximately inversely proportional to the maximum tensile stress $(\sigma - p)$. However, from their results it can be seen that neither the maximum tensile stress $(\sigma - p)$ or the effective stress σ was a satisfactory rupture criterion. Crussard and Friedel [15] discussed the interaction between a crack and a vacancy both in the presence and in the absence of a tensile stress. Without stress, they showed that the crack will shrink and finally collapse. In the presence of tensile stress they considered a flat elliptical crack (pancake shaped crack) in two orientations:

1. The plane of the crack is parallel to the tensile axis.
2. The plane of the crack is normal to the tensile axis.

Considering the effect of the stress distribution around these cracks on their interaction with vacancies, they deduced that in the first case the crack will collapse in the direction of its smallest axis (normal to the tensile axis) while in the second case the crack will grow at the edges. From these studies it seems reasonable to assume as a first approximation that under multiaxial stresses the intensity of the maximum tensile stress will control the rate of growth of cracks.

Phenomenological studies of creep do not as yet shed much light on the question of creep–rupture under multiaxial stress. One approach [16] often followed in predicting life under constant loads is based on the fact that the cross-section decreases more and more rapidly as the test continues. Thus, the creep rates increase rapidly towards the end of the test and the rupture-life is merely the time at which the strains become infinite. Such an estimate of rupture life may be useful for materials which are very ductile, but unfortunately it is an upper bound and may greatly overestimate the life of materials which fracture at low strains. In addition, it is not possible to use this approach to predict life when stress, or strain rate, is kept constant rather than the load. Another approach has been suggested by Kachanov [17] who interprets the fracture process as a damaged region spreading into undamaged material where the usual creep laws apply. In an interesting discussion of Kachanov's method by Odqvist and Hult [18], it is pointed out that his procedure leads to predictions which are equivalent to the linear cumulative creep damage equation of Robinson. Other phenomenological theories were proposed earlier by Siegfried and by Kochendorfer. These were considered by Dr Johnson during his work on creep–rupture and shown to be not generally applicable.

Another possible source of information on the criterion for rupture under combined stresses is the notched bar tension test. Although no complete analysis of the effect of notch geometry on life has yet been obtained, it can be seen qualitatively that if the linear cumulative damage rule of Robinson is accepted, notch strengthening cannot be explained if life is determined by the maximum tensile stress. On the other hand, notch strengthening can be rationalised if the effective stress is accepted as the criterion for failure. However, notch strengthening has been observed in practice in materials which, following Dr Johnson, we would expect to follow the maximum tensile stress criterion for rupture life. A tentative explanation of this contradiction has been offered by Voorhees and Freeman [19] who suggested that the maximum tensile stress controls crack initiation while the effective stress governs crack propagation. They explain Dr Johnson's observation that the maximum tensile stress controls rupture by pointing out that his specimens had a wall thickness of only 0·030 in (0·8 mm) and hence a crack once initiated would propagate very rapidly.

However, based on the metallurgical studies we have discussed, it appears that a more reasonable simple hypothesis would be to assume that the effective stress governs crack initiation while the maximum tensile stress controls crack propagation. In any event, there is evidence to indicate that both shear stresses and the maximum tensile stress are involved in the process of creep–rupture. Rabotnov [9] points out that in some cases an equivalent stress $\lambda\sigma_{max} + (1 - \lambda)\bar{\sigma}$ is more accurate for rupture prediction than the maximum tensile stress criterion alone, while Goldhoff and Brothers [20] have discussed the role of shear stresses and normal stresses in the initiation and propagation of cracks in notched bars.

In studying certain types of fracture, for example 'fatigue', the separation of the process into its initiation and growth phases has proved useful. It appears from the literature that this approach may be profitable also in studying creep–rupture. However, as the initiation and growth phases of creep–rupture may interact, it is by no means clear that simple criteria can be established to relate crack initiation and propagation to the state of stress in the specimen.

EXPERIMENTAL STUDY OF CREEP

The material selected for study was electrolytic tough pitch copper, which was received in the form of a 6 in (150 mm) diameter bar.

In preparing the specimens, the following procedure was adopted in an attempt to minimise anisotropy. Cubes of $4\frac{1}{8}$ in (105 mm) side were cut from the bar. These cubes were plastically deformed by compression to a strain of about 5 per cent in the x direction. Barrelling was minimised by providing shallow spirals on the faces of the cubes, and lubricating with molybdenum disulphide. The process was repeated in the y and z directions to restore the original geometry. This cycle was repeated four times. From each cube, twelve tubular specimens (0·7 in (17·8 mm) o.d., 0·6 in (15·25 mm) i.d. $3\frac{7}{8}$ in (98·5 mm) long and 2·5 in (63·5 mm) gauge length) were taken as shown in the figure. This procedure leaves 2 cubes of 2 in (50 mm) side and from these, twelve tensile specimens 0·187 in (4·75 mm) diameter, $1\frac{1}{2}$ in (38 mm) gauge length were prepared in a manner similar to that shown in Fig. 1. Again, from each of the 2 in cubes, two 1 in (25 mm) cubes remain and these were used to produce compression specimens 0·187 in (4·75 mm) diameter, 0·625 in (15·9 mm) length also in the x, y and z directions. The remaining $\frac{1}{2}$ in (13 mm) cubes were used for metallographic examination. The specimens were then sealed in Pyrex tubes under vacuum and annealed for 1 hr at 700°F (372°C), which leads to a grain size of 0·035 to 0·045 mm, *i.e.* about 30 grains across the thickness of the tubular specimens.

All of the creep tests reported in the present work were nominally at constant stress rather than constant load. This was achieved by adjusting axial loads and internal pressure during the tests, based on strain measurements and the assumption that no volume changes occur during creep.

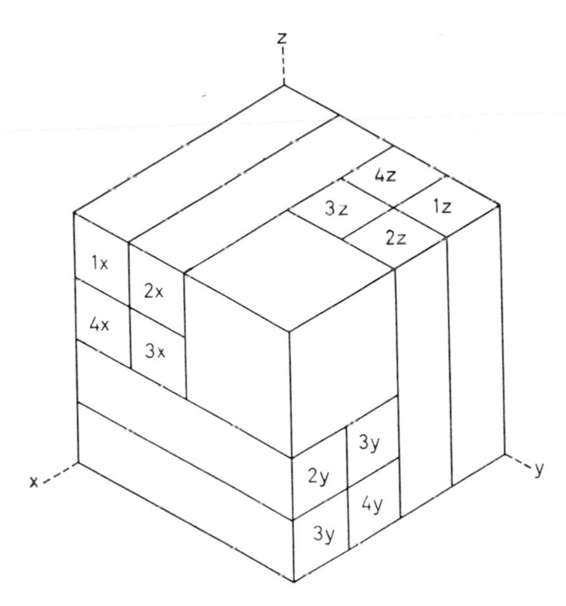

FIG. 1. Schematic representation of the way in which tubular specimens are cut from a cube in three perpendicular directions.

True strains are used throughout, but contrary to the practice often followed of reporting stress and strain for the outside diameter of tubular specimens we have given values corresponding to the mean tube diameter. This was done for convenience in subsequent studies of creep–rupture. The axial and hoop stresses were obtained from the usual statically determinate thin walled cylinder formulae while the radial stress, which changes during the test but is small, was taken as $-\frac{1}{4}$ (initial internal pressure + final internal pressure). Hoop strains measured at the outside diameter were converted to strains at the mean diameter by assuming that no volume changes occur during creep. At least until tertiary creep occurs, this assumption appears to be a reasonable one and it was checked in a few cases by sectioning specimens and measuring radial as well as hoop and axial strains.

Following Dr Johnson, the test temperature was taken as 250°C. However, instead of measuring creep rates at a given time, in this chapter we will merely compare the total strains in the different tests. As elastic strains are negligible in the present tests, this measurement is essentially the sum of the initial plastic strain and creep strain. Since it is difficult to separate the early stages of primary creep from initial plastic strain, we choose to compare the tests without making this distinction and at least in the beginning without making any assumptions about the stress or time dependence of the different strain components.

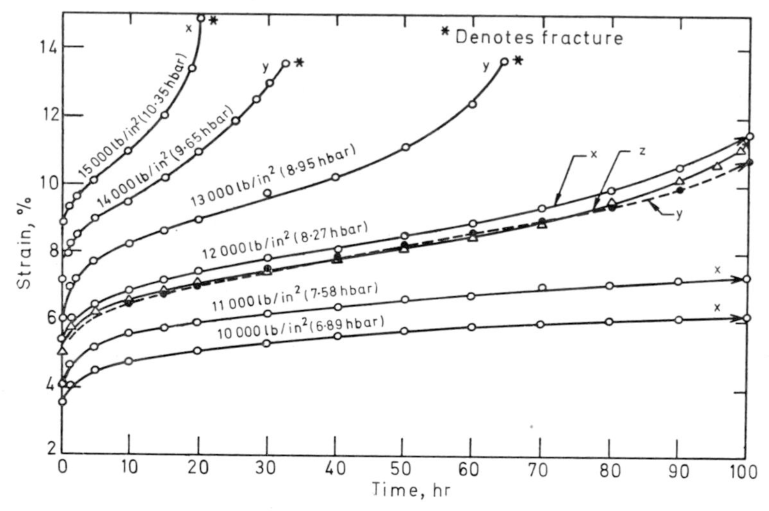

FIG. 2. Creep curves for tension tests on solid specimens. The direction x, y or z in which the specimen was taken from the cube is denoted by a letter on the curve.

Figure 2 shows the result of a series of constant stress tension tests. The specimens taken from the x, y and z directions and tested at 12 000 lb/in² (8·28 hbar) show a close agreement in strain values. The maximum strain variation between the three tests of about 6 per cent is within the scatter to be expected due to specimen tolerance limits, temperature variation, load measurement and strain measurement. Figure 3 presents a comparison of tension and compression tests and until tertiary creep intervenes the results from the two types of tests are strikingly similar. Seven torsion tests were performed and the effective strain values ($\bar{\varepsilon} = \gamma/\sqrt{3}$) at mean radius are shown in Fig. 4. The tests performed at a shear stress of 7100 lb/in^2 (4·9 hbar)—(effective stress 12 300 lb/in^2 (8·48 hbar)—on tubes taken in three perpendicular directions show approximately the same variation as the three tensile tests already discussed. Two torsion tests at $\bar{\sigma} = 11\,720$ lb/in^2 (8·1 hbar) on tubes from the y and z directions gave results which essentially coincide. Also, the single pressure test at $\bar{\sigma} = 11\,660$ lb/in^2 (8·05 hbar) on a tube in the x direction agrees reasonably well with the two torsion tests at $\bar{\sigma} = 11\,700$ lb/in^2 (8·08 hbar).

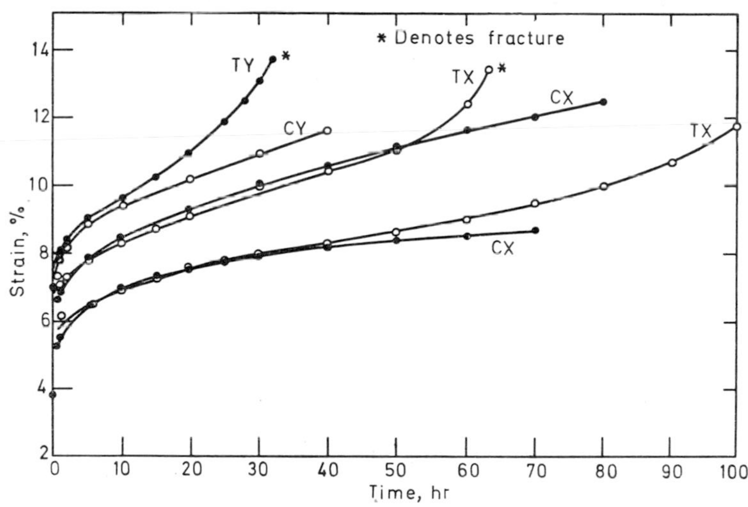

FIG. 3. Comparison of tension (T) and compression (C) creep curves at stress levels 12 000, 13 000 and 14 000 lb/in^2 (8·27, 8·95 and 9·65 hbar). The second letter on the curve denotes the direction in which specimens were taken from the cube.

By superimposing Figs. 2 and 4 it is seen that at a given time considerably greater effective strains occur in torsion or pressure tests than in tension tests at the same effective stress. Out to about 60 hr, the torsion test at $\bar{\sigma} = 11\,200$ lb/in^2 (7·74 hbar) coincides with tension at 12 000 lb/in^2 (8·28 hbar) and the pressure test at $\bar{\sigma} = 9350$ lb/in^2 (6·45 hbar) lies close to the results for tension at 10 000 lb/in^2 (6·90 hbar). Out to about 40 hr, the torsion tests at $\bar{\sigma} = 12\,310$ lb/in^2 (8·49 hbar) lie just slightly below tension at 13 000 lb/in^2 (8·96 hbar).

The results of Fig. 4 have been replotted in Fig. 5 to show maximum shear strain as a function of time for given values of twice the maximum shear stress. Since in a tension test $\gamma = \frac{3}{2}\varepsilon$, a direct comparison of Figs. 2 and 5 may be made if the ordinate in Fig. 2 is multiplied by 1·5. On this basis, considerably better agreement is found than when Figs. 2 and 4 were compared. The pressure test at $2\tau = 10\,780$ lb/in^2 (7·44 hbar) falls into

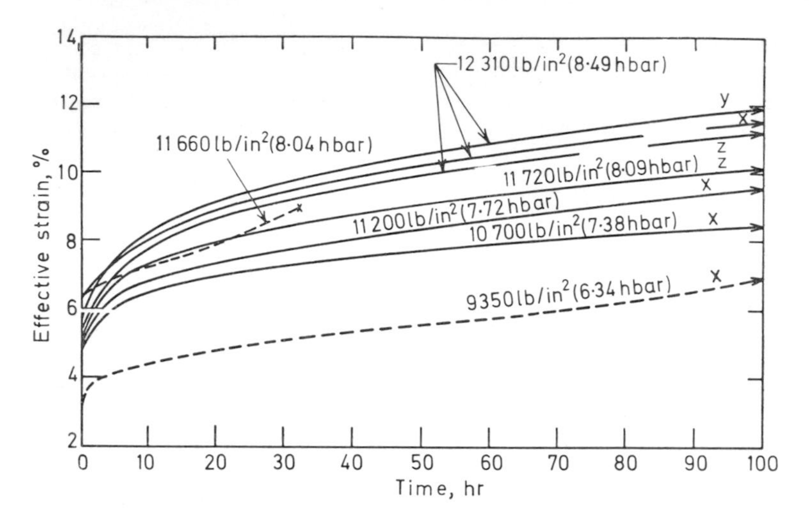

FIG. 4. Effective strain for torsion and internal pressure tests on tubes. The numbers shown on the curves are effective stress values and the letter is the direction in which specimens were taken from the cube. Data points have been omitted for clarity.

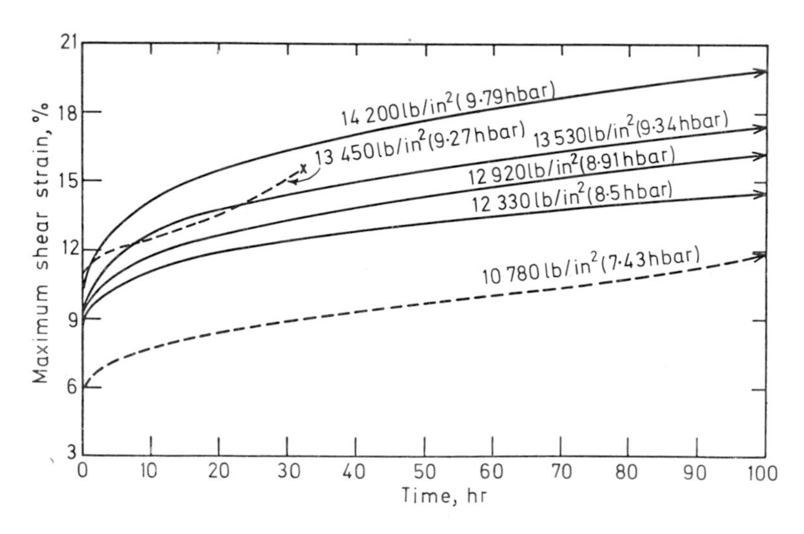

FIG. 5. Replot of data of Fig. 4. The numbers shown on the curves are twice the maximum shear stress.

place between tension at 10 000 and 11 000 lb/in^2 (6·89 and 7·59 hbar) while torsion at 2τ = 14 200 lb/in^2 (9·8 hbar) agrees well with tension at 14 000 lb/in^2 (9·65 hbar) until tertiary creep occurs. For the intermediate range of stresses the agreement between tension and torsion is less satisfactory, but is still considerably better than when they are compared on the basis of effective stress–effective strain.

Another possibility for the comparison of strain values in different tests is obtained by generalising eqn (7) into the form

$$\varepsilon_1 = \frac{F(\tau_{max})[\sigma_1 - \frac{1}{2}(\sigma_2 + \sigma_3)]G(t)}{\bar{\sigma}}$$

Substituting the expressions for ε_1, ε_2, ε_3 into the definition of effective strain leads to $\bar{\varepsilon} = F(\tau_{max})G(t)$. This approach may be tested by comparing Figs. 2 and 4 with the stress values on the curves in Fig. 4 replaced by those in Fig. 5. As with the effective stress–effective strain approach, the agreement is less satisfactory than that obtained using maximum shear stress–maximum shear strain.

To obtain another comparison of the torsion–tension and internal pressure tests we will assume that the total strain may be fitted by an expression $\varepsilon = \sigma^n G(t)$ which will enable us to present the data in a normalised form $\varepsilon/\sigma^n = G(t)$. This procedure cannot be exact because the stress dependence of initial plastic strain and creep strain are different. However, by plotting tension and torsion–pressure test data as logarithm ε versus logarithm σ for various times up to 100 hr we find that $\varepsilon \simeq \sigma^{2\cdot2}G(t)$. On this basis, the normalised effective strain $\bar{\varepsilon}(11\,000/\bar{\sigma})^{2\cdot2}$ is shown for the three types of tests in Fig. 6. The fairly narrow bands enclosing the torsion–pressure tests and the tension tests provide additional justification for the normalising procedure. In Fig. 6 the centre of the torsion–pressure band lies, on the average, about 20 per cent higher than that for tension at 100 hr, 16 per cent higher at 50 hr and 12 per cent higher at 10 hr. Replotting on the basis of maximum shear stress–maximum shear strain brings the bands into coincidence at 50 hr with the centre of torsion–pressure band being a few per cent higher than that for tension at 100 hr and a few per cent lower at 10 hr. We are reluctant to draw any conclusions for the times below 10 hr because the loading techniques vary from one kind of test to the other. Small variations in the period taken to apply the load may perturb the creep curve out to quite long times. For example, a 10 sec loading period may in some cases lead to 10 per cent error in strain recording at about 3 hr [21].

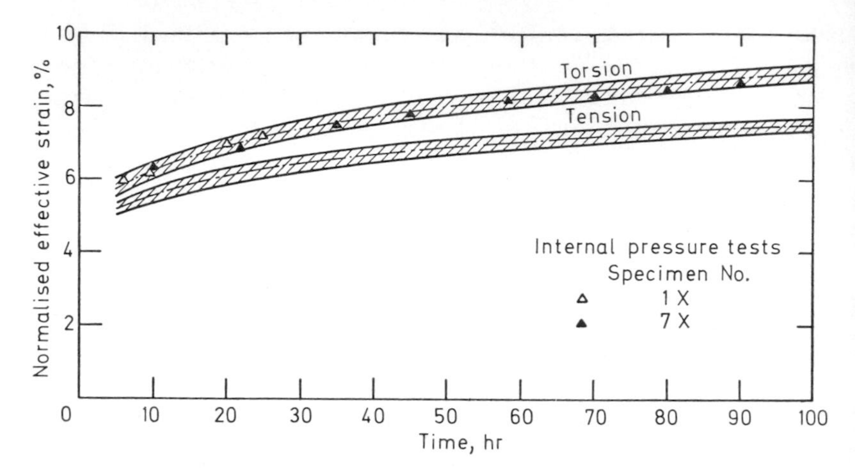

FIG. 6. Normalised effective strain $\bar{\varepsilon}(11\,000/\bar{\sigma})^{2\cdot 2}$ for tension, torsion and internal pressure tests.

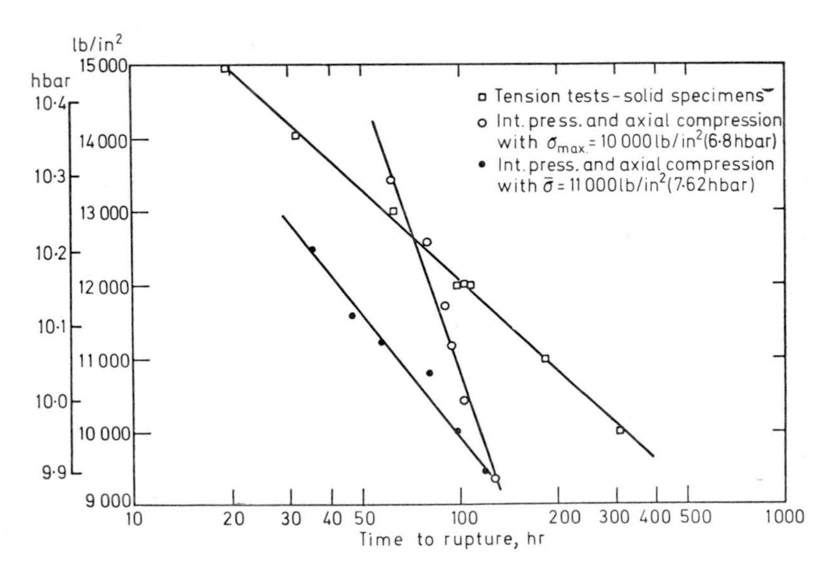

FIG. 7. Rupture life as a function of stress for combined stress tests on tubes and tension tests on solid specimens.

Now, we turn to examine the 'flow rule' that should be used to predict the magnitudes of the individual principal strains. For this purpose, a number of tests were made on tubular specimens under combined axial load and internal pressure. Measurements were made of axial and hoop strains in these tubes and radial strains were calculated on the assumption that no volume changes occur. Although the flow rule should, strictly, be written in terms of strain increments, as we work here with constant stresses and elastic strains are negligible, it will be expressed in terms of total strains.

Typical values for the observed and the predicted strain ratios in a number of tests are shown in Table 1. For each test, the ε_θ and ε_a strains were taken from the creep curve at a time just prior to tertiary creep. The predicted values for the more obvious flow rules, the Tresca and Levy–von Mises and the equations proposed by Bailey are also shown. A difficulty arises with the Tresca criterion when $\sigma_2 = \sigma_3$, as in the tension test, for the strain ratios must then be taken as $-0{\cdot}5$, $-0{\cdot}5$ for an isotropic material. However, even with this modification of the values shown in the table for cases in which $\sigma_a \simeq \sigma_r$, it is clear that the Tresca flow rule leads to unrealistic prediction while the Levy–von Mises flow rule in most cases agrees relatively well with experiment. Another approach to testing the various flow rules is to plot the Lode variables $\mu = (2\sigma_2 - \sigma_1 - \sigma_3)/(\sigma_1 - \sigma_3)$ and $v = 3\varepsilon_2/(\varepsilon_1 - \varepsilon_3)$. For the present tests, no clear trend could be detected by this method of plotting the data. However, if the values for tests 1Z, 8Y, 8Z are excluded, the remaining six points show behaviour very similar to that observed by Taylor and Quinney in their classic study of time independent plastic deformation of tubes loaded in combined tension–torsion. It has been pointed out by Prager that the results of Taylor and Quinney could be fitted by taking a plastic potential analogous to D of eqn (7), of the form

$$J'_2\left(1 - 0{\cdot}73\,\frac{J'^{\,2}_3}{J'^{\,3}_2}\right)$$

where J_2' and J_3' are invariants of the stress deviator tensor [22]. On the other hand, Dr Johnson in noticing similar deviations from the Levy–von Mises relations ($\mu = v$) when plotting the Lode variables, used eqn (2) to describe material behaviour.

In the case of the present tests, the best overall prediction of multiaxial behaviour, based on tension, creep data, appears to be given by eqn (4), but there are many other possibilities for predicting creep strains. For example, the Bailey equations, eqn (5), with $n = 2{\cdot}2$ and the choice of

TABLE 1

Comparison of observed strain ratios with predictions of Levy–von Mises, Tresca and Bailey flow rules

Specimen number	Stress lb/in² (hbar) σ_θ	σ_a	σ_r	Time for strain (hr)	Strain ratios: Measured $\varepsilon_a/\varepsilon_\theta$	Measured $\varepsilon_r/\varepsilon_\theta$	Tresca $\varepsilon_a/\varepsilon_\theta$	Tresca $\varepsilon_r/\varepsilon_\theta$	Levy-von Mises $\varepsilon_a/\varepsilon_\theta$	Levy-von Mises $\varepsilon_r/\varepsilon_\theta$	Bailey, $m = -1{\cdot}4$ $\varepsilon_a/\varepsilon_\theta$	Bailey, $m = -1{\cdot}4$ $\varepsilon_r/\varepsilon_\theta$
3X	11580 (8·00)	1735 (0·12)	−910 (−0·063)	20	−0·21	−0·79	0	−1	−0·32	−0·68	−0·23	−0·77
1Y	10820 (7·47)	0 (0)	−850 (−0·059)	30	−0·39	−0·61	0	−1	−0·44	−0·56	−0·39	−0·61
3Y	9450 (6·52)	−2330 (−0·161)	−730 (−0·050)	60	−0·63	−0·37	−1	0	−0·61	−0·39	−0·67	−0·33
1Z	10000 (6·89)	−1480 (−0·102)	−800 (−0·055)	50	−0·48	−0·52	−1	0	−0·55	−0·45	−0·56	−0·44
3Z	11200 (7·74)	810 (0·056)	−875 (−0·060)	20	−0·29	−0·71	0	−1	−0·39	−0·61	−0·31	−0·69
7Y	10000 (6·89)	0 (0)	−810 (−0·056)	30	−0·44	−0·56	0	−1	−0·44	−0·56	−0·40	−0·60
7Z	10000 (6·89)	−5000 (−0·345)	−770 (−0·053)	10	−0·80	−0·20	−1	0	−0·75	−0·25	−0·84	−0·16
8Y	10000 (6·89)	−3780 (−0·261)	−785 (−0·054)	40	−0·65	−0·35	−1	0	−0·68	−0·32	−0·77	−0·23
8Z	10000 (6·89)	−2500 (−0·173)	−780 (−0·0535)	40	−0·56	−0·44	−1	0	−0·61	−0·39	−0·67	−0·33

$m = -1{\cdot}4$ (based on results from tests in combined axial load and internal pressure) give reasonable predictions for the flow rule, as can be seen in Table 1, and yield good agreement between tension and torsion tests. However, unless the additional parameter m is relatively constant for different materials, this type of equation is not convenient since it requires test data in addition to tension before predictions can be made.

EXPERIMENTAL STUDY OF CREEP–RUPTURE

The two rupture criteria proposed by Dr Johnson, $\bar{\sigma}$ and σ_{max}, were examined by testing thin walled tubes under various combinations of axial compression and internal pressure. With this type of loading it is possible to hold either the effective stress or the maximum tensile stress constant and vary the other quantity. In this respect the compression–pressure test on tubes is more attractive than the combined tension–torsion test. However, in both types of tests creep buckling may precede rupture and for this reason the present tests had to be limited to stress ratios $\sigma_a/\sigma_\theta \geq -\frac{1}{2}$.

One difficulty in studying the different criteria for creep–rupture is that the effective stress has its greatest value at the inside surface of the tube while, in general, the maximum value of the tensile stress is at the outer wall. Since the combination of stresses that leads to the most severe conditions of crack initiation and growth is not known, it seems reasonable to quote all stresses for the mean diameter of the tube. On this basis, Table 2 shows the rupture time for five tests made at a constant maximum tensile stress of 10 000 lb/in^2 (6·89 hbar) and effective stress values ranging from 9350 to 13 400 lb/in^2 (6·45–9·25 hbar). Also shown are the results of six tests made with essentially constant effective stress values $\bar{\sigma} = 11\,300 \pm 300$ lb/in^2 (7·79 $\pm$ 0·21 hbar) and maximum tensile stresses ranging from 9450 to 12 480 lb/in^2 (6·51–8·60 hbar). The slight variation in the effective stress is due to the fact that when the tests were carried out, the radial stress was neglected leading to a value $\bar{\sigma} = 10\,820$ lb/in^2 (7·45 hbar) for all six tests. As shown in Table 2 we have made a slight correction to the observed rupture times so that they will all correspond to $\bar{\sigma} = 11\,065$ lb/in^2 (7·61 hbar). The correction is based on the observed dependence of life on effective stress when tests are made at constant σ_{max}.

The rupture life values from the table are shown in Fig. 7 along with tensile creep–rupture data. It is striking that rupture-life varies if either σ_{max} or $\bar{\sigma}$ is held constant and hence neither criterion for rupture is

TABLE 2

Rupture-life under combined axial compression and internal pressure

Specimen number	Stress lb/in² (hbar) $\sigma_\theta = \sigma_{max}$	σ_a	σ_r	$\bar{\sigma}$	Rupture-life, t (hr) Observed	Corrected
7X	10000	5000	−790	9350	126·1	
	(6·89)	(3·45)	(−0·0545)	(6·44)		
7Y	10000	0	−810	10430	102·6	
	(6·89)	(0)	(−0·056)	(7·20)		
7Z	10000	−5000	−770	13395	60·95	
	(6·89)	(−3·45)	(−0·053)	(9·23)		
8Y	10000	−3780	−785	12550	80·1	
	(6·89)	(−2·61)	(−0·054)	(8·65)		
8Z	10000	−2500	−780	11735	89·9	
	(6·89)	(−1·73)	(−0·0535)	(8·09)		
1X	12480	6240	−970	11660	32·4	35·6
	(8·50)	(4·3)	(−0·67)	(8·04)		
3X	11580	1735	−910	11400	44·13	46·8
	(7·98)	(1·195)	(−0·63)	(7·86)		
1Y	10820	0	−850	11270	77·8	80
	(7·46)	(0)	(−0·59)	(7·76)		
3Y	9450	−2325	−730	11065	118·0	118·0
	(6·51)	(−1·605)	(−0·50)	(7·61)		
1Z	10000	−1480	−800	11160	97·6	99·2
	(6·89)	(−1·02)	(−0·55)	(7·69)		
3Z	11200	+ 810	−875	11330	55·2	57·7
	(7·72)	(0·056)	(−0·060)	(7·82)		

adequate for this material. Since the data may be fitted by the exponential relations

$$t_{(\sigma_{max}=10\,000)} = A_1 \exp -b_1\bar{\sigma} \quad \text{and} \quad t_{(\bar{\sigma}=11\,065)} = A_2 \exp -b_2\sigma_{max}$$

we hypothesise that in the general case the rupture life may be written as

$$t = A \exp -(+b_1\bar{\sigma} + b_2\sigma_{max}) \tag{9}$$

where $A = A_1 \exp +10\,000b_2 = A_2 \exp +11\,065b_1$ and by using the method of least squares the following values are obtained:

$$A_1 = 5680{\cdot}0 \text{ hr}, \quad A_2 = 648{\cdot}5 \text{ hr}, \quad b_2 = 4{\cdot}09 \times 10^{-4} \text{ (lb/in}^2)^{-1}$$
$$b_1 = 1{\cdot}73 \times 10^{-4} \text{ (lb/in}^2)^{-1}$$

Applying eqn (9) to the case of uniaxial tension we obtain

$$t = 38\,500 \exp -5{\cdot}82\sigma \times 10^{-4} \text{ hr}$$

whereas a direct least squares fit of the tension data for solid specimens yields

$$t = 83\,000 \exp -5{\cdot}57\sigma \times 10^{-4} \text{ hr}$$

It is intriguing that the values of the exponents in the two equations agree so closely. The numerical factors in front of the exponential differ greatly, which is not surprising in view of the difference in the shape and size of the two types of specimens. The difference between the rupture life of solid and tubular specimens can also be seen in Dr Johnson's work and this appears to be an aspect of creep–rupture which deserves further examination.

A few tests were made with a constant maximum tensile stress $\sigma_{max} =$ 10 000 lb/in^2 (6·89 hbar) and with an effective stress value of $\bar{\sigma} =$ 13 400 lb/in^2 (9·23 hbar) for the first part of the test and $\sigma =$ 9350 lb/in^2 (6·44 hbar) for the remainder of the test. After unloading from $\bar{\sigma} =$ 13 400 to 9350 lb/in^2 (9·23 to 6·44 hbar), creep deformation almost stopped and these tests showed a considerably greater life than the specimen tested at $\sigma_{max} =$ 10 000 lb/in^2 (6·89 hbar) and a constant value of $\bar{\sigma} =$ 9350 lb/in^2 (6·44 hbar). Clearly, the linear cumulative damage rule does not hold under these conditions and in future work the linear cumulative octahedral strain criterion suggested by Dr Johnson will be examined.

Metallographic examination of failed specimens showed cracks present all over the cross-section and predominantly oriented normal to the maximum tensile stress. Cracking appears to start at the bore of the tube, which tends to support the hypothesis that crack initiation is governed by the effective stress. This observation also argues against the view [23] that creep–rupture in copper is caused by stress corrosion since oxidation of the outer surface is much greater than that of the inside. The inside surface is exposed only to the pressurising liquid, methyl phenyl silicone, and after rupture it is fairly clean. It is interesting to speculate on the differences in the amount of cracking observed in various specimens. For example, specimen 3Y $\bar{\sigma} =$ 11 065 lb/in^2 (7·61 hbar), $\sigma_{max} =$ 9450 lb/in^2 (6·51 hbar) was seen to be very extensively cracked at fracture while specimen 1X $\bar{\sigma} =$ 11 660 lb/in^2 (8·04 hbar), $\sigma_{max} =$ 12 480 lb/in^2 (8·50 hbar) had a relatively small number of isolated cracks in addition to the main crack. Evidently, in this latter case, the first crack to nucleate grew faster due to the relatively high value of σ_{max} and therefore caused final fracture before the bulk of the material was appreciably damaged as in the case of specimen 3Y.

CONCLUSIONS

In the tests reported, the initial plastic strains are of the same order as the creep strain to fracture. Such a situation may arise in practice, for example, in the localised region near a notch or other discontinuity. Even at the rather high strain levels used in the present study, isotropic behaviour was observed and the assumption that tension and compression creep are equivalent is seen to be satisfied. The coincidence of torsion and pressure creep tests shows that the hydrostatic stress has a negligible effect, at least until the beginning of tertiary creep. Hence, the material conforms to the assumptions usually employed in formulating equations for creep under multiaxial stress. Following these assumptions, the stress–strain (or strain rate) relations are usually obtained by differentiating a potential which is a function of J_2' and J_3'. Since the Odqvist equations correspond to the case in which the potential is a function of J_2' alone, the fact that they do not give a precise correlation of tension and torsion tests indicates that the potential must be a function of J_3' as well as J_2'. The difficulty then lies in selecting an appropriate functional relation from the infinite number that are available. The Bailey equations fall into this category and indeed enable fairly precise predictions to be made once the additional parameter is determined. However, in the present tests, equally satisfactory strain prediction could be obtained by departing from the formalism of the flow potential and combining the Levy–von Mises flow rule with the maximum shear stress as in eqn (4).

Creep–rupture studies show that life under multiaxial stress depends on the two criteria suggested by Dr Johnson, effective stress and maximum tensile stress. It is suggested, as a first approximation, that these aspects of the stress state may govern crack initiation and crack growth respectively.

ACKNOWLEDGEMENT

This work was supported by the US Army Research Office, Durham.

REFERENCES

1. ODQVIST, F. K. G. (1966). 'Mathematical theory of creep and creep-rupture.' TEMPLE, G. and JAMES, I. M. (Eds.). *Oxford Mathematical Monographs.* Oxford: Clarendon Press.
ODQVIST, F. K. G. (1935). 'Creep stresses in a rotating disc.' *Proc. 4th Int. Congr. Appl. Mech.*, p. 228. London: Cambridge University Press.
2. SODERBERG, C. R. (1936). 'Interpretation of creep tests for machinery design.' *Trans. Amer. Soc. Mech. Engrs*, **58**, 733–743.
3. BAILEY, R. W. (1935). 'The utilisation of creep test data in engineering design.' *Proc. Instn Mech. Engrs*, **131**, 131–349.
4. JOHNSON, A. E., HENDERSON, J. and KHAN, B. (1962). *Complex-stress Creep, Relaxation and Fracture of Metallic Alloys.* Edinburgh: HMSO.
5. WAHL, A. M. (1956). 'Analysis of creep in rotating discs based on the Tresca criterion and associated flow rule.' *J. Appl. Mech.*, **23**, 231–238.
WAHL, A. M. (1957). 'Stress distribution in rotating discs subjected to creep at elevated temperature.' *J. Appl. Mech.*, **24**, 299–305.
WAHL, A. M. (1958). 'Further studies of stress distribution in rotating discs and cylinders under elevated temperature creep conditions.' *J. Appl. Mech.*, **25**, 243–250.
WAHL, A. M., SANKEY, G. O., MANJOINE, M. J. and SHOEMAKER, E. (1954). 'Creep tests of rotating discs at elevated temperature and comparison with experiment.' *J. Appl. Mech.*, **21**, 225–235.
6. COFFIN, L. F., SHEPLER, P. R. and CHERNIAK, G. S. (1949). 'Primary creep in design of internal pressure vessels.' *J. Appl. Mech.*, **16**, 229–241.
7. DAVIS, E. A. (1961). 'The Bailey flow rule and its associated yield surface.' *J. Appl. Mech.*, **28E**, 310.
8. WAHL, A. M. (1962). 'Application of the modified Bailey equations under biaxial tension stress.' *Proc. 4th US Nat. Conf. Appl. Mech.*, Berkeley. New York: American Society of Mechanical Engineers.
9. RABOTNOV, YU. N. (1965). 'Experimental data on creep of engineering alloys and phenomenological theories of creep—a review.' *J. Appl. Mech. & Tech. Phys.*, **1**, 141–159.
10. BERMAN, I. and PAI, D. H. (1966). 'A theory of anisotropic steady state creep.' *Int. J. Mech. Sci.*, **8**(5), 341–352.
11. FINNIE, I. (1963). 'An experimental study of creep in tubes.' *Proc. Jnt. Int. Conf. on creep*, pp. 2–21. London: Institution of Mechanical Engineers.
12. JOHANSEN, F. P. (1963). *An experimental study of creep in aluminium under multiaxial stress.* M.Sc. Thesis. University of California, Berkeley.
13. BALLUFFI, R. W. and SEIGLE, L. L. (1957). 'Growth of voids in metals during diffusion and creep.' *Acta Metall.*, **5**(8), 449–454.
14. HULL, D. and RIMMER, D. (1959). 'The growth of grain boundary voids under stress.' *Phil. Mag.*, **4**(42), 673–687.
15. CRUSSARD, C. and FRIEDEL, J. (1956). 'Theory of accelerated creep and rupture.' *Creep and Fracture of Metals at High Temperatures.* London: HMSO.
16. HOFF, N. J. (1953). 'The necking and rupture of rods subjected to constant tensile loads.' *J. Appl. Mech.*, **20**, 105–108.

17. KACHANOV, L. M. (1961). 'Rupture time under creep conditions' (in Russian). *Akad. Nauk SSSR* 1*gv OTN*, **5**, 88–92.
18. ODQVIST, F. K. G. and HULT, J. (1961). 'Some aspects of creep rupture.' *Ark. Fysik*, **19**(26), 379.
19. VOORHEES, H. R. and FREEMAN, J. W. (1959). 'Notch-sensitivity of high temperature alloys.' *WADC Report* TR59-470. Ohio: Wright Air Development Centre, Wright-Patterson AFB.
20. GOLDHOFF, R. M. and BROTHERS, A. J. (1968). 'The influence of notches on mechanical behaviour at elevated temperature: some metallographic observations.' *J. Bas. Engng*, **90**, 37.
21. FINNIE, I. and HELLER, W. R. (1959). *Creep of Engineering Materials*, p. 169. New York: McGraw-Hill.
22. HILL, R. (1950). *The Mathematical Theory of Plasticity*, p. 174. London: Oxford University Press.
23. BLEAKNEY, H. H. (1965). 'The creep-rupture embrittlement of copper.' *Can. Metall. Q.*, **4**(1).

Chapter 17

CREEP STRAINS IN THICK RINGS SUBJECTED TO INTERNAL PRESSURE

H. FESSLER, P. A. T. GILL AND P. STANLEY

SUMMARY

Lead alloy rings were subjected to creep conditions under internal pressure exerted by axially compressed silicone rubber plugs. Surface strains were determined with the moiré technique; outside diameter measurements were taken with dial gauges. Ten rings of three different diameter ratios were tested at seven different pressures for 100 *hr at room temperature. Good repeatability of results was achieved.*

The experimental results are compared with the results from computations; these showed that, in the cases examined, stress redistribution occurs extremely quickly and that different stress–strain–time relationships and hardening behaviour assumptions lead to very similar results.

At the outside diameter the measured hoop total strain rates were generally smaller than predicted by the computations.

When the strains were expressed as multiples of the outside diameter hoop strain, the computed ratios were independent of pressure and of time within the test period. Computed results for different diameter ratios fell on the same two curves. The measured strain ratios differed little from these curves, except some of the hoop values near the bore.

NOTATION

d Diameter

K Outside diameter/bore.

p Internal pressure.

r Radius.
t Time.
A, B, b, m, n, s Constants.
ε Strain.
σ Stress.

Subscripts
r, θ Polar coordinates.
c Creep.
i Inner.
o Outer.
T Total.

INTRODUCTION

As the theoretical treatments for time dependent strains and the computational methods of handling them become more powerful, the need to examine their validity by careful, searching experiments increases. The point was made by Manning in the discussion of a well known paper by Dr Johnson and his co-workers [1], who expressed regret that the theoretical results had not been checked experimentally. In view of his very wide experience in the field, it is most unlikely that Manning did not know the reason for this, but the author's reply was explicit:

> . . . adequate experimental investigation of the theory involved technical problems in regard to strain measurement which were exceedingly difficult, and quite probably impossible to solve, and such investigation was accordingly not lightly, or even advisedly undertaken.

Although valuable experimental work has been carried out over the intervening years, the need for confirmatory tests has not been fully met. Experimental model studies at room temperature, in conjunction with new strain measuring techniques, offer wide scope for such testing in which the experimental difficulties usually associated with creep testing are considerably eased. With a suitable material, test loads and durations can be greatly reduced; in room temperature work, free access is possible which facilitates loading and strain measurement, and the temperature control problem is eased. The present chapter describes work of this kind in which the creep behaviour of a series of radially thick rings (*i.e.* short open ended cylinders), in a lead alloy, subjected to a constant internal pressure has been studied and compared with theory.

The criteria of choice and the calibration of the material used, an extruded 0·2 per cent Sb, 0·02 per cent As, lead alloy, have been described in detail elsewhere [2]. Briefly, the material crept readily under relatively small loads at room temperature without structural changes, its uniaxial tensile creep behaviour was satisfactorily represented by equations of the form $\varepsilon_c = B\sigma^n f(t)$ or $\varepsilon_c = D \sinh (\sigma/\sigma_o) f(t)$, (these and other material constants for the material are summarised in Appendix 1) in which the stress index n was the same as that for a structural steel, and its behaviour under varying stress tended to be of the strain hardening rather than the time hardening type.

As well as for the practical reasons of ease of manufacture and loading, the ring models were chosen because they contained a non-uniform, biaxial stress system in which appreciable stress redistribution was known to occur and strains could be measured over the entire section from the bore to the periphery. It was considered that the experimental study of strains under this relatively complex stress system would itself be worthwhile and additionally that their prediction from the available uniaxial data would provide a valuable and searching test for the theory and methods used in the computation.

PREVIOUS WORK

Theoretical

Bailey's work provided the first generally useful, basic design method for creep. In a later paper [3] he presented a method for the application of tensile data to the case of a pressurised cylinder. General equations are obtained from the three principal stresses with the assumptions that the creep at all points is in the same stage and elastic strains are negligibly small. The treatment requires that the principal stresses remain constant throughout any one stage of creep.

Dr Johnson, Henderson and Khan [1] give a theoretical solution for the problem of creep in thick cylinders which assumes plane strain conditions with zero axial creep strain and constant axial elastic strain at all times. These assumptions allow the derivation of a differential equation relating the principal stress difference $(\sigma_r - \sigma_\theta)$ at a point to the radius (r), the time (t) and a number of material constants. The equation may be solved graphically and values of σ_r and σ_θ can be readily obtained. The solution cannot be adapted [4] for the plane stress condition applicable in the case of rings. Furthermore, the assumption that the sum of the hoop and radial stresses at a point is constant, which follows from the axial strain assumptions, is incompatible with the equilibrium requirements when conditions of rapid stress redistribution occur.

Henderson has emphasised (in a private communication) that the elastic and creep strains must be similar in magnitude if the assumptions made in the theory are to be satisfied. This further limits the suitability of the solution for predicting strains in model tests since the reliable measurement of strains of only twice the elastic strain was not practicable.

It has been shown by Larke and Parker [5] that the equations due to Bailey and Dr Johnson *et al.* for the calculation of bore and outer diameter hoop strains are mathematically identical. They also derive that the ratio of bore hoop strain to od hoop strain ($\varepsilon_{\theta i}/\varepsilon_{\theta o}$) is equal to the square of the ratio of o.d. to bore (K^2).

A treatment of creep in cylinders has been published by Smith [6], which contains none of the previous axial strain assumptions and can deal with either the plane stress or plane strain conditions. The method has been extended and programmed by the Berkeley Nuclear Laboratories of the Central Electricity Generating Board [7]; further details are given in a later section.

Experimental

The earliest reference to creep strain measurements in a thick walled cylinder appears to be due to Bailey [8]. A lead tube ($K = 3$) was tested with an internal pressure of 1200 lbf/in^2 (82·75 bar) at 19·5°C. Outside diameter hoop strain measurements were taken up to 5 per cent at about 2 hr when the tube presumably ruptured. (This is not stated, but the o.d. hoop creep rate increased rapidly at this time.) No uniaxial creep data are given for the tube material.

Norton [9, 10] and Norton and Soderberg [11] reported tests on thinner walled cylindrical pressure vessels ($K = 1{\cdot}07$ and $1{\cdot}23$) in 0·5 per cent molybdenum steel. Outside diameter hoop strain measurements were taken at 800°, 900° and 1050°F (427°, 483° and 565°C) for a range of pressure up to 9240 lbf/in^2 (637 bar).

Nakahara [12] has tested lead cylinders ($K = 1{\cdot}39$ and $1{\cdot}56$) with pressures up to 450 lbf/in^2 (31 bar) at an unspecified temperature (presumably room temperature). Outside diameter hoop strains were measured, but few details are given.

Davis [13] has reported creep rupture tests on stainless steel tubular specimens, with $K = 2$ and 3, at 1200°F (648°C). The specimens were subjected to either simple internal pressure (up to 24 000 lbf/in^2; 1655 bar) or equal biaxial tension. Outside diameter hoop strains were measured and a favourable correlation with uniaxial test data was obtained.

EXPERIMENTAL WORK

Loading rig

A uniform and constant radial pressure was obtained in a ring model by axially compressing a soft rubber plug, fitted into the bore, by means of a deadweight load acting through a lever arm system. The arrangement is shown schematically in Fig. 1.

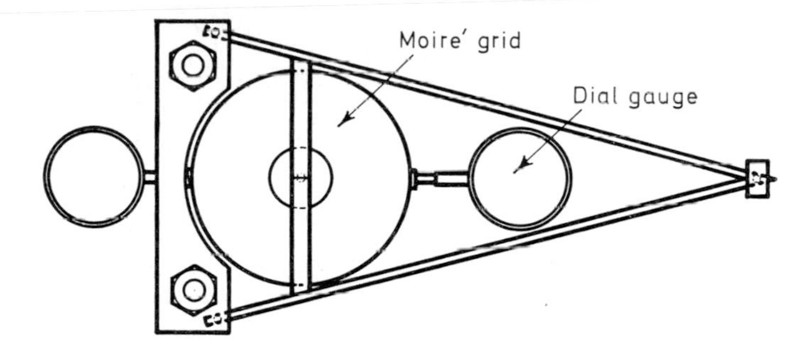

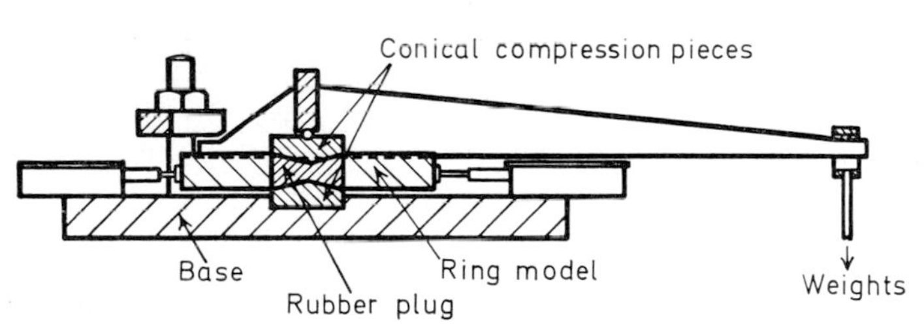

FIG. 1. Loading system for ring models.

The rubber plugs had re-entrant conical ends to facilitate location of the fitting steel compression pieces and to eliminate the barrelling effect which can be pronounced in plane ended compression members and which would have resulted in an axial variation in the pressure loading. In preliminary tests, plugs with cone angles of 120°, 150° and 180° (plane ended) had been fitted with a set of very thin Araldite rings and compressed between conforming plattens in a compression testing machine. Ring deformations, obtained from micrometer readings, showed there to be no appreciable axial variation in hoop strains in either the 120° or the 150° angled plugs. The latter was chosen for all the ring model testing since it allowed a shorter plug and, consequently, axially thinner ring models. The relationship between axial compression and radial pressure for a 150° angled plug was also obtained from these tests from the mean measured deformations of the Araldite rings using an independently determined value of Young's modulus for Araldite. The results for loading and unloading are shown in Fig. 2. The relationship was linear at least up to a hoop strain of 1·6 per

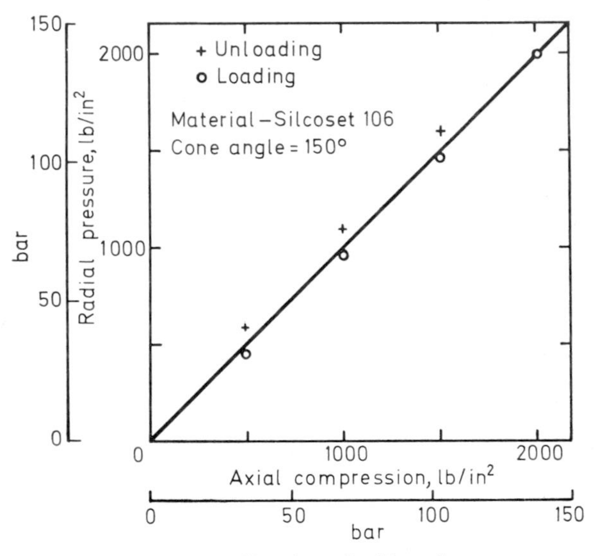

FIG. 2. Calibration of rubber plug.

cent, the maximum obtainable with the Araldite rings; the small hysteresis effect was not important for the present application; the behaviour was approximately 'fluid-like' in that axial and radial stresses were very nearly equal.

The plug material was Silcoset 106, a cold curing, easily mouldable silicone rubber. The ground steel compression pieces from the loading rig were incorporated in the mould to ensure a perfect fit.

In order to minimise the obstruction to viewing the model surface, a divided form of lever arm was used. Small steel balls, cemented to the lever, formed the pivots to ensure that a known constant force was exerted on the rubber plug. The balls could slide to minimise horizontal forces as the lever tilted.

Models

For the reasons given earlier, the 'specimen' chosen for this work was a thick, flat circular ring subjected to a constant internal pressure. The rings were turned from a single length of extruded tube, 1 in (25 mm) bore and 4 in (100 mm) o.d., using a single point tool; dimensions are given in Table 1. All models were stabilised after final machining by heating for 1 hr at 100°C followed by air cooling.

Four ring geometrics were studied. The 'A' rings, the first to be designed and tested, had the maximum o.d. and minimum bore to give the radially thickest possible models with an axial length of 0·5 in (12·5 mm). These rings were sufficiently robust to be machined and handled without difficulty, they were stable when subjected to an internal pressure and it was felt that the stress state would approximate closely to the plane stress condition assumed in the theoretical work. The dimensions of the 'B' rings were the same as those of the 'A' rings, except that the axial length was approximately doubled to 0·97 in (24·6 mm), the value common to rings 'C' and 'D'. In the 'C' rings, K was reduced to 2·00, a typical 'thick cylinder' value frequently used as an example in theoretical work. Since the maximum surface area was required for maximum resolution in the strain measurements, the greatest practicable o.d. of 3·875 in (98·4 mm) was retained and the bore diameter was increased. Because calibration of the plugs was only carried out for one shape (that of the 'A' rings), this shape was preserved as nearly as possible in rings 'C' and resulted in an axial length of 0·97 in (24·6 mm). In the 'D' rings, the bore to o.d. ratio was further reduced to 1·40, at the lower end of the range of practical interest. Since the radial thickness was too small in any event to allow the measurement of surface strains, the bore of the 'C' rings was retained to avoid making another plug mould.

Strain measurement

The moiré fringe technique [14, 15] was used to measure strains on the upper surface of all models except A1, D1 and D2. A strain sensitivity of $\pm 0{\cdot}1$ per cent was obtained. Full details of the apparatus, procedure and analysis will be published elsewhere.

In addition, 0·0001 in (2·5 μm) dial gauges were included in the loading rig (Fig. 1) to provide an independent measurement of hoop strains at the outer diameter. The bore and outer diameter were measured independently at the completion of the test when the model had been taken from the loading frame. Model thickness changes were not considered important and were therefore not measured.

Table 1

Model details

Ring model	Geometry: o.d./Bore K	Geometry: Length/Bore	Geometry: Bore d_i in (mm)	Experimental: Pressure p lbf/in² (bar)	Experimental: Initial equivalent stress at bore lbf/in² (bar)	Experimental: Remarks	Time for which computed results are available (hr): σ^n [a] sh[c]	σ^n [a] pl sh	σ^n [a] pl th	σ^n [a] th	$\sinh(\sigma/\sigma_o)$ [b] sh	$\sinh(\sigma/\sigma_o)$ [b] th
A1	3·44	0·44	1·125 (28·6)	1120 (77·2)	2120 (146)	No moiré	30	–	(3)	25(75)	–	–
A2	3·44	0·44	1·125 (28·6)	1000 (69)	1900 (132)		100	10	–	75	–	–
A3	3·44	0·44	1·125 (28·6)	750 (51·7)	1420 (97·9)		100	–	–	100	–	–
B1	3·44	0·86	1·125 (28·6)	1000 (69)	1900 (132)		As for model A2[d]					
B2	3·44	0·86	1·125 (28·6)	750 (51·7)	1420 (97·9)		As for model A3[d]					
C1	2·00	0·50	1·938 (49·3)	650 (44·8)	1520 (104·9)		75	–	–	75	100	100
C2	2·00	0·50	1·938 (49·3)	650 (44·8)	1520 (104·9)	Repeat of C1	75	–	–	75	100	100
C3	2·00	0·50	1·938 (49·3)	450 (31)	1050 (72·4)		100	–	–	100	–	–
D1	1·40	0·50	1·938 (49·3)	350 (24·1)	1290 (88·8)	No moiré	30	–	–	100	–	–
D2	1·40	0·50	1·938 (49·3)	250 (17·2)	920 (63·5)	No moiré	100	–	–	100	–	–

Notes

1 The material behaviour relationships are:
  [a] $\varepsilon_c \propto \sigma^n(t^s + bt)$
  [b] $\varepsilon_c \propto \sinh(\sigma/\sigma_o)(t^s + bt)$
  sh Strain hardening
  th Time hardening
  pl Including effect of instantaneous plasticity.
2 [c] This is defined as the 'standard' computation.
3 [d] The computations assumed a plane stress state and were independent of axial thickness.
4 Times in parentheses for mode A1 are based [illegible]

Model pressures

Model A1 was tested at a high stress level without a moiré grid, to determine the practical limitations of the rig; the rubber plug split when the bore hoop strain of the ring was 7 per cent. Maximum pressure levels in subsequent tests were therefore chosen to give a bore hoop strain of about 5 per cent after a test period of 100 hr. Each ring geometry was also tested at a lower pressure level to give a bore hoop strain of about 1 per cent after 100 hr.

The required pressure values were estimated using a result due to Larke and Parker [5] who showed that, for a material with uniaxial creep behaviour represented by the expression

$$\varepsilon_c = A\sigma^n f(t)$$

the bore hoop strain under plane strain conditions is given by

$$\varepsilon_{c\theta} = K^2 \frac{A}{2} 3^{(n+1)/2} \left[\frac{p}{n(K^{2/n} - 1)} \right]^n f(t)$$

Loads were derived from the calculated radial pressures assuming perfect 'fluid-like' behaviour, Fig. 2. The use of the plane strain assumption was considered justified for this purpose. Pressures are tabulated in Table 1; material constants are given in Appendix 1.

Test procedure

The tests were conducted in a temperature controlled (±1 deg C) laboratory. The test period was 100 hr. The deformed moiré grid was photographed immediately on application of the load and after 0·1, 0·3, 1·0, 3·0, 10, 30 and 100 hr, for subsequent strain analysis; the two dial gauges were read at the same times for o.d. hoop strains. The photographing and the dial gauge readings took about 15 sec on each occasion.

Moiré grids were not used on rings D1 and D2, these being too thin to give a sufficiently extensive moiré fringe pattern for worthwhile analysis.

The bores and outside diameters were measured before and after the tests.

Note: With rings B1 and B2, where the axial length to bore ratio was greatest, it was found at the completion of each test that the bore had become tapered, with the bore hoop strain on the lower surface approximately twice as great as that at the upper surface which carried the moiré grid. As this effect was not found in any of the other model rings, surface reinforcement due to the moiré grid was not suspected. It was assumed that the loading system became unsatisfactory if the diameter of the rubber plug was about equal to its length. In all other tests the plug diameter was at least twice the plug length.

COMPUTED RESULTS

All the computed results have been obtained from the CEGB Berkeley Nuclear Laboratories [7] programme, based on Smith's [6] finite difference solution of the equilibrium and compatibility equations.

Deformation is considered to occur in a series of steps, each consisting of a short period of steady state creep followed by instantaneous stress readjustment to re-satisfy the equations of equilibrium and compatibility. The creep strain increments are derived from the mean values of stress during the interval, which are estimated from the stresses at the beginning of each step. Cumulative changes in radius are allowed for, but small strain theory is used over each interval.

The programme is quite general. Any separable form of stress and time functions may be used, with either the strain or time hardening hypothesis. The 'standard' computation, *see* Table 1, incorporated a uniaxial creep law $\varepsilon_c = B\sigma^n(t^s + bt)$ with the strain hardening hypothesis and did not include effects due to instantaneous plasticity. The use of an alternative stress function (sinh (σ/σ_0)) and the time hardening hypothesis, and the inclusion of instantaneous plasticity were investigated.

The 'test times' for the different computations are summarised in Table 1. It was not practicable to obtain computed results up to 100 hr for all the shapes and pressures used for the five different cases listed in Table 1, nor was this necessary. In spite of the gaps and short runs, the computations are sufficiently extensive to allow valuable conclusions to be drawn from them independent of the experimental work; these results will be discussed in this section.

Instantaneous plasticity

The effect of instantaneous plasticity on the computed total hoop strains may be seen in Fig. 3 where values obtained for model A1 with and without its inclusion are compared. An early estimate of the yield stress of the material of 200 lbf/in^2 (13·8 bar) was used in the computations. (The yield stress used in the computation for instantaneous plasticity for models A2 and B1 in Table 1 was 1000 lbf/in^2 (68·9 bar); *see* Appendix 1). Figure 3 shows the largest effects of including instantaneous plasticity which are

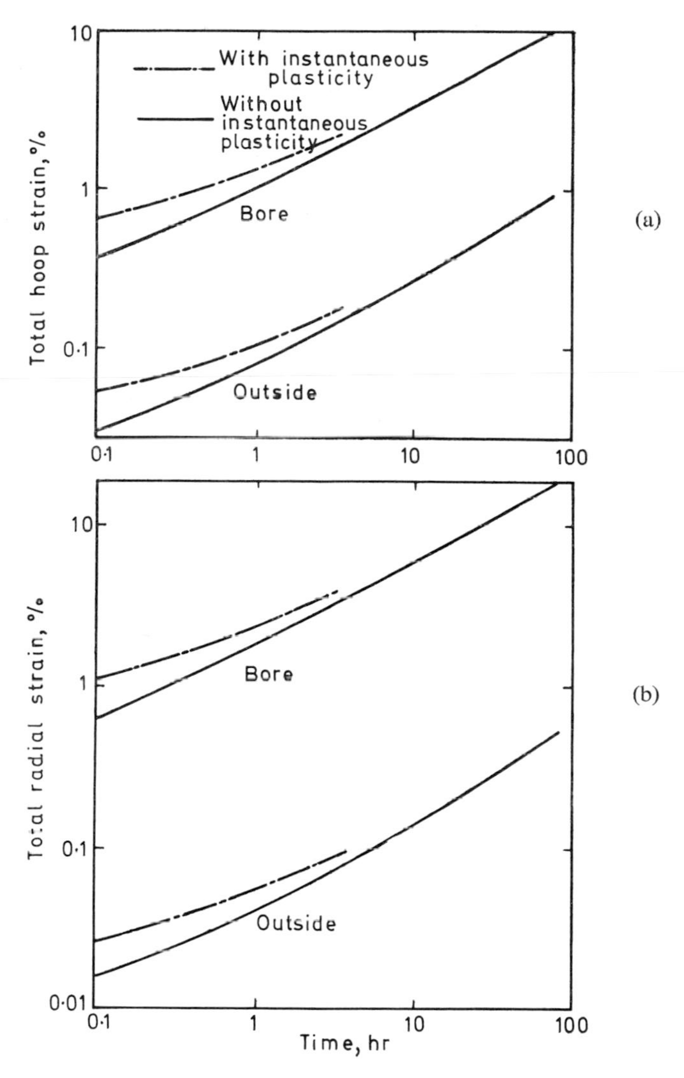

FIG. 3(a). Total hoop strains.

FIG. 3(b). Total radial strains.

FIG. 3. Model A1: effect of instantaneous plasticity on computed strains. $K = 3{\cdot}44$, $p = 1120$ lbf/in^2 (77·4 bar). Based on provisional creep data with yield stress $= 200$ lbf/in^2 (13·8 bar) and a creep law $\varepsilon_c = 3{\cdot}1 \times 10^{-17}\sigma^{5\cdot0}t^{0\cdot5}$, broken line with instantaneous plasticity, continuous line without instantaneous plasticity.

likely to occur within the range of this work because it is based on the greatest diameter ratio K, the greatest internal pressure and the smallest estimate of yield stress. As may be expected, the differences in computed hoop strains due to the inclusion of instantaneous plasticity decrease with time. It is noteworthy that the percentage difference at the outside diameter should be the same as at the bore and approximately the same as in both the bore and od radial strains, which are also shown in the figure. With the higher yield stress, the effect of instantaneous plasticity as computed for model A2 was negligible.

TABLE 2

Percentage differences in total strains computed with time and strain hardening hypotheses

Corresponding values of $(\varepsilon_t - \varepsilon_s)/\varepsilon_s \times 100$

Ring	o.d./Bore	Pressure lbf/in² (bar)	Time (hr)	Hoop at: Bore	Hoop at: Mid-thickness	Hoop at: Outside	Radial at: Bore	Radial at: Mid-thickness	Radial at: Outside
A1	3·44	1120	0·0001	−19	−15	−15	−26	−16	−16
		(77·2)	0·1	+1	−1	−1	+5	0	−1
			10	0	+1	+1	+1	+1	+1
A2, B1	3·44	1000	0·0001	−15	−11	−12	−23	−12	−13
		(69)	0·1	+3	0	0	+7	+1	−1
			10	+1	0	0	+2	0	0
A3, B2	3·44	750	0·0001	−8	−6	−8	−15	−7	−5
		(51·7)	0·1	+5	+2	+1	+11	+3	+2
			10	+4	+2	+2	+6	+2	+2
			100	0	0	0	+1	0	0
C1 & 2	2·00	650	0·0001	−12	−11	−10	−18	−12	−11
		(44·8)	0·1	−3	−4	−3	−2	−3	−3
			10	−1	−1	−1	0	−1	−1
C3	2·00	450	0·0001	−4	−3	−4	−8	−4	−2
		(31)	0·1	+1	0	0	+3	0	−1
			10	+1	+1	+1	+2	+1	+1
			100	0	0	0	0	0	0
D1	1·40	350	0·0001	−14	−13	−13	−17	−15	−14
		(24·1)	0·1	−5	−5	−5	−5	−5	−5
			10	−2	−2	−3	−2	−2	−2
D2	1·40	250	0·0001	−12	−13	−13	−10	−11	−12
		(17·2)	0·1	−4	−3	−4	+1	−3	−4
			10	0	0	0	0	0	0
			100	−1	0	−1	0	0	−1

Time or strain hardening

The question whether time or strain hardening is the more realistic description of the behaviour of a material creeping under varying stresses has been widely debated [16] and the opportunity was taken here of trying to clarify the point for the present cases in planning the computations.

The results (Table 2) show that, for thick rings subjected to internal pressure, the percentage differences in the computed total strains derived with the two hypotheses are small. The differences are zero at the start of the test, $t = 0$. It is appreciated that times of 0·0001 hr are of no practical significance, but the computed values at these times are given as examples of the differences at very small times. It may be concluded from these results that:

1. Strain hardening appears always to predict higher strains than time hardening at the early times.
2. After creep times of 0·1 hr or greater the differences between time and strain hardening are less than the likely experimental error in the tests.
3. In the majority of cases the differences at the bore are greater than the differences at the outside.
4. At the bore the differences in radial strains are greater than the differences in hoop strains.
5. The difference is in no case greater than 3 per cent after 10 hr.

The above conclusions have been obtained from calculations using the power stress relationship, but there is no reason to assume that they do not also apply when the sinh stress relationship is used.

The stress functions

The sinh stress relationship was used to compute the strains for the rings with diameter ratio $K = 2{\cdot}00$ only, as a typical value. The percentage differences between the total strains obtained from the two stress functions are shown in Table 3.

It may be concluded from these results that:

1. The differences are small but increase with time and pressure at all points for both hardening hypotheses.
2. The differences between the tabulated ratios in any one row are trivial; at any time, the hoop and radial values are almost the same at bore, mid-thickness and outside.

TABLE 3

Percentage differences in total strains predicted with power and sinh stress functions for K = 2·00

Corresponding values of $(\varepsilon_{\text{sinh}} - \varepsilon_n)/\varepsilon_n \times 100$

Ring	Hypothesis	Pressure lbf/in²	(bar)	Time (hr)	Hoop at: Bore	Hoop at: Mid-thickness	Hoop at: Outside	Radial at: Bore	Radial at: Mid-thickness	Radial at: Outside
C1 & 2	Time hardening	650	(44·8)	0·0001	+1	+1	+1	+4	+1	+1
				0·1	−9	−9	−9	−8	−8	−9
				10	−12	−13	−13	−11	−12	−12
C1 & 2	Strain hardening	650	(44·8)	0·0001	+1	0	+1	+3	0	0
				0·1	−10	−10	−10	−9	−10	−8
				10	−13	−13	−13	−11	−13	−13
C3	Strain hardening	450	(31·0)	0·0001	0	0	0	0	0	0
				0·1	+2	+2	+3	0	+1	+3
				10	+4	+5	+5	+2	+4	+5
				100	+5	+5	+5	+2	+5	+6

3. As may have been expected from the previous sub-section, the changes in the differences due to the use of strain hardening instead of time hardening are insignificant.
4. The greatest differences computed were 13 per cent which can be extrapolated to approximately 15 per cent in 100 hr.

Stress redistribution

The manner in which the computed stress distributions change with time is illustrated for model A2 in Fig. 4, for which stresses were calculated at 13 equispaced radial positions. The radial stresses at the inner and outer surfaces are, of course, constant and the greatest changes in radial stresses

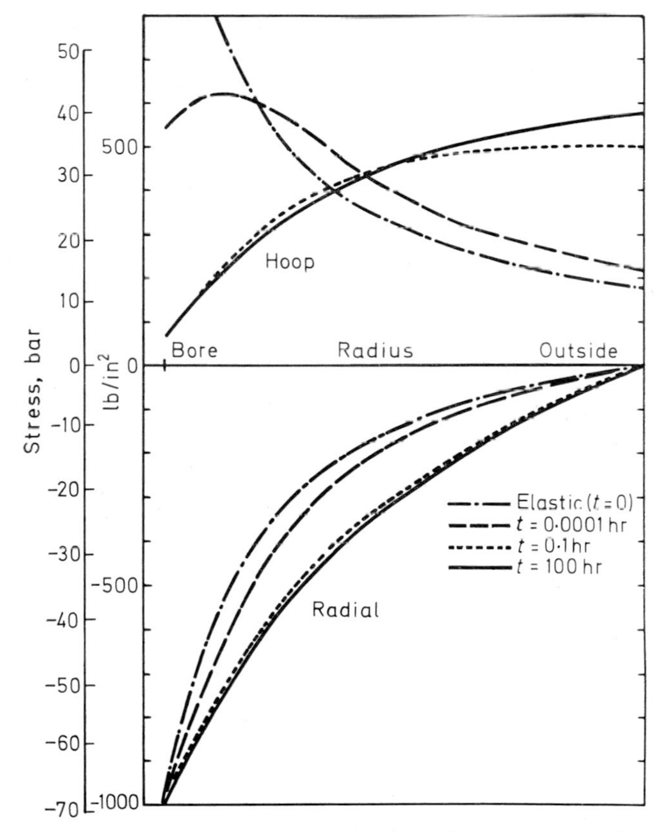

FIG. 4. Model A2: Computed stress distributions, $K = 3{\cdot}44$, $p = 1000$ lbf/in^2 (68·9 bar). Based on strain hardening without instantaneous plasticity, using the power stress function.

TABLE 4

Computed stresses
(based on strain hardening, without instantaneous plasticity, using the power stress function)

Ring model	K	Pressure lbf/in² (bar)	Time (hr)	Bore Value lbf/in² (bar)	Bore Change (%)	Hoop stress Mid-thickness Value lbf/in² (bar)	Hoop stress Mid-thickness Change (%)	Outside Value lbf/in² (bar)	Outside Change (%)	Radial stress Mid-thickness Value lbf/in² (bar)	Radial stress Mid-thickness Change (%)
A1	3·44	1120	0	1327		351		206		−145	
		(77·2)		(91·5)	−61	(24·2)	+28	(14·2)	+31	(−10)	+30
			0·0001	514		451		269		−188	
				(35·4)	−34	(31·1)	+20	(18·5)	+159	(−12·9)	+87
			0·1	61		520		596		−314	
				(4·2)	0	(35·8)	+1	(41·1)	+21	(−21·6)	+7
			10	60		525		639		−324	
				(4·15)		(36·2)		(44·0)		(−22·3)	
A2, B1	3·44	1000	0	1185		313		184		−130	
		(69)		(81·7)	−54	(21·6)	+21	(12·7)	+22	(−8·95)	+22
			0·0001	546		380		225		−158	
				(37·6)	−40	(26·2)	+28	(15·5)	+152	(−10·9)	+89
			0·1	68		467		505		−274	
				(4·7)	−2	(32·2)	−1	(34·8)	+33	(−18·9)	+8
			10	49		465		565		−285	
				(3·38)	+1	(32·1)	+4	(39·0)	+9	(−19·6)	+6
			100	65		477		581		−293	
				(4·48)		(32·9)		(40)		(−20·2)	

A3, B2	3·44	750	0	888		235		138		−97	
		(51·7)		(61·3)	−35	(16·2)	+9	(9·5)	+9	(−6·7)	+9
			0·0001	579		257		151		−106	
				(40)	−52	(17·7)	+45	(10·4)	+91	(−7·3)	+73
			0·1	114		362		276		−177	
				(7·85)	−9	(24·9)	−6	(19)	+100	(−12·2)	+38
			10	37		348		414		−214	
				(2·55)	0	(24·0)	+2	(28·5)	+11	(−14·7)	+4
			100	39		352		429		−218	
				(2·69)		(24·2)		(29·6)		(−15·0)	
C1, C2	2·00	650	0	1083		602		433		−168	
		(44·9)		(74·8)	−32	(41·5)	+10	(29·8)	+16	(−11·6)	+14
			0·0001	735		662		504		−192	
				(50·7)	−24	(45·6)	+2	(34·7)	+53	(−13·2)	+26
			0·1	478		672		734		−236	
				(33)	0	(46·4)	+1	(50·5)	+5	(−16·3)	+2
			10	478		680		754		−239	
				(33)		(46·9)		(52·0)		(−16·5)	
C3	2·00	450	0	750		416		300		−116	
		(31)		(51·6)	−14	(28·7)	+4	(20·7)	+4	(−8·0)	+4
			0·0001	641		431		313		−121	
				(44·1)	−37	(30·4)	+9	(21·6)	+51	(−8·35)	+31
			0·1	362		468		467		−157	
				(25·6)	−5	(32·2)	0	(32·2)	+16	(−10·8)	+7
			10	327		466		514		−165	
				(22·5)	0	(32·1)	+1	(35·4)	+3	(−11·4)	0
			100	330		471		522		−165	
				(22·7)		(32·5)		(36)		(−11·4)	

TABLE 4—*continued*

Computed stresses

(*based on strain hardening, without instantaneous plasticity, using the power stress function*)

Ring model	K	Pressure lbf/in² (bar)	Time (hr)	Bore		Hoop stress Mid-thickness		Outside		Radial stress Mid-thickness	
				Value lbf/in² (bar)	Change (%)	Value lbf/in² (bar)	Change (%)	Value lbf/in² (bar)	Change (%)	Value lbf/in² (bar)	Change (%)
D1	1·40	350	0	1078		860		728		−132	
		(24·1)		(74·3)	−16	(59·3)	+3	(50·2)	+12	(− 9·1)	+8
			0·0001	910		882		816		−142	
				(62·6)	−9	(60·8)	0	(56·3)	+14	(−9·8)	+6
			0·1	810		881		919		−150	
				(55·8)	+2	(62·1)	+2	(63·4)	+3	(−10·3)	+1
			10	828		901		942		−151	
				(57)		(62·1)		(65)		(−10·4)	
D2	1·40	250	0	770		615		520		−94	
		(17·2)		(53)	−24	(42·4)	+5	(35·9)	+13	(−6·5)	+10
			0·0001	586		648		585		−103	
				(40·4)	0	(44·6)	−3	(40·3)	+12	(−7·1)	+4
			0·1	585		629		649		−107	
				(40·3)	−1	(43·4)	0	(44·7)	+2	(−7·4)	+1
			10	579		631		660		−108	
				(39·9)	+2	(43·5)	+2	(45·5)	+3	(−7·45)	0
			100	593		645		675		−108	
				(40·9)		(44·5)		(46·5)		(−7·45)	

occur at approximately one fifth of the wall thickness from the bore. The large initial elastic hoop stresses at the inner surface are redistributed extremely rapidly. Between 0·1 and 100 hr (the time interval over which measurements were taken, henceforth called the 'test period') there are no appreciable changes except in the hoop stresses near the outside.

It was not considered necessary to present the complete stress distributions for the other six cases. Since neither a logarithmic nor a linear time scale is satisfactory for the complete representation of the variation of stress with time at a particular position, the stresses at three radial positions and at four times are recorded in Table 4. The changes in these stresses during each tubulated time interval, expressed as a percentage of the corresponding initial elastic stress, are also shown. The rings, therefore, creep under an almost stable, fully redistributed stress system. (It was unfortunate that so little stress redistribution occurred during the actual test period. The test series was too advanced for changes in experimental techniques to be made when this became apparent.)

In the majority of cases the changes in these stresses during the practicable test period, as detailed in Table 4, are less than 10 per cent. The greatest absolute changes during the test period are always the increases in outside hoop stress. The biggest percentage change during the practicable test period based on the corresponding elastic stress is 111 per cent for models A3 and B2, the thickest rings, at the lowest pressure. The changes during the test period appear in most cases to be greatest for the lower pressure cases; for the higher pressure cases appreciable stress redistribution will have occurred before the 'test period' has begun.

At mid-thickness the radial compressive stresses always increase. The bore hoop stresses decrease to very small values in the thickest rings but decrease to only about three quarters of their elastic values in the thinnest rings. In the latter, the hoop stresses become fairly constant across the section and increase everywhere, owing presumably, in part, to the increasing mean hoop stress caused by increase in bore and decrease in radial thickness. The programme allows for changes in geometry during the test period.

EXPERIMENTAL RESULTS

Variation of o.d. hoop strain with time

It is convenient to describe the experimental results in two parts: firstly the variation of the outer diameter hoop strains with time and secondly the strain distributions across the rings at different times. The work described in this first section is based on dial gauge measurements whereas the strain distributions were obtained from moiré fringe patterns.

The measured o.d. hoop strains are plotted to natural scales in Fig. 5 and compared with the computed values based on the standard computation (*see* Table 1). This method of presentation is convenient for assessing

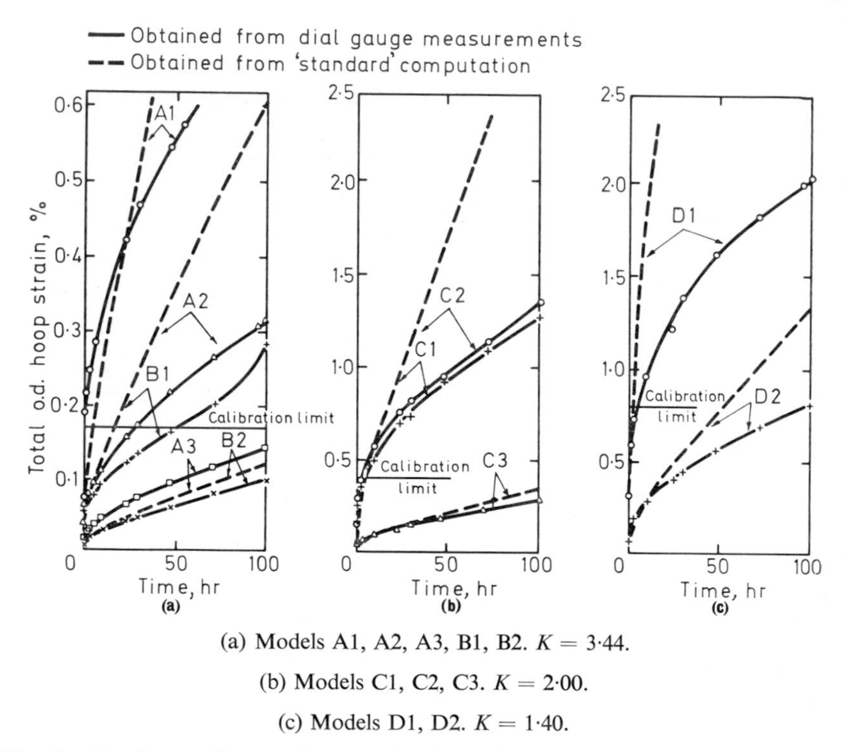

(a) Models A1, A2, A3, B1, B2. $K = 3{\cdot}44$.

(b) Models C1, C2, C3. $K = 2{\cdot}00$.

(c) Models D1, D2. $K = 1{\cdot}40$.

FIG. 5. Total outer diameter hoop strains (natural scales). Broken line obtained from 'standard' computation, continuous line from dial gauge measurements.

the secondary creep behaviour. It may be seen that there is close agreement between the measurements for models C1 and C2 which are nominally identical rings tested under the same nominal pressure. This gives a good indication of the overall repeatability of these measurements. In contrast with this close agreement, the differences between computed and measured strains in many cases are large.

The primary creep behaviour, to which increasing practical importance is being attached, is best shown in the log strain–log time curves in Fig. 6.

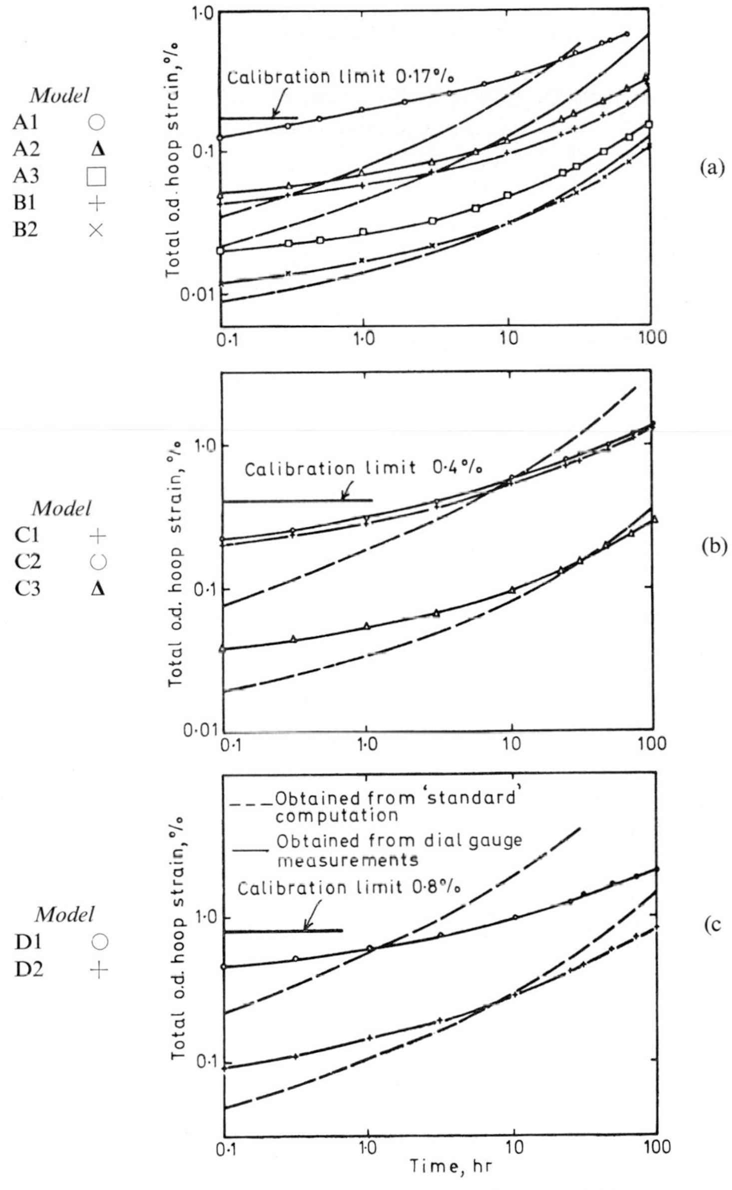

FIG. 6(a). Models A1, A2, A3, B1, B2. $K = 3{\cdot}44$.

FIG. 6(b). Models C1, C2, C3. $K = 2{\cdot}00$.

FIG. 6(c). Models D1, D2. $K = 1{\cdot}40$.

FIG. 6. Total outer diameter hoop strains (logarithmic scale). Broken line obtained from 'standard' computation, continuous line from dial gauge measurements.

There is a measure of agreement between calculated and measured values shown by the fact that each calculated curve intersects the experimental one within the test period. It should be noted that the material constants were evaluated at 10 hr. The differences due to the use of the final material data instead of the provisional are small compared with the differences between measured and computed strains. (Compare the A1 curves in Figs. 3a and 6a.)

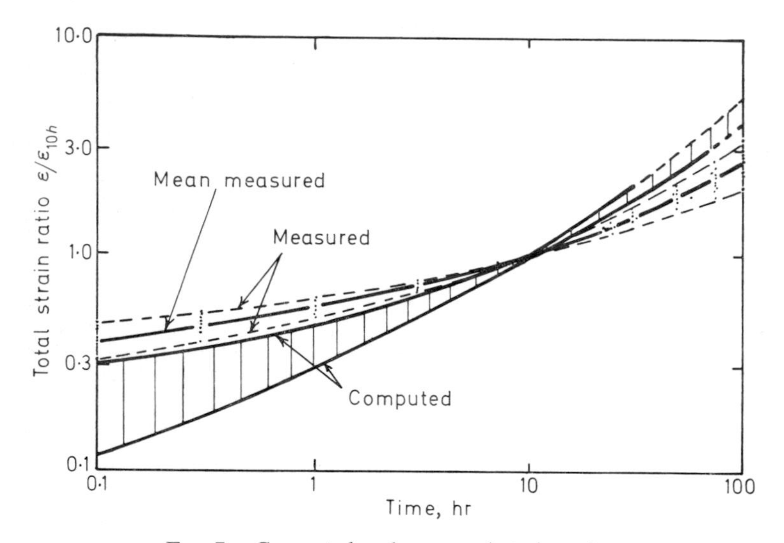

FIG. 7. Computed and measured strain ratios.

Because the pressures exerted by the rubber plugs may be less than the nominal values at large strains, the o.d. hoop strain corresponding to the bore hoop strain calibration limit of 1·6 per cent is shown on Figs. 6a to c. These points indicate the extent of the calibration tests; it is not implied that the pressure was less than the nominal at greater values, but this possibility must be considered as an explanation for some of the measured strains being less than the computed ones.

The measured values in Fig. 6 lie on curves of similar shape; this is apparent in Fig. 7 in which all strains are expressed as multiples of the corresponding strain at 10 hr. At times less than 10 hr the upper and lower limits are defined by points from models D1 and D2, respectively. At 0·1 hr the range of strain ratios is 40 per cent of the mid-value; neglecting model D2 reduces this to 20 per cent, with the C2 value as the new lower limit. For times greater than 10 hr the upper and lower limits are derived from models B2 and D1, respectively. Results from models in which the maximum bore hoop strain did not exceed the calibration limit of the rubber plug (A3, B2, C3 and D2) were greater than the results from the other models (A1, A2, B1, C1, C2 and D1).

The mean curve in Fig. 7 may be represented by the expression

$$\frac{\varepsilon}{\varepsilon_{10_h}} = 0{\cdot}56(t^{0\cdot19} + 0{\cdot}025t)$$

where t is measured in hours.

Figure 7 also shows the computed results expressed as multiples of the corresponding strain at 10 hr. Below 10 hr the spread of computed results is greater than for the measured values, with the upper and lower limits defined by models A3 (B2) and D1 which were subjected to the smallest and greatest mean hoop stresses, respectively. Although the magnitude of the effect of including instantaneous plasticity would be expected to depend on both mean hoop stress and stress gradient, it is of interest to note that the lower limit model, D1, is subjected to the greatest mean hoop stress in the tests. Based on the result shown in Fig. 3, it may be deduced that the inclusion of instantaneous plasticity would reduce the slope of the curves below the mean curve for times less than 10 hr and consequently reduce the spread of the computed strain ratios at the earlier times.

Above 10 hr the range of computed results, which would be little affected by instantaneous plasticity, is slightly less than the measured range.

In addition to the *total* o.d. hoop strains discussed above, the o.d. hoop *creep* strains were also studied. These were obtained by subtracting the first readings, obtained less than 15 sec after application of the loads, from all subsequent values. The log creep strains varied linearly with log time as shown in Fig. 8 for $K = 2{\cdot}00$. The creep strains after 10 hr and the slopes of the log–log plots are recorded in Table 5 with the total measured hoop strains after 10 hr. It should be noted that the slopes varied little from the mean value of 0·46.

In an attempt to relate the measured o.d. hoop strains from the three different diameter ratios tested, the total and creep strains after 10 hr were plotted against the mean hoop stress, $p/(K - 1)$, in each ring in Fig. 9. With the exception of model A1, these results show approximately linear relationships which, for the lead alloy used and defined in Appendix 1, lead to the empirical expressions for total o.d. hoop strain

$$\varepsilon_T = 0{\cdot}019 \left(\frac{p}{K-1}\right)^{3\cdot3} f(t)$$

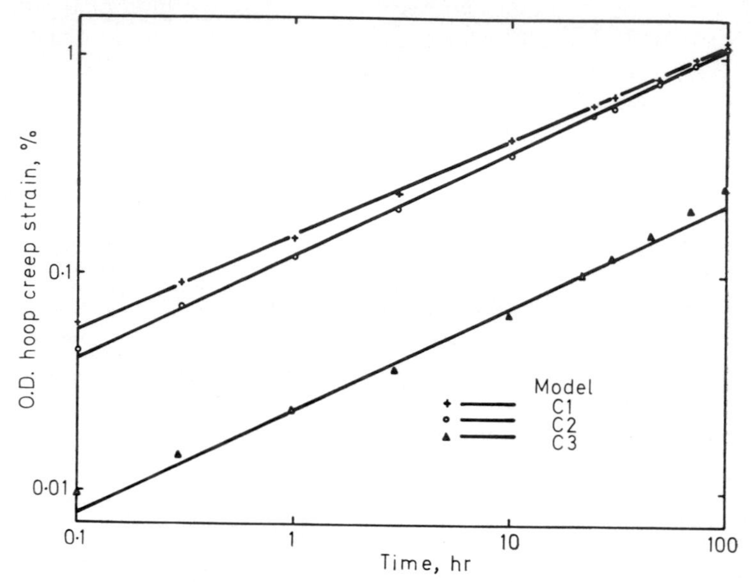

FIG. 8. Measured o.d. hoop creep strain–time relationships for $K = 2{\cdot}00$.

TABLE 5

Constants for experimental outside hoop strain–time curves

Ring model	Mean hoop stress $p/(K-1)$ lbf/in² (bar)	Strain after 10 hr Total (%)	Creep (%)	Slope of log creep strain–log time plot
A1	460 (31·7)	0·316	0·239	0·36
A2	410 (28·2)	0·114	0·072	0·48
B1	410 (28·2)	0·094	0·058	0·46
A3	308 (21·2)	0·046	0·030	0·50
B2	308 (21·2)	0·030	0·020	0·52
C1	650 (44·8)	0·571	0·416	0·44
C2	650 (44·8)	0·510	0·354	0·48
C3	450 (31·0)	0·095	0·067	0·48
D1	875 (60·3)	0·970	0·655	0·37
D2	625 (43·1)	0·281	0·216	0·48

and for the outer diameter hoop creep strain

$$\varepsilon_c = 0{\cdot}012\left(\frac{p}{K-1}\right)^{3\cdot3} f(t)$$

In these expressions p is measured in thousands of pounds per square inch and t in hours.

The time function $f(t)$ obtained from the mean curve of the experimental values in Fig. 7 ($t^{0\cdot19} + 0{\cdot}025t$) differs from that obtained from the slopes

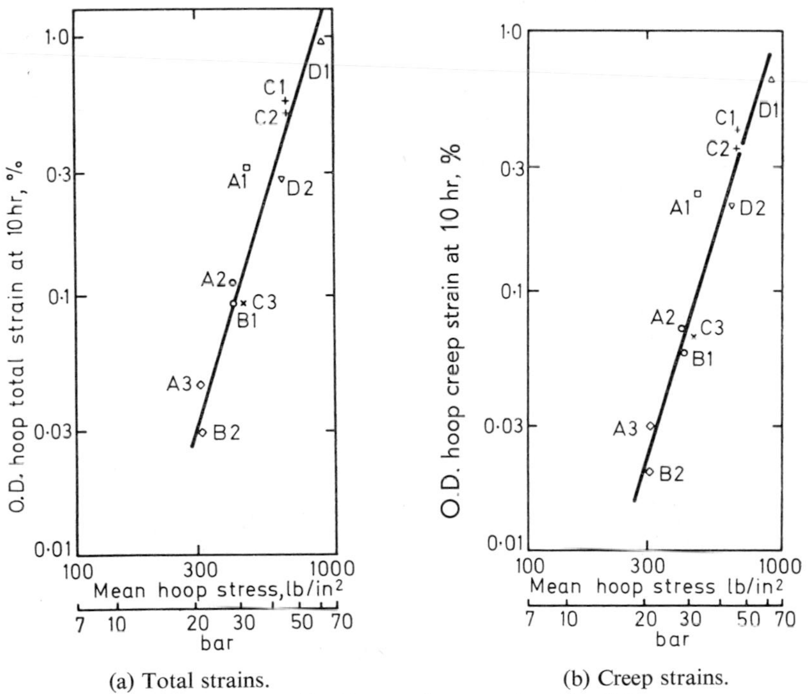

(a) Total strains. (b) Creep strains.

FIG. 9. Outside diameter hoop strains after 10 hr.

of the creep strain–time curves ($t^{0\cdot46}$). The former predicts higher values than the latter at times less than 10 hr. Further tests would be required to determine which of these functions gives the better representation of the model behaviour.

Radial distribution of strain

It was found convenient to express all strains as multiples of the corresponding (same model, same time, experimental with experimental, theoretical

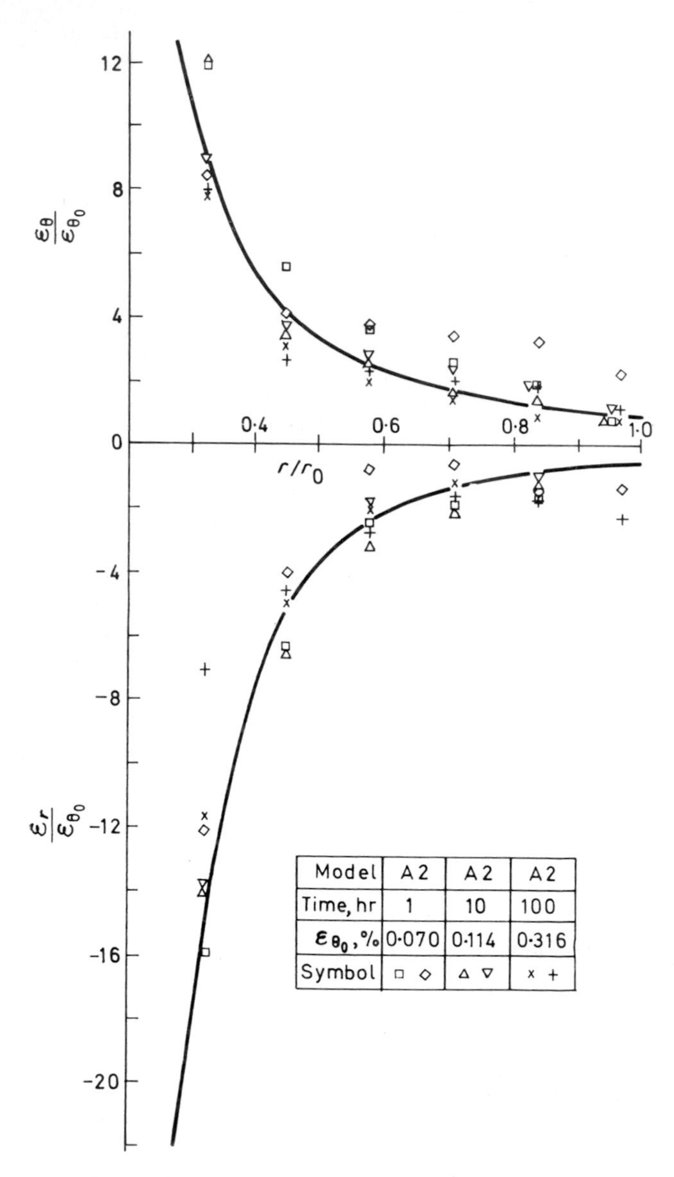

Model	A2	A2	A2
Time, hr	1	10	100
ε_{θ_0}, %	0·070	0·114	0·316
Symbol	□ ◇	△ ▽	× +

FIG. 10(a). Model A2, $K = 3{\cdot}44$, $p = 1000$ lbf/in^2.

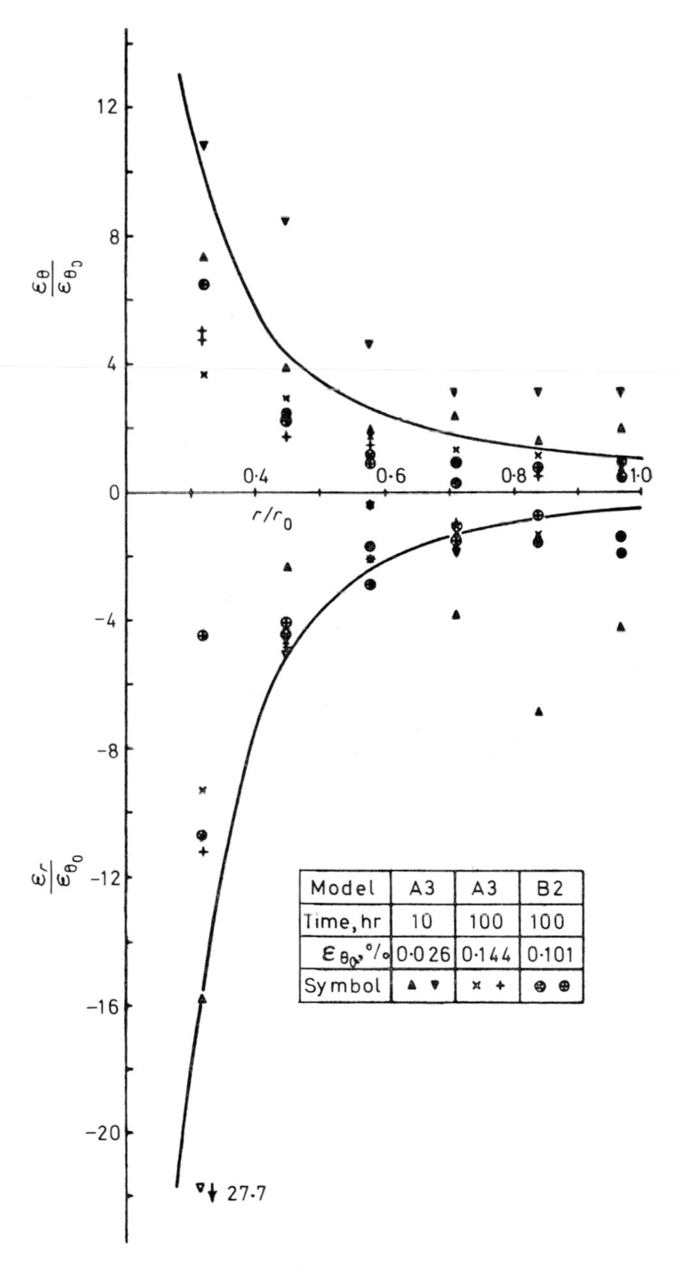

Model	A3	A3	B2
Time, hr	10	100	100
ε_{θ_0}, %	0·026	0·144	0·101
Symbol	▲ ▼	× +	⊕ ●

FIG. 10(b). Models A3 and B2, $K = 3{\cdot}44$, $p = 750$ lbf/in².

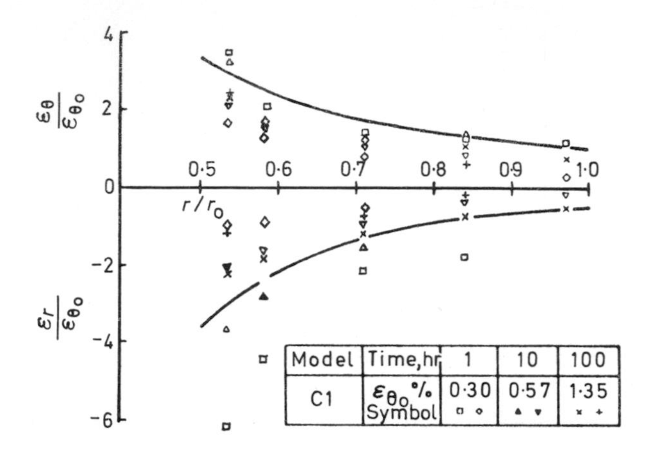

FIG. 10(c). Model C1, $K = 2{\cdot}00$, $p = 650$ lbf/in^2.

with theoretical) o.d. hoop strains. For the experimental ratios, the o.d. hoop strains were obtained from dial gauge readings. The resulting strain ratios are shown in Fig. 10a to e for the strains obtained from moiré measurements and computations plotted against the radius ratio r/r_0. As the magnitude of the outer diameter hoop strain varied considerably for the different models and test times, it is stated in the legend to assist the reader in assessing the significance of the scatter of the various experimental ratios.

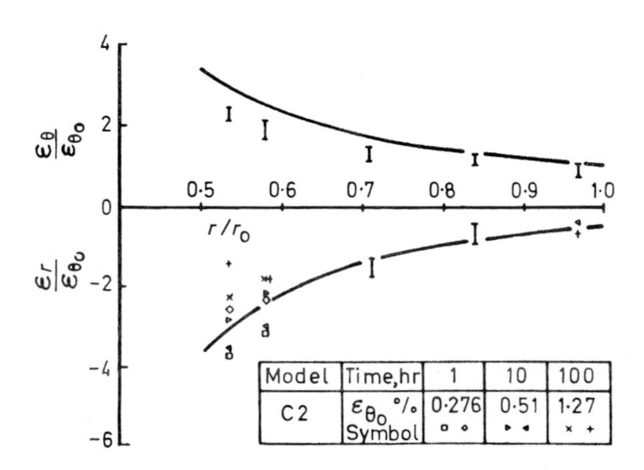

FIG. 10(d). Model C2, $K = 2{\cdot}00$, $p = 650$ lbf/in^2. I—range of all six results.

For each of the ring geometries it was found that, within the test period, computed hoop strain ratios fell very close to one single curve, computed radial strain ratios on another. This was regardless of time and test pressure. Ratios were obtained only for the times at which the moiré patterns were evaluated, but there is no reason to doubt the generality of this statement in respect of time. The greatest differences between computed strain ratios at all times and pressures for a given K ratio occurred at the bore; the ranges of values at this position are shown in Table 6. The

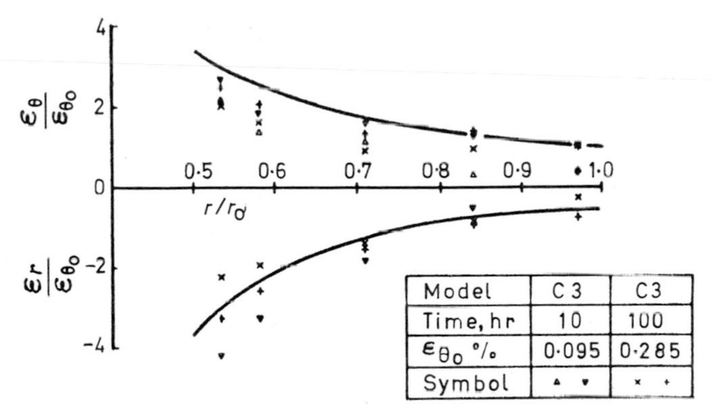

FIG. 10(e). Model C3, $K = 2{\cdot}00$, $p = 450$ lbf/in^2.

FIG. 10. Distribution of total strain in rings. Curves are (computed strains)/(computed o.d. hoop strains). Experimental points are (measured strains)/(measured o.d. hoop strains from dial gauge measurements).

changes with increasing test time showed no systematic trends. It should be noted that the variations in the computed ratios were much smaller than the differences between them and the predictions based on Larke and Parker's [5] analysis, but the latter treated the plane strain case whereas the computations were for plane stress.

The computed radial strain ratio–radius ratio curve may be expressed as:

$$\frac{\varepsilon_r}{\varepsilon_{\theta o}} = 0{\cdot}5\left(\frac{r}{r_o}\right)^{-3}$$

The value 0·5 is consistent with the assumption of incompressible deformation at the outer edge of the rings.

The measured strain ratios show the same type of distribution as the computed ones, except for the hoop strain ratios near the bores which are generally lower than the computed values.

TABLE 6

Range of bore strain ratios

K	$\varepsilon_\theta/\varepsilon_{\theta o}$ Larke and Parker	Computed	$\varepsilon_r/\varepsilon_{\theta o}$ Computed
3·44	11·8	12·6 to 13·0	−21·8 to −24·8
2·00	4·00	3·38 to 3·42	−3·58 to −3·76
1·40	1·96	1·70 to 1·78	−1·27 to −1·31

No significant trends were apparent in the differences between the experimental strain ratios at any one radial position and the differences were possibly due to experimental error. As may be expected, the effects of experimental errors are generally greater for the smaller test times.

No results are presented for model B1 because the bore of this model did not expand uniformly. Some other moiré patterns were unsuitable for analysis. The computed results for Fig. 10c and d are only available for 1 hr and 10 hr but they can be compared with experimental values at 100 hr if it is accepted that for this ring geometry also computed strain ratios do not differ significantly between 10 hr and 100 hr.

DISCUSSION AND CONCLUSIONS

The model material has, in general, two possible roles, namely:

1. The prediction of prototype deformation characteristics.
2. The confirmation of computer work.

The first role is clearly of major importance to designers but the main objective of the work described has been to compare computed and experimental creep data.

The running of a comprehensive set of computations, even for such a simple specimen, was unusual and a number of valuable observations can be made on this part of the work alone, without reference to the experimental work:

1. The inclusion of the effects of instantaneous plasticity in the computations proved extremely time-consuming. In the two cases in which a comparison was made, the differences between strains computed with and without instantaneous plasticity were appreciable only for the lower value of the yield stress, during the first few hours, and conditions in this case were such that the differences obtained were likely to be greater than in other models.

2. For the cases studied, the predicted strains are not sensitively dependent on which hardening rule is used in the computation.
3. In view of the small differences in computed strains obtained with the sinh and power stress functions, it seems unnecessary in many cases to establish which is the better 'fit' for uniaxial creep data.
4. The greater part of the stress redistribution occurs extremely quickly.

It would be premature at the present time to attempt to base any broader generalisations on these observations; they are nevertheless useful pointers in this complex field in which work aimed at providing working simplifications should not be neglected.

Successful methods for exerting constant internal pressure and for measuring surface strains on lead rings have been developed.

Experimental and computed o.d. hoop strains are presented together in Figs. 5 and 6. In most cases, except at the very early times, there is a clearly defined tendency for the computed results to be greater than the experimental ones. The good agreement between the results for the nominally identical models C1 and C2 is some indication of the consistency of the experimental work. It has been shown [17] that an extremely small spread in creep data can be obtained if certain experimental precautions are taken. Special attention was not paid to these points in the present work but excessive random errors are not suspected. Relevant factors in comparing the experimental and computed data are:

1. The omission of effects due to instantaneous plasticity in the calculations. This would increase the early predicted strains.
2. A possible slight reduction in model loads with increasing strain due to the non-fluidlike behaviour of the rubber plugs as the strains became greater than the calibration limit.

Even with these factors in mind, the remaining impression is that, with one possible exception (A3), the observed decrease in strain rate with time is consistently more rapid than predicted. Further work is necessary to explain why this should be so. It could be that neither the conventional time nor strain hardening hypotheses are adequate for the relatively large strains in these models or that the use of separable time and stress functions to describe the creep behaviour is not appropriate for large strains.

The most useful generalisations to be drawn from the work are presented in Fig. 10. When the computed data are plotted in the form of strain ratios versus radius ratios, a single curve is sufficient for all the hoop strains, a second for all radial strains. This applies for:

1. All times within the test period.
2. All pressures used.
3. All diameter ratios used.

The measured strain ratios followed the trend of these curves, except some of the hoop values nearer the bore.

For the lead alloy used, for any values of K and p, the entire strain distribution at any time in the test period may be obtained from the o.d. hoop strain measurement and one computation based on the uniaxial creep function of the material. There is no reason to suppose that this would not apply for other materials. It would be valuable to determine whether similar results could be obtained for cylinders subjected to internal pressure.

NOTE: *In discussions with CEGB representatives, following the completion of this investigation, it has been pointed out that outstanding importance attaches to the form of representation of the uniaxial data, over the* whole *range of stresses, and for the complete time scale, relevant to the model tests. In these model tests, the uniaxial data were extrapolated to cover the high initial stresses present for very short times at the beginning of some of the model tests. Work carried out at the Berkeley Nuclear Laboratories has shown good agreement between computed and measured outside diameter creep strains, obtained from steel cylinders, for much smaller strains than those of the model tests.*

ACKNOWLEDGEMENTS

This work was supported by CEGB, Berkeley Nuclear Laboratories, which also carried out the computer work. Mr R. Fidler of CERL, Leatherhead gave valuable information regarding his use of the moiré technique. The authors wish to thank staff of CEGB and technicians at Nottingham University for their assistance.

REFERENCES

1. JOHNSON, A. E., HENDERSON, J. and KHAN, B. (1961). 'Behaviour of metallic thick-walled cylindrical vessels or tubes subject to high internal or external pressures at elevated temperatures.' *Proc. Instn Mech. Engrs*, **175,** 1043–1069.

2. FESSLER, H., GILL, P. A. T. and STANLEY, P. (1968). 'A material for accelerated creep testing with models.' *JBCSA Conf. Recent Advances in Stress Analysis*. London: Royal Aeronautical Society.
3. BAILEY, R. W. (1951). 'Creep relationships and their application to pipes, tubes and cylindrical parts under internal pressure.' *Proc. Instn Mech. Engrs*, **164,** 425–431.
4. GILL, P. A. T. (1968). Ph.D. Thesis. University of Nottingham.
5. LARKE, E. C. and PARKER, R. J. (1966). 'Circumferential creep strain of cylinders subjected to internal pressure: a comparison of the theories of Johnson and Bailey.' *J. Mech. Eng. Sci.*, **8,** 22.
6. SMITH, E. M. (1965). 'Analysis of creep in cylinders, spheres and thin discs.' *J. Mech. Eng. Sci.*, **7,** 82–92.
7. FREDERICK, C. O., CHUBB, E. J. and BROMLEY, W. P. (1965/6). 'Cyclic loading of a tube with creep, plasticity and thermal effects.' *Proc. Instn Mech. Engrs*, **180** (Pt 3I); Part 5, 448–461.
8. BAILEY, R. W. (1930). 'Thick-walled tubes and cylinders under high pressure and temperature.' *Engineering*, **129,** 772–773; 785–786; 818–819.
9. NORTON, F. H. (1939). 'Creep in tubular pressure vessels.' *Trans. Amer. Soc. Mech. Engrs*, **61,** 239–244.
10. NORTON, F. H. (1941). 'Progress report on tubular creep tests.' *Trans. Amer. Soc. Mech. Engrs*, **63,** 735–736.
11. NORTON, F. H. and SODERBERG, C. R. (1942). 'Report on tubular creep tests.' *Trans. Amer. Soc. Mech. Engrs*, **64,** 769–777.
12. NAKAHARA, M. (1939). 'New theories of creep of metals.' *Jap. Soc. Mech. Engrs*, **5** (18), 12–20.
13. DAVIS, E. A. (1960). 'Creep rupture tests for design of high pressure steam equipment.' *Trans. Amer. Soc. Mech. Engrs, J. Basic Engng*, **82D**, 453–461.
14. FIDLER, R. and NURSE, P. (1965). *CERL Report*, No. RD/L/R 1289. Leatherhead, Surrey: Central Electricity Research Laboratories.
15. RILEY, W. F. (1967). 'Moiré method of strain analysis.' *Exp. Mech.*, **7,** 19A.
16. Institution of Mechanical Engineers (1963/64). 'Thermal loading and creep in structures and components.' *Proc. Instn Mech. Engrs*, **178** (Pt 3L).
17. BERRY, D. A. and ANSTEE, R. F. W. (1968). 'A preliminary study of the creep and recovery behaviour of DTD 5070 A aluminium alloy.' *JBCSA Conf.: Recent Advances in Stress Analysis*. London: Royal Aeronautical Society.

Appendix 1

DETAILS OF MODEL MATERIAL

Composition	Antimony	0·2 per cent
	Arsenic	0·02 per cent
	Lead	remainder
Young's modulus		$3{\cdot}05 \times 10^6$ lbf/in^2 (210 kbar)

Poisson's ratio	0·43
Yield stress (provisional)	200 lbf/in^2 (13·8 bar)
Yield stress (final)	1000 lbf/in^2 (69 bar)
Uniaxial creep behaviour	

i The 'standard' function

$$\varepsilon_c = B\sigma^n(t^s + bt)$$

where

$B = 3{\cdot}02 \times 10^{-19}$ lbf in hr units
$n = 5{\cdot}46$
$s = 0{\cdot}39$
$b = 0{\cdot}106$

ii The alternative function

$$\varepsilon_c = D \sinh(\sigma/\sigma_0)(t^s + bt)$$

where

$D = 1{\cdot}355 \times 10^{-5}$ lbf in hr units
$\sigma_o = 143$ lbf/in^2
$s = 0{\cdot}39$
$b = 0{\cdot}106$

iii The provisional function (used for Fig. 3 only)

$$\varepsilon_c = A\sigma^{n'}t^m$$

where

$A = 3{\cdot}1 \times 10^{-17}$ lbf in hr units
$n' = 5{\cdot}0$
$m = 0{\cdot}5$

Chapter 18

STRESS REDISTRIBUTION CAUSED BY CREEP IN A THICK WALLED CIRCULAR CYLINDER UNDER AXIAL AND THERMAL LOADING

J. M. CLARKE AND J. F. BARNES

SUMMARY

A thick walled tube was subjected to an axial load and a radial temperature distribution which caused thermal stresses. The creep strains and the rupture times were observed and compared with results of conventional creep tests and theoretical analysis. Theory suggested and experiments confirmed that stress redistribution caused the overall strain behaviour to approach that for the mean axial stress and the mean radial temperature.

The experimental technique and apparatus includes a novel and simple optical extensometer. Appendices contain a complete analytical treatment of the triaxial stress problem in a long thick tube in the presence of an arbitrary distribution of non-elastic strains, and a treatment of some conditions under which stress redistribution calculations can lead to a 'steady state' or 'fully redistributed' stress pattern.

A less rigorous theoretical treatment, which ignores radial constraints, is shown to lead to an underestimate of the thermal stresses and of the time required for stress redistribution to occur.

NOTATION

A	Area.
a, b	Arbitrary constants.

c	Non-elastic strain.
E	Young's modulus.
e	Total strain (elastic + thermal + creep).
$F(r)$, $G(r)$, $H(r)$	Function of radius
K	Constant in eqn (A48).
P	Pressure.
p, q	Arbitrary constants.
r	Radius.
R, S	Constants of integration.
s	Mean axial stress.
T	Temperature.
t	Time.
u	Radial displacement.
v, w	Stress and temperature sensitivities of strain rate.
x, y	Coordinates of $\ln(\varepsilon/\lambda)$ versus $\ln(t/\tau)$ curve.
ε	Creep strain.
$\bar{\varepsilon}$	Effective creep strain.
λ	A reference strain.
ν	Poisson's ratio.
σ	Stress.
$\bar{\sigma}$	Effective stress.
τ	A reference time.

Subscripts

1, 2	Before and after a time increment.
1, 2, 3	Axial radial and hoop directions.
i, o	Inside and outside surfaces.
m	Area mean.
s	Strain hardening hypothesis.
t	Time hardening hypothesis.

INTRODUCTION

The application of the digital computer to stress analysis allows solutions to be obtained easily to a large number of problems which have to be tackled by numerical methods. One example is the redistribution of stress in an internally cooled gas turbine rotor blade [1]. This redistribution is caused by the accumulation of creep strains at different rates which depend on the temperature, stress and previous history of each element of the blade. The most important result of the stress redistribution process is that the stress pattern becomes more uniform than that calculated simply on a basis of the thermal strains and a perfectly elastic material. This is

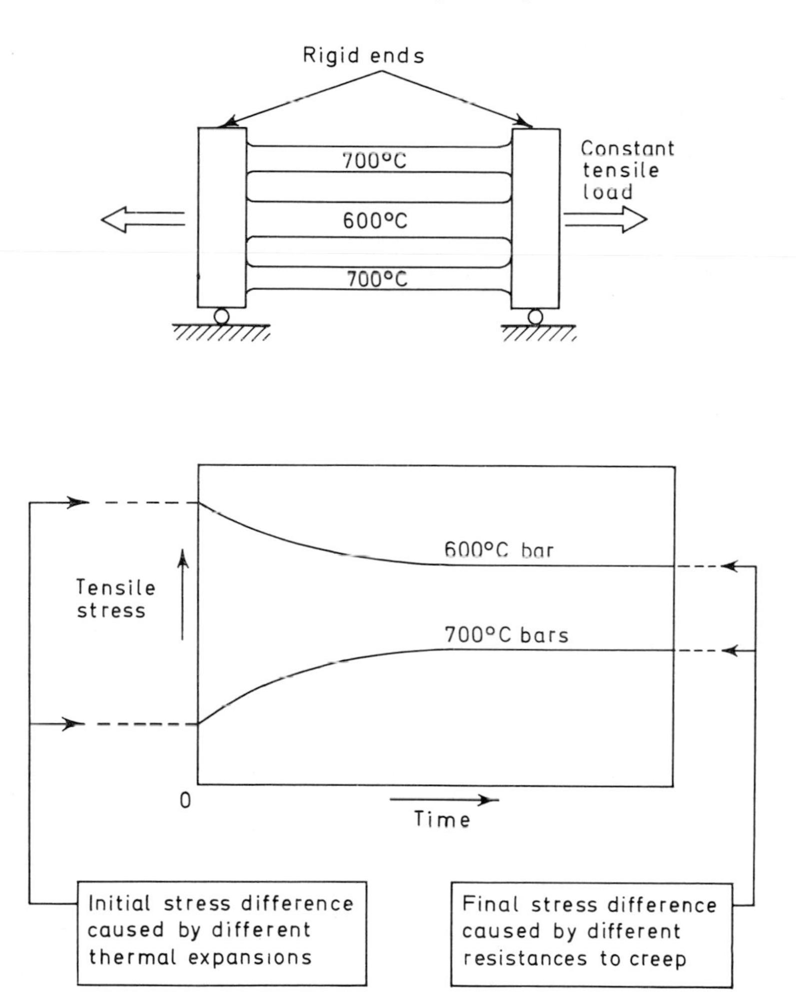

FIG. 1. Stress redistribution in a simple tensile system.

illustrated for a very simple tensile system in Fig. 1. By consideration of rupture times, based on the most severe conditions of stress and temperature within a blade, the stress pattern can be shown to become more favourable as redistribution proceeds; furthermore, the redistribution is predicted to be complete before any part of the blade would be expected to fail. The time to rupture of the blade should therefore be more closely related to the mean stress and mean temperature at a spanwise position than to the most severe combination of stress and temperature at the beginning of redistribution.

In order to obtain some guidance on the validity of the assumptions and to compare the predicted rupture life with that actually observed, an experimental and theoretical investigation has been conducted on a thick walled circular cylinder subjected to an axial tensile load and a radial temperature gradient. Apart from being worthy of study in its own right, the choice of a cylindrical geometry made the problem amenable to analysis while corresponding closely to the geometrically more complex situation within a cooled turbine rotor blade. It also enabled an experiment to be conducted under carefully controlled laboratory conditions and strain measurements to be taken. Confirmation that the accumulation of axial creep strain and the observed rupture lives agree well with those based on the conditions of mean stress and temperature are the most important results from this experimental investigation.

DETAILS OF EXPERIMENTAL WORK

Figure 2 shows a schematic arrangement of the apparatus used. The axial load was applied by a 5 ton (50 kN) Denison creep machine and the radial temperature gradient was produced by passing an electric current axially along the test specimen and allowing the latter to radiate freely from its outside surface. A number of experiments were repeated to determine scatter, and some control experiments were made to separate variables, the whole programme involving about 5000 hr of creep testing.

The dimensions, temperatures and stresses used for the experiment were chosen to satisfy:

1. A modest electric power requirement.
2. Tight manufacturing tolerances, especially in the bore of the test specimen.
3. Mean stress and temperature sufficient to cause rupture in about 400 hr if redistribution occurred.
4. A combination of local stress and temperature severe enough to initiate rupture in a much shorter time if redistribution did not occur.
5. The temperature to be high enough to cause the required radiation intensity from the outside surface.

A number of exploratory calculations using manufacturer's data [2] for Nimonic alloy 105 led to the choice of the following values:

Tube length	3 in (76 mm)
Inside diameter	$\frac{5}{16}$ in (8 mm)
Outside diameter	$\frac{5}{8}$ in (16 mm)
Mean axial stress	4 ton/in^2 (6·2 hbar)
Maximum axial stress	7 ton/in^2 (10·8 hbar)
Mean temperature	950°C

For these conditions, the rupture times for the most severe (initial) stress and temperature, and the mean stress and temperature were estimated to be 40 hr and 400 hr, respectively. The calculations showed that there was very little freedom to depart from the chosen values.

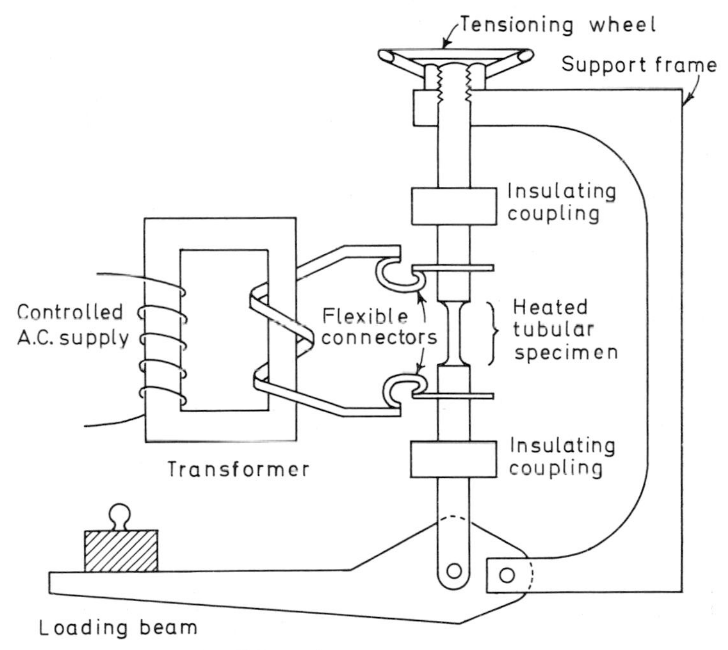

FIG. 2. Schematic arrangement of apparatus.

The tests on tubular specimens containing a radial temperature gradient (Type 1 tests) were supported by four other test arrangements intended to act as controls and to provide creep data for the same batch of material. They are shown on Fig. 3.

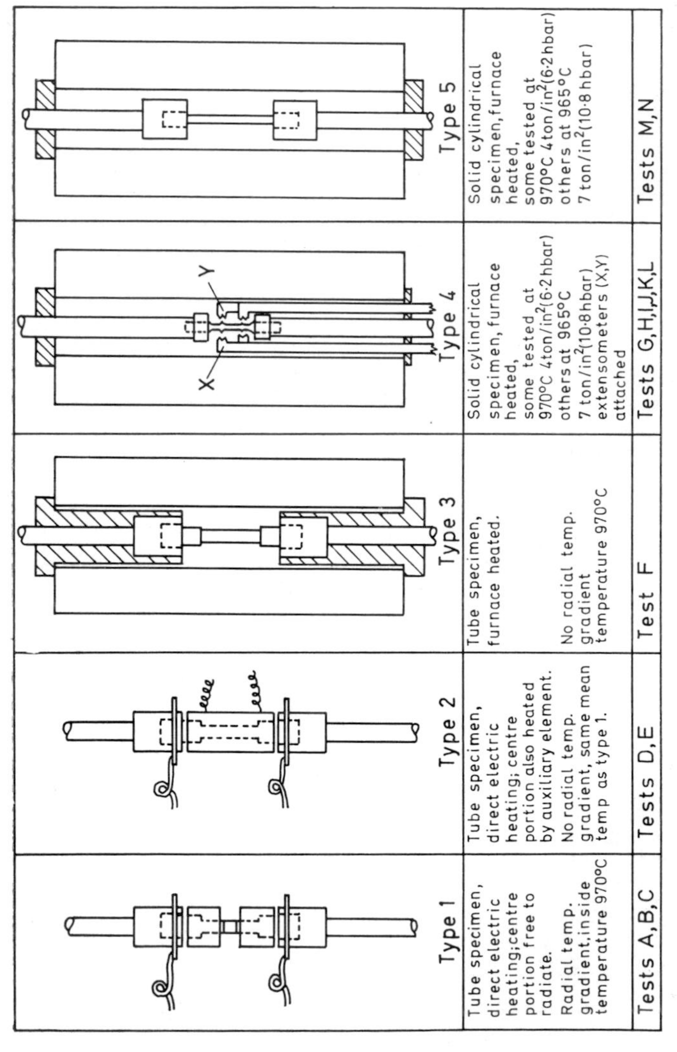

FIG. 3. Experimental arrangements.

Type 2 tests simulated Type 1 except for the provision of external insulation and an auxiliary heater to ensure a uniform radial temperature. These tests were therefore free from thermal stresses.

The single Type 3 test also used a tubular specimen but heated it in a conventional creep test furnace.

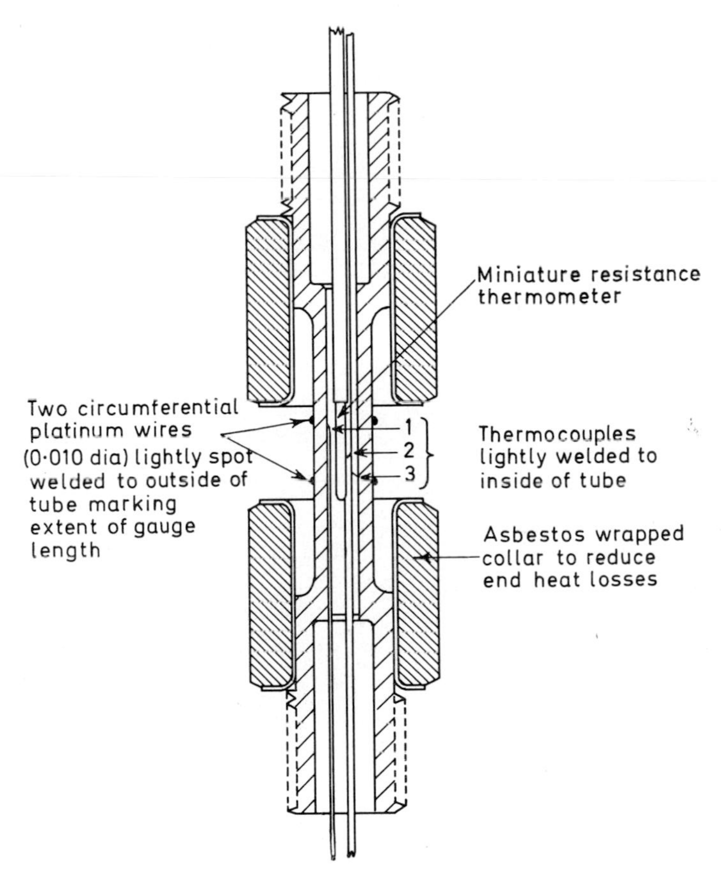

FIG. 4. Arrangement No. 1 showing instrumentation.

Type 4 and 5 tests used solid cylindrical specimens of two different types to establish the creep strain behaviour of the material. The test technique conformed with orthodox creep test methods.

The temperature sensitivity of rupture times, at the chosen mean stress and temperature levels, was 4·5 per cent reduction in life per deg C rise in temperature. The temperature of the furnace heated specimens was generally controlled within $\pm\frac{1}{2}$ deg C and that of the tube tests within approximately ± 1 deg C. Platinum resistance thermometers were used as sensors for the proportional temperature controllers but temperature measurements were made using platinum/platinum rhodium thermocouples.

The calculation of temperature difference referred to in the last section assumed no loss of heat from the gauge length (the central $\frac{5}{8}$ in) by conduction along the tube axis. It was therefore necessary to use the insulating collars shown in Fig. 4 to ensure a uniform temperature for about an inch in the centre of the tube length.

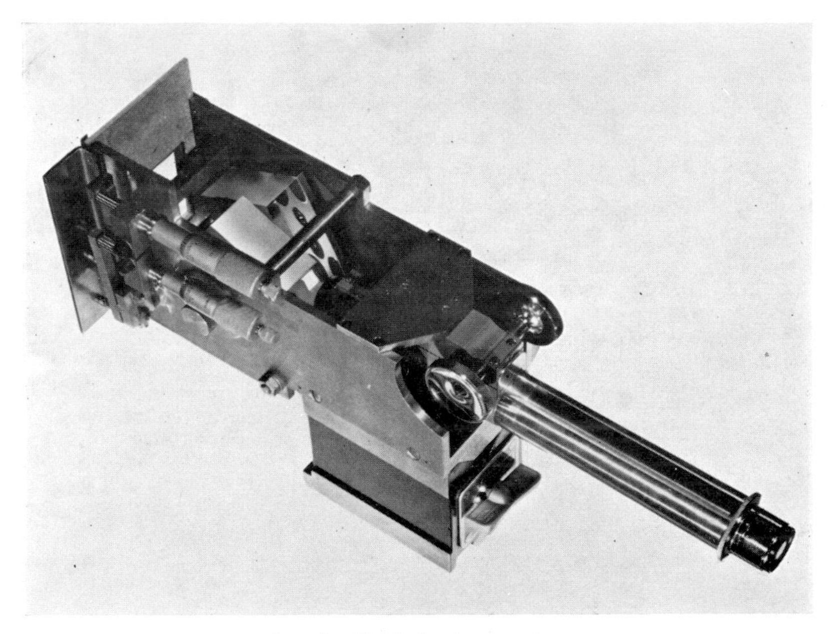

FIG. 5. Optical extensometer.

It was considered desirable to measure strains on the radiating specimens if this could be done without interfering with the symmetry of heat flux. For this purpose, a simple and novel optical extensometer was constructed (*see* Fig. 5). The principle employed was the displacement of a light path by the interposition of inclined glass blocks as shown in Fig. 6. One pair of fixed blocks served to superimpose the images while a second pair of thinner blocks were rotated to accommodate and measure strains.

The results for individual tests are distinguished by code letters. The key to these letters is given in Table 1 which should be used with Figs. 7 to 10.

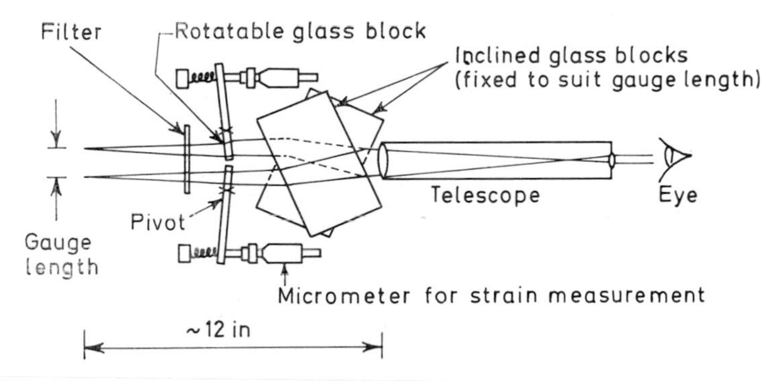

FIG. 6. Schematic arrangement showing principle of optical extensometer.

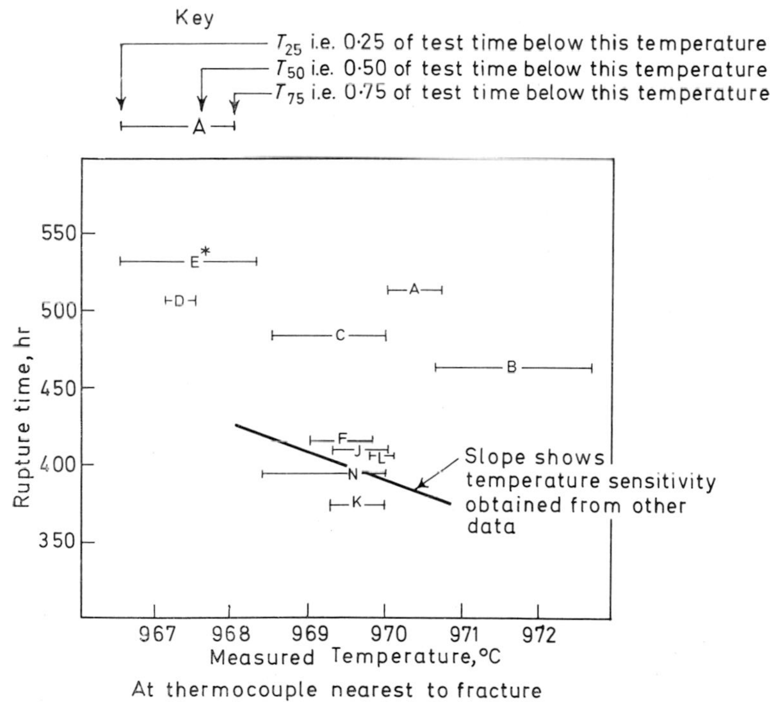

FIG. 7. Rupture times for specimens stressed to 4 ton/in² (6·2 hbar). (For letter code see Table 1.)

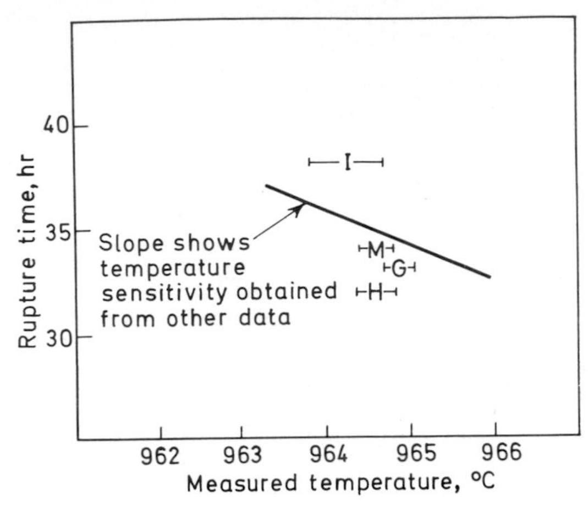

FIG. 8. Rupture times for specimens stressed to 7 ton/in^2 (10·8 hbar).

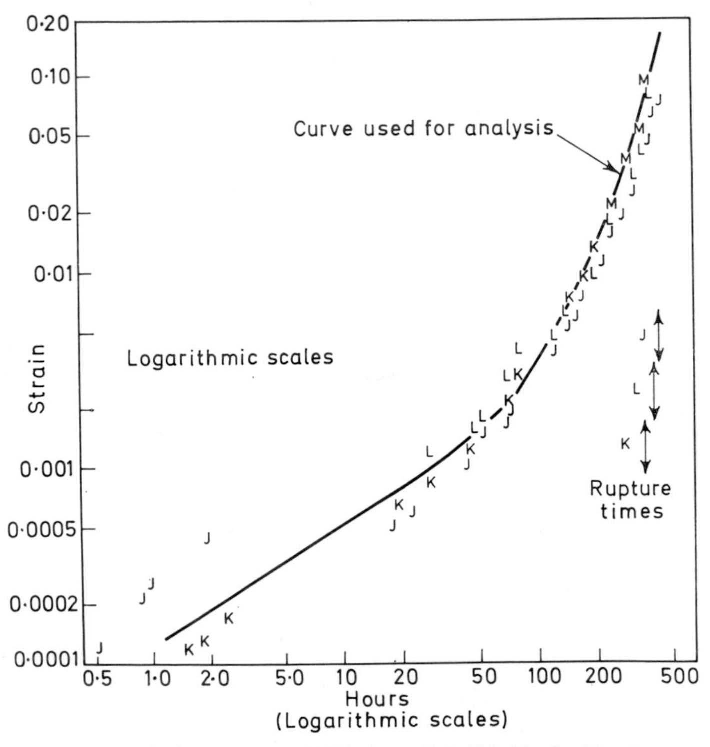

FIG. 9. Strain records for tests at 970°C, 4 ton/in^2 (6·2 hbar). (For letter code see Table 1.)

The rupture times and reductions of area are given in Table 1 together with the test conditions. The same rupture time information is shown graphically on Figs. 7 and 8 with the temperatures measured by the thermocouple nearest to the eventual rupture line.

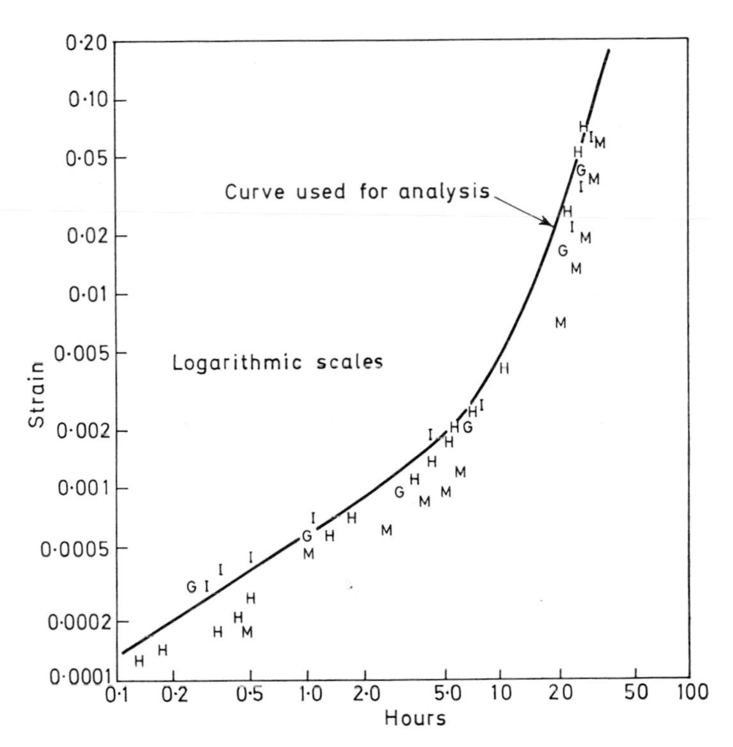

FIG. 10. Strain records for tests at 965°C, 7 ton/in² (10·8 hbar). (For letter code see Table 1.)

Only two strain records were obtained from the three Type 1 tests because the first was completed before the construction of the optical extensometer. Figures 11 and 12 show the results. For the small strains the scale has been magnified. The short vertical lines on the figures show the range between the greatest and least strains measured on each occasion.

Test pieces in type 2 tests were enclosed by lagging so that strain measurement using the optical extensometer was not possible.

Figures 9 and 10 show the creep strain readings for the Type 4 and 5 tests, respectively, at the two different conditions of stress and temperature.

TABLE 1

Rupture times, reduction of area and test conditions

Test	Stress ton/in^2 (hbar)	T_{25}(°C)	T_{50}(°C)	T_{75}(°C)	Rupture time (hr)	Reduction of area (%)
A	4·0 (6·2)	970·0	970·2	970·7	515	18·4
B	4·0 (6·2)	970·6	971·6	972·7	464	18·5
C	4·0 (6·2)	968·5	969·4	970·0	485	18·4
D	4·0 (6·2)	967·1	967·3	967·5	508	17·4
E	3·96 (6·11)	966·5	967·5	968·3	553	17·0
F	4·0 (6·2)	969·0	969·4	969·8	415	16·5
G	7·0 (10·8)	964·7	964·9	965·0	33	29·1
H	7·0 (10·8)	964·5	964·6	964·8	$31 < t < 32{\cdot}5$	(19·5)
I	7·0 (10·8)	963·8	964·3	964·7	38	33·7
J	4·0 (6·2)	969·3	969·7	970·0	415	21·6
K	4·0 (6·2)	969·3	969·8	970·0	374	20·9
L	4·0 (6·2)	969·8	969·9	970·1	408	19·5
M	7·0 (10·8)	964·5	964·6	964·8	34	31·7
N	4·0 (6·2)	968·4	969·6	970·0	396	13·5

[a] This test was terminated by a power failure after 31 hr but the strain record showed that rupture would have occurred before 32·5 hr. The percentage reduction of area was for a low temperature fracture caused by the specimen's contraction while cooling.

Note: T_{25} = Temperature below which 25 per cent of test duration was spent.
T_{50} = Temperature below which 50 per cent of test duration was spent.
T_{75} = Temperature below which 75 per cent of test duration was spent.

T_{25}, T_{50}, T_{75} were derived from measurements by the thermocouple nearest to the rupture line.

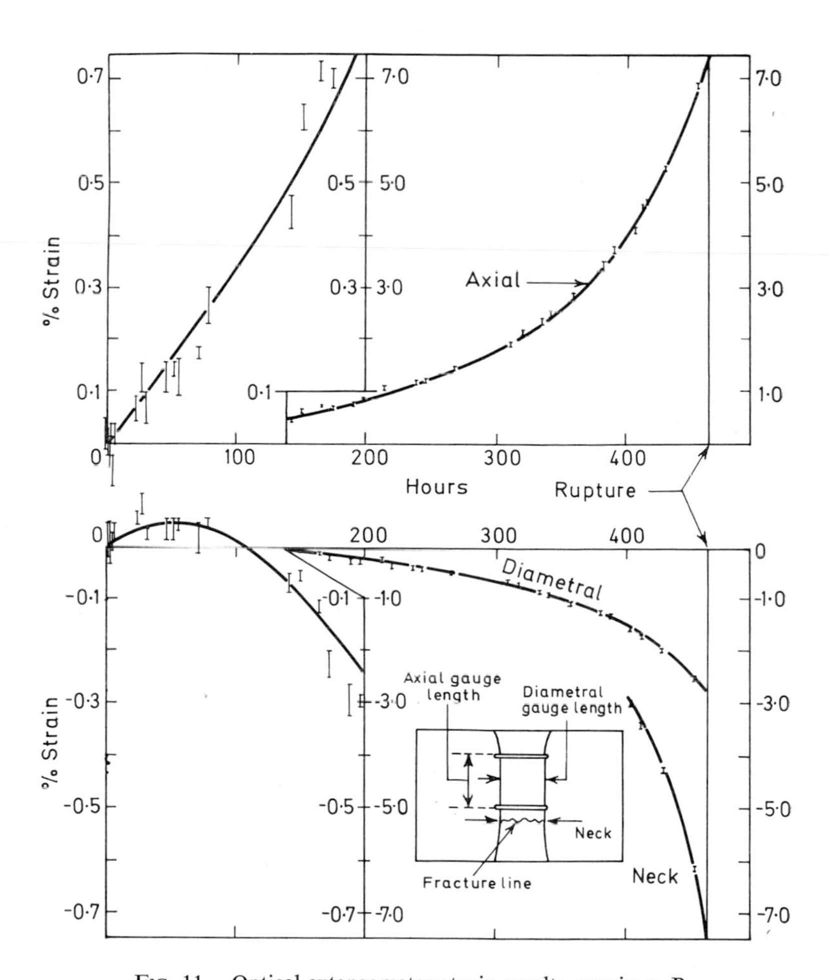

FIG. 11. Optical extensometer strain results, specimen B.

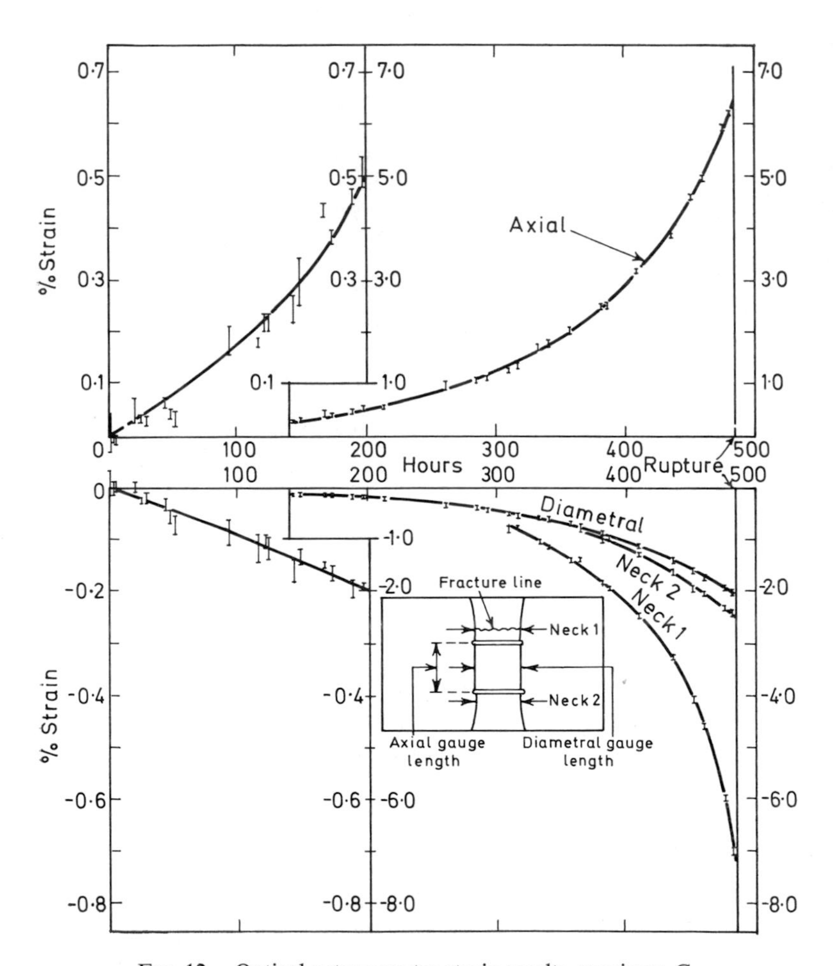

FIG. 12. Optical extensometer strain results, specimen C.

THEORETICAL ANALYSIS

Some order of magnitude considerations

Using manufacturer's data for Nimonic alloy 105 at 970°C it may be shown that a strain of 0·06 per cent may be caused by:

either elastic strain from a uniaxial stress of 5 ton/in^2 (7·7 hbar)

or creep strain after about 5 hr at 5 ton/in^2 (7·7 hbar)

or a temperature difference of 15°C.

The equal strains caused by the conditions listed above, all of which are fairly easily realised, show that the behaviour of a hot component is typified by a strong interaction between elastic, thermal expansion and creep effects. At lower temperatures it would also be necessary to consider plastic flow.

Stress analysis

The elastic thermal stress analysis of a long thick cylinder with an arbitrary radial temperature distribution is available from standard texts [3]. The temperature influences the stress pattern by introducing thermal strains. Whereas thermal strains are concerned with volume changes, creep strains are concerned with shape changes and it was therefore necessary to extend the solution. Appendix 1 gives details of the extended analysis. The approach is analytical rather than numerical but completely closed forms of solution for the stresses are obtainable only if analytic functions are assumed for the distributions of the various non-elastic strains. In practice, these distributions were represented by arrays of numbers within a digital computer and the accuracy of solution then depends on the extent of these arrays and the numerical method used for their integration.

The stress analysis does not need to distinguish the various (*e.g.* thermal, plastic, creep) components of non-elastic strain and the procedure adopted in the theoretical work described here was to add together the terms caused by thermal expansion and creep at any instant prior to the calculation of the corresponding stress pattern. Implicit assumptions in the analysis are that the non-elastic strains should be small enough to permit the usual tensor representation and that the principal directions of non-elastic strain must coincide with those of the elastic stresses and strains, namely the hoop, radial and axial directions. Furthermore, the material is assumed to be elastically and plastically isotropic.

Calculation of creep strain increments

Creep behaviour is still described in empirical terms and comprehensive results from creep tests of materials such as nickel base alloys are only available for tests at constant stress and temperature under uniaxial tensile conditions. In stress redistribution studies, involving a changing pattern of multiaxial stresses, it is necessary therefore to augment the data by using certain hypotheses in order to:

1 Permit estimates of behaviour under changing uniaxial conditions.

2 Permit predictions of behaviour under multiaxial stress situations including compression.

The hypotheses used in the work described here are not new and are reproduced simply for completeness.

Time and strain hardening hypotheses

These are two of the many alternative hypotheses currently being considered. For more complete descriptions of them see, for instance, refs. 4 to 8.

Suppose creep strain data from tests at constant stress and temperature are available in the form

$$\varepsilon = f(\sigma, T, t) \tag{1}$$

where

ε = Creep strain

σ = Stress (uniaxial tensile)

T = Temperature

t = Time measured from application of the load

f = A function with σ, T and t as variables

Then in principle the same information can also be expressed as

$$t = g(\sigma, T, \varepsilon) \tag{2}$$

where

g = A different function

According to the 'time hardening' hypothesis, the increment of creep strain caused by a period (Δt) of constant stress (σ_1) and temperature

(T_1), following the instant at which the strain was ε_1 and the time was t_1, is given by

$$\begin{aligned}\Delta\varepsilon_t &= f(\sigma_1, T_1, t_1 + \Delta t) - f(\sigma_1, T_1, t_1)\\ &= f_t(\sigma_1, T_1, t_1, \Delta t)\end{aligned} \tag{3}$$

where the subscript t refers to time hardening.

According to the 'strain hardening' hypothesis the increment of creep strain caused by the same conditions would be

$$\begin{aligned}\Delta\varepsilon_s &= f[\sigma_1, T_1, g(\sigma_1, T_1, \varepsilon_1) + \Delta t] - \varepsilon_1\\ &= f_s(\sigma_1, T_1, \varepsilon_1, \Delta t)\end{aligned} \tag{4}$$

where subscript s refers to strain hardening.

It is apparent that the influence of previous loading and straining is being introduced by the single values of time (t_1) and strain (ε_1), respectively, and a convenient (non-iterative) use of the strain hardening hypothesis depends on the existence of both 'f' and 'g' functions. Suitable expressions are given in ref. 9. It should be noted that many forms of 'f', for example sums of terms containing powers of t, do not lend themselves to conversion to 'g' functions.

Effective stress and strain

Equations (3) and (4) above, which refer to uniaxial tensile data, can be extended to multiaxial situations by postulating effective stresses ($\bar{\sigma}$) and strains ($\bar{\varepsilon}$) so that

$$\Delta\bar{\varepsilon}_t = f_t(\bar{\sigma}_1, T_1, t_1, \Delta t) \tag{5}$$

$$\Delta\bar{\varepsilon}_s = f_s(\bar{\sigma}_1, T_1, \bar{\varepsilon}_1, \Delta t) \tag{6}$$

where $\bar{\varepsilon}$, $\Delta\bar{\varepsilon}$ and $\bar{\sigma}$ are expressed in terms of the principal strains ($\varepsilon_1, \varepsilon_2, \varepsilon_3$) and principal stresses ($\sigma_1, \sigma_2, \sigma_3$) by

$$\bar{\varepsilon} = \frac{\sqrt{2}}{3}\sqrt{[(\varepsilon_1 - \varepsilon_2)^2 + (\varepsilon_2 - \varepsilon_3)^2 + (\varepsilon_3 - \varepsilon_1)^2]} \tag{7}$$

$$\Delta\bar{\varepsilon} = \frac{\sqrt{2}}{3}\sqrt{[(\Delta\varepsilon_1 - \Delta\varepsilon_2)^2 + (\Delta\varepsilon_2 - \Delta\varepsilon_3)^2 + (\Delta\varepsilon_3 - \Delta\varepsilon_1)^2]} \tag{8}$$

and

$$\bar{\sigma} = \frac{1}{\sqrt{2}}\sqrt{[(\sigma_1 - \sigma_2)^2 + (\sigma_2 - \sigma_3)^2 + (\sigma_3 - \sigma_1)^2]} \tag{9}$$

The multiplying constants are chosen so that for plastic flow under uniaxial stress,

$$\bar{\sigma} = \sigma_1 \quad \text{when} \quad \sigma_2 = \sigma_3 = 0$$

and

$$\bar{\varepsilon} = \varepsilon_1 \quad \text{when} \quad \varepsilon_2 = \varepsilon_3 = -\tfrac{1}{2}\varepsilon_1$$

Levy–von Mises flow rule

This concerns the direction of shape change and in its incremental flow form [10, 11] it ensures that the increment of effective shear strain corresponds in its direction with the current direction of effective shear stress. It may be stated in the following form:

$$\frac{\sigma_1 - \sigma_2}{\Delta\varepsilon_1 - \Delta\varepsilon_2} = \frac{\sigma_2 - \sigma_3}{\Delta\varepsilon_2 - \Delta\varepsilon_3} = \frac{\sigma_3 - \sigma_1}{\Delta\varepsilon_3 - \Delta\varepsilon_1} \tag{10}$$

It follows from the definitions of effective stress and strain that the ratio in eqn (10) also equals $2\bar{\sigma}/3\Delta\bar{\varepsilon}$.

Constant volume condition

Equations (5) or (6) with (7), (8), (9) and (10) suffice to determine differences between strain increments in the principal directions. In order to determine completely the actual values, a further condition is required and this is taken to be the constant volume condition,

$$\Delta\varepsilon_1 + \Delta\varepsilon_2 + \Delta\varepsilon_3 = 0 \tag{11}$$

Simultaneous use of the above conditions is assured by calculating strain increments using the following expressions:

$$\begin{aligned}
\Delta\varepsilon_1 &= \frac{\Delta\bar{\varepsilon}}{2\bar{\sigma}}(2\sigma_1 - \sigma_2 - \sigma_3) \\
\Delta\varepsilon_2 &= \frac{\Delta\bar{\varepsilon}}{2\bar{\sigma}}(2\sigma_2 - \sigma_1 - \sigma_3) \\
\Delta\varepsilon_3 &= \frac{\Delta\bar{\varepsilon}}{2\bar{\sigma}}(2\sigma_3 - \sigma_2 - \sigma_1)
\end{aligned} \tag{12}$$

Representation and evaluation of creep properties

The formulae adopted for the description of creep properties (as measured during conventional creep strain tests at constant tensile stress and temperature) were developed in 1963 for use in the calculation of stress redistribution effects in cooled gas turbine blades, and have been found to provide an accurate and convenient method for all the Nimonic alloys [9]. They have the particular advantage, which is unusual in formulae used for this purpose, that while permitting a representation of both primary and tertiary creep they allow explicit evaluation of strain from time and also time from strain.

Because the experimental programme was concerned with a limited number of tests and a relatively small range of stresses and temperatures, it was decided to concentrate on only two different stress/temperature combinations and to do several tests at each to indicate the scatter. Other more extensive creep strain data for Nimonic alloy 105 were available and were slightly modified in order to comply with the observed strain versus time results for this particular batch. The modified data are summarised in Table 2 using the rotation from ref. 9.

TABLE 2

Derived creep properties for Nimonic 105 *fully heat treated. Cast AE* 148

Symbol	Value						Units
θ	27·8						Degrees of arc
φ	75·0						Degrees of arc
A	0·863						Natural log cycles
B	−6·430						Natural log cycles
C	0						Natural log cycles/ (ton/in^2)
at	900°C	920°C	940°C	960°C	980°C	1000°C	
D	10·45	9·81	9·29	8·81	8·39	7·97	Natural log cycles
E	−0·62	−0·680	−0·758	−0·850	−0·960	−1·10	Natural log cycles/ (ton/in^2)

A computer program was written in Elliott 803 Autocode to calculate, using either time or strain hardening hypotheses, the redistribution of stresses within a thick tube with non-uniform radial temperature pattern.

The loading conditions on the tube are specified by:

1 Mean axial stress.
2 Inside and outside pressures (maintained zero in this investigation).
3 Inside and outside radii.
4 Temperatures at a number of equispaced positions through the thickness.

DISCUSSION OF EXPERIMENTAL AND THEORETICAL RESULTS

The rupture time results fall into three categories:

1 Electrically heated tube specimens, both with and without radial temperature gradients and loaded by a mean axial stress of 4 tons/in^2 (6·2 hbar) at a mean temperature of 967·4°C broke after about 500 hr. (Tests A, B, C, D and E.)
2 Furnace heated specimens with an axial stress of 4 tons/in^2 (6·2 hbar) and a temperature of 970°C broke after about 400 hr. (Tests F, J, K, L and N.)
3 Furnace heated specimens with a stress of 7 tons/in^2 (10·8 hbar) and a temperature of 965°C broke after about 35 hr. (Tests G, H, I and M.)

The strain records for the two Type 1 tests, shown in Figs. 11 and 12, reveal a number of features:

1 A very high diametral strain rate at the neck immediately before fracture.
2 Necking occurred on each side of the gauge length.
3 Examination and comparison of the diametral and axial strain rates suggests that there was a volumetric expansion in both tests for approximately the first 100 hr. The magnitude of this effect was much greater for Test B than C but its duration was similar. Subsequently, with the onset of necking (and non-uniform strain in the axial gauge length) the comparison between axial and diametral strains becomes less meaningful. It is of interest to note that Levy and Barody [12] have observed an apparent volume increase in the early stages of creep tests on aluminium sheet material.

The strain records for the normal creep tests are shown on Figs. 9 and 10. The continuous lines show the hyperbolic relation between log (strain) and log (time) which was used for the theoretical work.

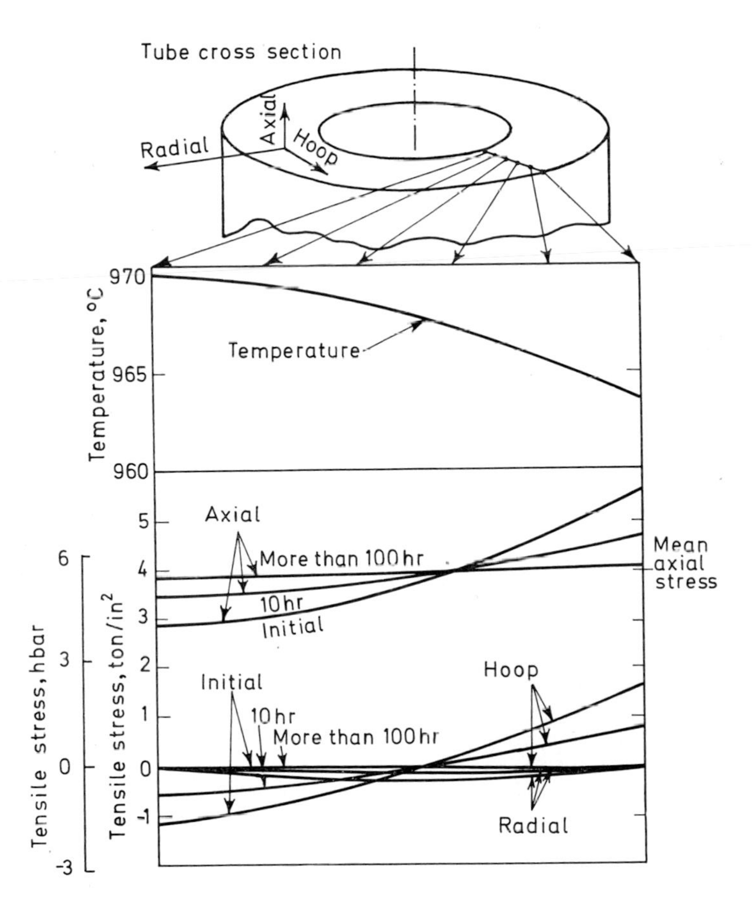

FIG. 13. Calculated temperature and stress profiles before, during and after redistribution.

Calculated stress redistribution

Figure 13 shows the radial temperature pattern and also the calculated stresses in the tube before, during and after stress redistribution. Figure 14 shows the time variation of axial stress, according to both time and strain hardening hypotheses. The small difference between the hypotheses as shown on Fig. 15 arises from the relatively small stress changes involved,

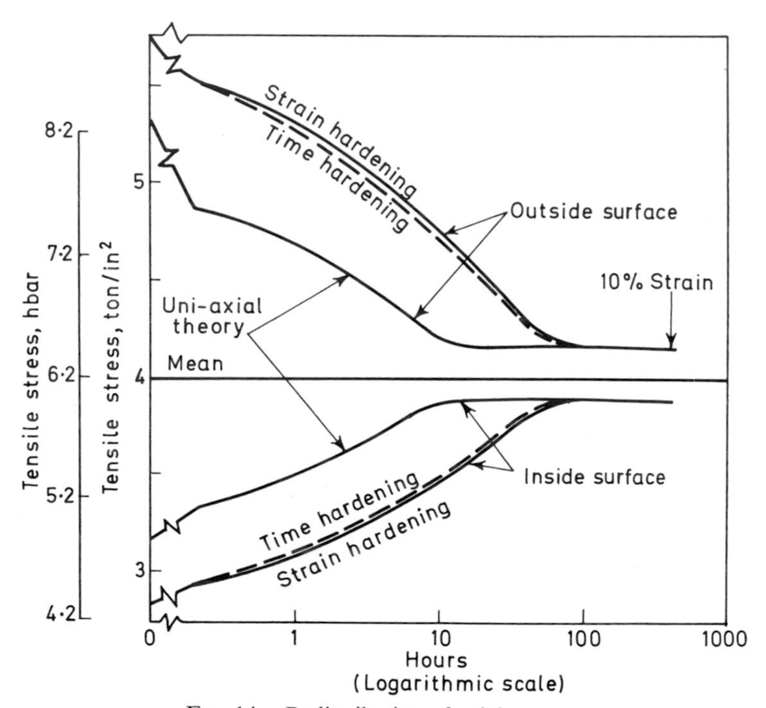

FIG. 14. Redistribution of axial stresses.

and the fact that some stresses are increasing while others are decreasing [13]. Figure 14 also shows results calculated more simply by ignoring the radial constraints, that is by envisaging the tube as a set of separated concentric tubes each with its own uniform temperature and uniaxial stress but having the same total lengths. It is evident that ignoring the constraints causes an underestimate of the thermal stresses and the time required for redistribution. The final distribution is unaffected because in this example, and in the case of turbine blades, the final distribution happens to be uniaxial.

Figure 15 shows the time variation of hoop and radial stresses, and Fig. 16 the directions (in the octahedral plane) of the effective stress and strain components. It is apparent that the strain direction lags behind the stress direction as a consequence of the incremental flow rule, but they approach the uniaxial condition.

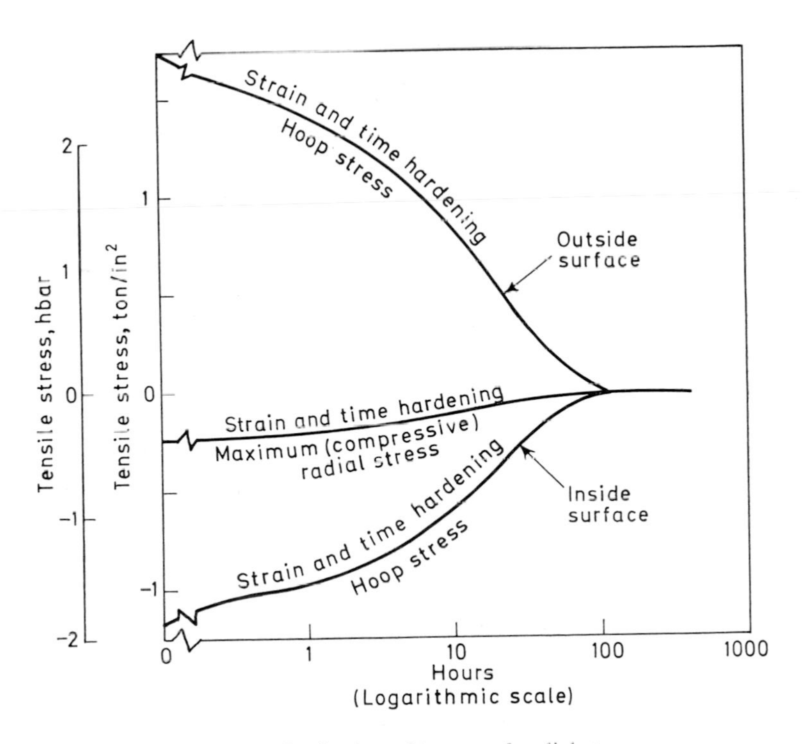

FIG. 15. Redistribution of hoop and radial stresses.

Measured and calculated strain results

Figure 17 compares the theoretical predictions and measured values of axial strain. The upper line shows theoretical values of axial elongation according to the theory (strain and time hardening hypotheses being indistinguishable). The dashed line at the lower end depicts the results of a calculation for a cylinder without thermal stressses but having the same mean axial stress and mean radial temperature. The other continuous lines are the measured strains (transferred from Figs. 11 and 12) for the tubular specimens. The lower strain level of the observed readings is associated

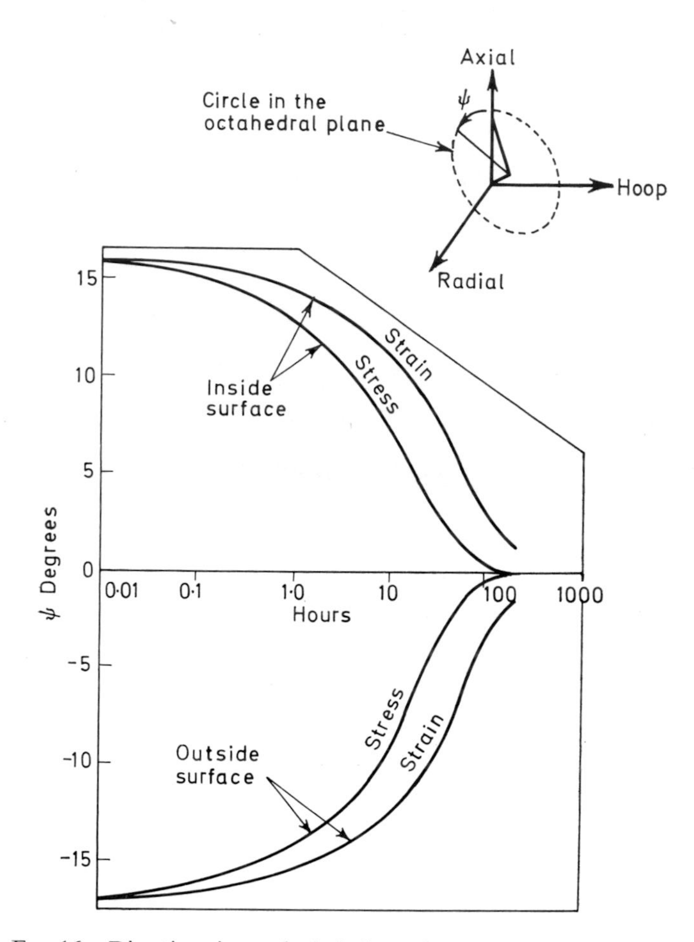

FIG. 16. Directions in octahedral plane of stresses and creep strains.

with the direct electric heating method which seems to give consistently longer rupture times and, in part, to the theoretical values used to describe the creep behaviour. Figure 9 shows the latter to give rather high creep strains in the strain range from about 1 to 10 per cent.

Examination of Fig. 17 does show that application of creep data, obtained under conditions of constant stress and temperature, to a stress redistribution problem gives a strain behaviour which agrees reasonably well with

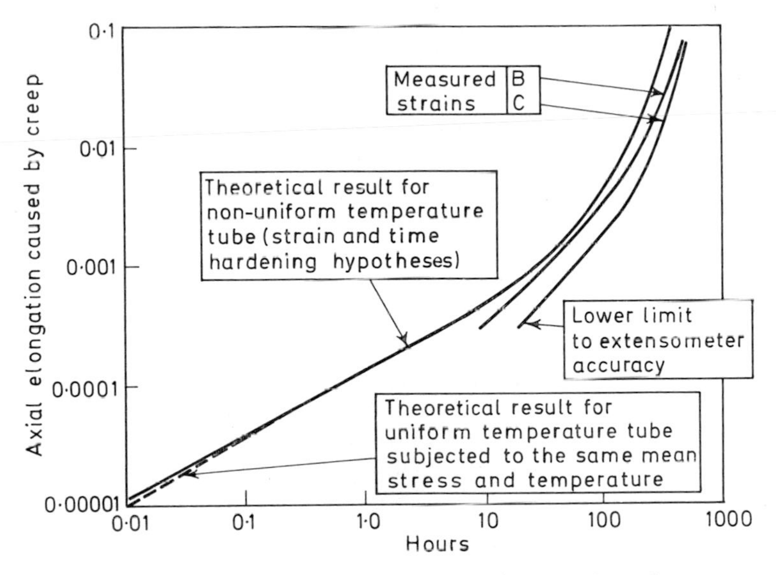

FIG. 17. Comparison of measured and calculated elongations.

that actually observed. Moreover the values of the rupture life, predicted on the basis of the fully redistributed stress pattern, agree very well with the observed values. This result is important because it implies that redistribution occurs as the result of very small creep strains and hence with little damage—thus confirming an assumption inherent in the present methods of calculating turbine blade lives.

CONCLUSIONS

The following conclusions seem justified for the particular alloy (Nimonic 105) and the test condition (970°C and 4 tons/in^2; 6·2 hbar).

There is no significant difference between rupture times for specimens with and without modest initial thermal stresses, provided that the mean stress and temperature levels are identical and do not vary with time.

A comparison of axial and diametral strains on the tube experiments with a temperature gradient indicates that some volume increase occurred during the first 100 hr and was particularly significant in the initial stages.

There seemed to be no significant difference between the rupture times for the two shapes and three sizes of test specimen when all were furnace tested using the same type of testing machine.

An analytical solution has been derived for the stress distribution in a long thick cylinder containing an arbitrary distribution of non-elastic strains. The non-elastic strains may include any combination of thermal, creep and plastic strains, while the boundary conditions may include any combination of internal and external pressures and any axial mean stress.

The analysis has been incorporated into an Elliott 803 Autocode computer program. The program has been used, together with creep data derived from creep tests on the same batch of Nimonic alloy 105, to predict the behaviour of the tube during the process of stress redistribution arising from the initial thermal stress situation.

The results of the calculation show themselves relatively insensitive to the 'hardening hypothesis' used because stress changes are relatively small and because some stresses increase as others decrease [13].

A simpler analysis ignoring radial and circumferential stresses (and therefore radial continuity of displacement) is shown to lead to an underestimate of both the magnitude of the thermal stresses and the time required for them to redistribute. The simpler analysis does however give correct values for the redistributed stress pattern because in this particular case (no internal or external pressure) the radial and circumferential stresses disappear.

The predicted shape of strain/time curve corresponds, except at extremely small strains and times, to that of a uniformly stressed specimen having the same mean stress and temperature. Thus, the theory and experiment both indicate an overall creep behaviour and rupture life consistent with mean stress/mean temperature conditions.

ACKNOWLEDGEMENTS

The author gratefully acknowledges the assistance given by Mr J. E. Northwood and Mr G. R. M. Jackson, who undertook metallographic examinations, and by the staff of the Instrument Section of NGTE who constructed the optical extensometer and carefully welded thermocouples at 'impossible' positions.

REFERENCES

1. Barnes, J. F. and Clarke, J. M. (1964). 'The significance of creep in cooled gas turbine blades.' *Conf. on Thermal Loading and Creep in Structures and Components. Proc. Instn Mech. Engrs*, **178** (3L), 43–52.
2. *The Nimonic Series of High Temperature Alloys*. Birmingham: Henry Wiggin & Co. Ltd.
3. Timoshenko, S. (1934). *Theory of Elasticity*, p. 366. London: McGraw-Hill.
4. Johnson, A. E., Henderson, J. and Mathur, V. (1958). 'Creep under changing complex stress systems.' *Engineer*, **206**, 209–216.
5. Mendelson, A., Nirschberg, M. H. and Namson, S. S. (1960). 'A general approach to the practical solution of creep problems.' *Trans Amer. Soc. Mech. Engrs, Series D, J. Basic Engng*, **5** (20).
6. Rabotnov, Yu. N. (1963). 'On the equation of state of creep.' *Jt Int. Conf. on Creep. Proc. Instn Mech. Engrs*, **178** (3A), 2-117–2-122.
7. Graham, A. (1953). 'The phenomenological method in rheology.' *Research*, **6**, 92–97.
8. Graham, A. and Walles, K. F. A. (1955). 'Relationships between long and short-time creep and tensile properties in a commercial alloy.' *J. Iron and Steel Inst.*, **179**, 105–120.
9. Clarke, J. M. (1966). 'A convenient representation of creep strain data for problems involving time-varying stresses and temperatures.' *NGTE Report No. R*284.
10. Phillips, A. and Kaechele, L. (1956). 'Combined stress tests in plasticity.' *J. Appl. Mech.*, **23**, 43–48.
11. Marin, J. and Hu, L. W. (1956). 'Biaxial plastic stress-strain relations of a mild steel for variable stress ratios.' *Trans. Amer. Soc. Mech. Engrs*, **78**, 499–509.
12. Levy, J. C. and Barody, I. I. (1964). Poisson's ratio in creep using the strain-replica method. *Conf. on Thermal Loading and Creep in Structures and Components. Proc. Instn Mech. Engrs*, **178** (3L), 193–197.
13. Clarke, J. M. (1964). Discussion at session 4. *Conf. on Thermal Loading and Creep in Structures and Components. Proc. Instn Mech. Engrs*, **178** (3L), 169.

Appendix 1

CALCULATION OF STRESSES IN A LONG THICK WALLED TUBE WITH ARBITRARY MEAN AXIAL STRESS, INSIDE AND OUTSIDE PRESSURES AND RADIAL DISTRIBUTION OF NON-ELASTIC STRAINS

Hooke's law in the presence of non-elastic strains may be written:

$$
\begin{aligned}
e_1 &= \frac{1}{E}\sigma_1 - \frac{v}{E}(\sigma_2 + \sigma_3) + c_1 \\
e_2 &= \frac{1}{E}\sigma_2 - \frac{v}{E}(\sigma_1 + \sigma_3) + c_2 \\
e_3 &= \frac{1}{E}\sigma_3 - \frac{v}{E}(\sigma_2 + \sigma_1) + c_3
\end{aligned}
\tag{A13}
$$

Elimination of σ_1 from the above and some rearrangement gives:

$$
\sigma_2 = \frac{Ev}{(1-2v)(1+v)}[(e_1 - c_1) + (e_2 - c_2) + (e_3 - c_3)] + \frac{E}{1+v}[e_1 - c_2] \tag{A14}
$$

$$
\sigma_2 - \sigma_3 = -\frac{E}{1+v}[(e_2 - e_3) - (c_2 - c_3)] \tag{A15}
$$

The radial equilibrium condition in the absence of body forces is:

$$
\frac{\mathrm{d}\sigma_2}{\mathrm{d}r} + \frac{\sigma_2 - \sigma_3}{r} = 0 \tag{A16}
$$

and the compatibility condition under the assumed axisymmetric and longitudinally uniform conditions is:

$$
\begin{aligned}
e_1 &= \text{const} \\
e_2 &= \frac{\mathrm{d}u}{\mathrm{d}r} \\
e_3 &= \frac{u}{r}
\end{aligned}
\tag{A17}
$$

Substitution from eqn (A17) into eqn (A14) and (A15) and then from eqns (A14) and (A15) into eqn (A16) leads to the following second order non-homogeneous ordinary differential equation for the displacement:

$$\frac{\mathrm{d}}{\mathrm{d}r}\left(\frac{\mathrm{d}u}{\mathrm{d}r} + \frac{u}{r}\right) = F(r) \tag{A18}$$

where

$$F(r) = \frac{v}{1 - v}\frac{\mathrm{d}}{\mathrm{d}r}(c_1 + c_2 + c_3) + \frac{1 - 2v}{1 - v}\left(\frac{\mathrm{d}c_2}{\mathrm{d}r} + \frac{c_2 - c_3}{r}\right) \tag{A19}$$

Two integrations of eqn (A18) lead to the form of the solution for displacement:

$$u = Rr + S\frac{r_\mathrm{i}^2}{r} + \frac{1}{r}\int_{r_\mathrm{i}}^{r} r\int_{r_\mathrm{i}}^{r} F(r)\,\mathrm{d}r\,\mathrm{d}r \tag{A20}$$

and for convenience the notation

$$G(r) = \int_{r_\mathrm{i}}^{r} F(r)\,\mathrm{d}r$$

$$H(r) = \frac{1}{r^2}\int_{r_\mathrm{i}}^{r} rG(r)\,\mathrm{d}r$$

is introduced so that eqn (A20) can be written

$$u = Rr + Sr_\mathrm{i}^2\frac{1}{r} + rH(r) \tag{A21}$$

The problem is therefore reduced to finding the constants R, S and e_1, such that the following boundary conditions apply:

$$\begin{array}{lll} 1 & \sigma_{2\mathrm{i}} = -P_\mathrm{i} & \text{the inside pressure} \\ 2 & \sigma_{2\mathrm{o}} = -P_\mathrm{o} & \text{the outside pressure} \\ 3 & 2\int_{r_\mathrm{i}}^{r_\mathrm{o}} \sigma_1 r\,\mathrm{d}r/(r_\mathrm{o}^2 - r_\mathrm{i}^2) = s & \text{the mean axial stress} \end{array} \tag{A22}$$

Substitution from eqn (A17) into eqn (A14) expresses σ_2 in terms of u and its derivative so that eqn (A20) is then available to express σ_2 in terms of the constants R, S and e_1.

$$\begin{aligned} \sigma_2 = {} & \frac{E}{(1 - 2v)(1 + v)}(R + ve_1) - \frac{E}{(1 + v)}\left[S\frac{r_\mathrm{i}^2}{r^2} + H(r)\right] \\ & + \frac{E(1 - v)}{(1 - 2v)(1 + v)}[G(r) - c_2] - \frac{Ev}{(1 - 2v)(1 + v)}(c_1 + c_3) \end{aligned} \tag{A23}$$

Substitution for $\sigma_2 = -P_i$ at $r = r_i$ and $\sigma_2 = -P_o$ at $r = r_o$ leads to values for $(R + ve_1)$ and S

$$S = \frac{r_o^2}{r_o^2 - r_i^2}\left\{\frac{1+v}{E}(P_i - P_o) - \frac{1-v}{1-2v}G(r_o) + H(r_o) + \frac{v}{1-2v}\right.$$

$$\left.\times\,[(c_{1o} - c_{1i}) + (c_{3o} - c_{3i})] + \frac{1-v}{1-2v}(c_{2o} - c_{2i})\right\} \quad \text{(A24)}$$

$$R + ve_1 = S(1-2v) - \frac{(1-2v)(1+v)}{E}P_i + (1-v)c_{2i} + v(c_{1i} + c_{3i}) \quad \text{(A25)}$$

The last two equations permit σ_2 to be found using eqn (A23) then σ_3 can be found from:

$$\sigma_2 - \sigma_3 = \frac{E}{1+v}\left[2S\frac{r_i^2}{r^2} + 2H(r) - G(r) + (c_2 - c_3)\right] \quad \text{(A26)}$$

The radial and loop stresses have now been determined but the third boundary condition has not been introduced. This can now be done by integrating the first equation in (A13).
Hence

$$e_1 = \frac{2}{r_o^2 - r_i^2}\left[\frac{1}{E}\int_{r_i}^{r_o} r\sigma_1\,\mathrm{d}r - \frac{v}{E}\int_{r_i}^{r_o}(\sigma_2 + \sigma_3)r\,\mathrm{d}r + \int_{r_i}^{r_o} c_1 r\,\mathrm{d}r\right]$$

i.e.

$$e_1 = \frac{S}{E} - \frac{2}{r_o^2 - r_i^2}\left[\frac{v}{E}\int_{r_i}^{r_o}(\sigma_2 + \sigma_3)r\,\mathrm{d}r - \int_{r_i}^{r_o} c_1 r\,\mathrm{d}r\right] \quad \text{(A27)}$$

Sufficient information is now available to calculate σ_1 from the first equation in (A13), and the displacement function u from eqn (A20).

Appendix 2

SOME CONDITIONS REQUIRED FOR THE ESTABLISHMENT OF A STEADY STATE STRESS DISTRIBUTION IN A UNIAXIAL AXISYMMETRIC SITUATION

The analysis which follows illustrates a number of properties of the 'fully redistributed' or 'steady state' stress pattern which is generally achieved after stress redistribution calculations.

It is shown that if the log (creep strain)/log (time) curves are of the same shape and simply displaced for different stresses and temperatures then the existence of a general steady state requires that the shape involved is a member of a family of curves containing three arbitrary constants and including as a sub-set all straight lines. A method is shown for calculating directly the redistributed solution.

It is assumed that the creep strain/time results of ordinary creep tests over a range of stresses and temperatures can be represented by a formula of the form:

$$\ln\left(\frac{\varepsilon}{\lambda}\right) = f_1\left[\ln\left(\frac{t}{\tau}\right)\right] \tag{A28}$$

or for convenience:

$$y = f_1(x) \tag{A29}$$

where the function f_1 does not depend on stress or temperature but λ and τ can each depend on both stress and temperature. The function f_1 is for the moment arbitrary and so are λ and τ which are the strain and time values for which y and x respectively are zero. A hyperbolic relation between x and y is used in ref. 9 and also in the theoretical work described in this chapter.

It can be shown that the corresponding expression for strain rate is:

$$\frac{d\varepsilon}{dt} = \frac{dy}{dx} \cdot \exp(y - x) \cdot \frac{\lambda}{\tau} \tag{A30}$$

or

$$\frac{d\varepsilon}{dt} = f_2(x) \cdot \frac{\lambda}{\tau} \tag{A31}$$

where

$$f_2(x) = \frac{dy}{dx} \cdot \exp(y - x) \tag{A32}$$

Now in a uniaxial axisymmetric stress situation in which neither the stress nor temperature are changing with time the creep strain rate (say $\dot{\varepsilon}_1$) must be the same at all positions, so that

$$f_2(x) = \dot{\varepsilon}_1 \tau/\lambda \tag{A33}$$

where in general τ/λ and x will be different at different positions.

Some time later although τ/λ remains constant (because stresses and temperatures are constant) the strain rates (say $\dot{\varepsilon}_2$) common to all positions might be different, so that

$$f_2(x + \Delta x) = \dot{\varepsilon}_2 \tau/\lambda \tag{A34}$$

Now

$$\Delta x = \ln\left(1 + \frac{\Delta t}{t}\right) \tag{A35}$$

which is independent of x and is the same at all points hence, for all x,

$$\frac{f_2(x + \Delta x)}{f_2(x)} = \frac{\dot{\varepsilon}_2}{\dot{\varepsilon}_1} \tag{A36}$$

i.e., the proportional change in f_2 with x does not depend on the instantaneous value of f_2. The most general function with this property is the exponential function which may be written:

$$f_2(x) = \exp(ax + b) \tag{A37}$$

where a, b are constants or equivalently, and more conveniently in this case:

$$f_2(x) = p \exp[(p - 1)x + q] \tag{A38}$$

where p, q are constants. Comparison with eqn (A32) yields the differential equation

$$\exp(y) \cdot \frac{\mathrm{d}y}{\mathrm{d}x} = p \exp(px + q) \tag{A39}$$

Integration of this equation gives:

$$y = \ln[\exp(px + q) + \exp(y_{-\infty})] \tag{A40}$$

where $y_{-\infty}$ is the value of y when $x = -\infty$. An important special case of eqn (A40) is the straight line

$$y = px + q \tag{A41}$$

which corresponds to a strain/time relation of the form

$$\varepsilon \propto t^p \tag{A42}$$

and assumes zero creep strain at zero time.

Substitution from eqn (A38) into eqn (A31) gives

$$\frac{\mathrm{d}\varepsilon}{\mathrm{d}t} = p \exp[(p - 1)x + q] \cdot \frac{\lambda}{\tau} \tag{A43}$$

i.e.,

$$\frac{\mathrm{d}\varepsilon}{\mathrm{d}t} = p \exp(q) \cdot t^{p-1} \cdot \frac{\lambda}{\tau^p} \tag{A44}$$

In the general case when both λ and τ vary with stress and temperature the steady state stress pattern must therefore satisfy (because the other term in eqn (A44) is the same for all elements):

$$\frac{\lambda}{\tau^p} = \text{constant} \tag{A45}$$

and the constant can be obtained from the requirement that the integrated stress be equal to the applied load. The solution therefore depends on p unless λ is constant.

SIMPLE EXAMPLE

Assuming the following variations of λ and τ with stress and temperature:

$$\begin{aligned} \lambda &= \text{constant} \\ \tau &= \tau' \exp\,[v(\sigma - \sigma') + w(T - T')] \end{aligned} \tag{A46}$$

Here τ', σ', T' are typical constant values for the time scale, stress and temperature of the problem while v and w represent sensitivities of τ to departures from σ' and T'.

Then for constant λ/τ^p as required by eqn (A45) for a steady state solution

$$\tau = \text{constant} \tag{A47}$$

and from eqn (A46)

$$v(\sigma - \sigma') + w(T - T') = \text{constant} = K \tag{A48}$$

Integrating with respect to area gives:

$$v \int \sigma \, \mathrm{d}A + w \int T \, \mathrm{d}A = (K + v\sigma' + wT') \int \mathrm{d}A \tag{A49}$$

i.e.,

$$v(\sigma_m - \sigma') + w(T_m - T') = K \tag{A50}$$

where σ_m and T_m are area mean values of stress and temperature.

From eqn (A50) and eqn (A46) it follows that the effective time scale factor τ for the non-uniform system corresponds to that obtainable from a uniform stress situation at the same mean stress and temperature.

It should be emphasised that the relatively simple expressions (A46) which led to this conclusion only apply for rather limited departures of stress and temperature from the σ', T' values. Nevertheless the analysis serves to illustrate that the strain accumulation after redistribution corresponds roughly with mean stress/mean temperature conditions and the solution of eqn (A45) with more accurate forms of eqn (A46) gives a useful technique for the direct calculation of steady state stress patterns from applied temperature patterns.

Chapter 19

CREEP-BUCKLING OF FLAT RECTANGULAR PLATES WHEN THE CREEP EXPONENT RANGES FROM 3 TO 7

N. J. HOFF, L. BERKE, T. C. HONIKMAN AND I. M. LEVI

SUMMARY

The accuracy of an earlier analysis of the creep deflections and of the critical time of flat, rectangular plates subjected to uniform uniaxial compression parallel to one pair of edges is examined and is found satisfactory. The analysis was carried out for a plate whose material creeps in accordance with the power law and the creep exponent n *was taken as* 3. *New closed form expressions are derived for* n = 5 *and* n = 7, *and it is shown that these are also sufficiently accurate for engineering purposes if the initial deviations of the plate from the ideal plane are small. Correction factors are proposed for the case when the initial imperfections of the plate are large.*

NOTATION

a, b	Width and length of the plate.
A_{ij}	Coefficients in the series for σ_y.
B_{ij}	Coefficients in the series σ_x.
C	Constant in Odqvist's creep law.
C_1, C_2	Constants in the general expression for the critical time.
C_{11}, D_{11}	Coefficients in the series for the shear stresses and twisting moments.
D	Distance between the faces of the sandwich.
E	Modulus of elasticity.

h	Thickness of the sandwich faces.
h_{real}	Thickness of the real plate.
J_2	Second invariant of the stress deviation tensor.
k	Constant in the uniaxial creep law.
m	Exponent in Odqvist's creep law.
n	Exponent in the uniaxial creep law.
P	Applied compressive load.
Q	Shear stress resultant (transverse).
S	Shear stress resultant (in-plane)
s_{ij}	Component of the stress deviation tensor.
t	Time.
V	Shear stress resultant (transverse).
w	Lateral deflection function.
W	Amplitude of the lateral deflection.
x, y	Coordinates in the plane of the plate.
α	Argument $\pi x/a$ of trigonometric function.
β	Argument $\pi y/b$ of trigonometric function.
δ_{ij}	Kronecker delta.
$\dot{\varepsilon}$	Uniaxial creep strain rate.
ε_{ij}	Component of the strain tensor.
κ	Curvature.
σ	Applied uniaxial stress.
τ_{xy}	Shear stresses in the sandwich faces.
ω	Dimensionless lateral displacement amplitude.

Derivative notation

$(\dot{})$	Derivative of ( ) with respect to time.

Subscripts

o	Outer face of the sandwich.
i	Inner face of the sandwich.
E	Euler quantity pertaining to the elastic buckling of a square plate of width a.
nom	Nominal.
equiv	Equivalent.
ave	Average.
crit	Critical.

Superscript

*	Coefficient of the series for a stress in the inner flange.

INTRODUCTION

It has been known for a long time that slender or thin walled structural elements are likely to buckle when the loads acting upon them give rise to compressive stresses. If the material of the element is perfectly elastic and the element has been manufactured with great care, this buckling takes place rather suddenly when the so-called critical load of the structural element is reached in the process of loading. If the material of the element exhibits creep deformations at the temperature at which the test is being carried out, the element buckles even at loads smaller than the critical load provided that the loads are allowed to act on the structure for a sufficiently long time. Thus, in the presence of creep, the calculation of the critical load is replaced by the calculation of the critical time.

It is possible to give a simple explanation of this phenomenon. A structural element is never perfect because it can be manufactured in the shop only within tolerances which cannot be reduced to zero. But a column which is not perfectly straight will increase its small initial curvature as time passes in consequence of the creep deformations. The increase in curvature means increased lateral displacements from the theoretical straight axis of the column, and these displacements serve as lever arms for the compressive load that acts along the axis. The increased lever arm causes an increase in the bending moment acting on the column which, in turn, results in an increased rate of change of the curvature of the column. When the creep law is the non-linear one characterising metals, the creep rate increases much more rapidly than the stress. Thus, the process is an accelerating one which leads to excessively large lateral displacements and collapse within a finite time. This finite time has been called the critical time.

The history of the development of our knowledge of creep-buckling was surveyed by the senior author [1] in his invited lecture delivered before the Third US National Congress of Applied Mechanics in 1958. He showed there that the analytical approach most likely to lead to agreement with experimental results was the one in which the structural element was assumed to exhibit small initial deviations from the exact shape, and he presented arguments why the elegant and mathematically correct theory of Rabotnov and Shesterikov [2], in which the element is assumed to have a perfect shape, is less likely to do so.

The assumption of initial imperfections underlies the analysis of the creep-buckling of flat plates in a report [3] prepared by the senior author for a conference held in Cambridge in honour of Sir John Baker in 1968.

In this report, a closed form solution was obtained for the critical time in spite of the highly non-linear character of the problem, but in the derivation of the solution a number of simplifying assumptions were made. One of the purposes of the present chapter is to investigate the effects of these assumptions in some detail. A second purpose is to extend the validity of the critical time formula obtained for the creep exponent $n = 3$ to other values of n.

RÉSUMÉ OF THE EARLIER SOLUTION

In the original solution of ref. 3 the following assumptions, whose effects will be investigated here, were made:

1 The exponent n in the uniaxial creep law $\dot{\varepsilon} = k\sigma^n$ was taken as 3; in this equation, σ is the uniaxial tensile stress, $\dot{\varepsilon}$ the corresponding strain rate and k and n are empirical constants.
2 In the expression for the second invariant J_2 of the deviatoric part of the stress tensor, which appears in the biaxial creep law, the effect of the shear stresses was disregarded.
3 A simplified expression was obtained for the lateral displacement rate of the plate by adding the expressions derived for small displacements to those obtained for large displacements.
4 The lateral displacements and the stress components were represented by trigonometric series, but of the infinitely large number of terms only a very restricted number was retained in the analysis.

The significance of these restrictive assumptions was discussed in the original paper, and estimates of the error incurred were given wherever possible. It is useful, however, to calculate more accurately the displacement rates and the critical times in order to establish confidence in the solution, and to extend the validity of the formulae by obtaining the values of the numerical coefficients for a sufficiently wide range of values of the creep exponent n.

A further assumption underlying the analysis of ref. 3 as well as the work presented in this chapter is that the deformations are caused only by steady, or secondary, creep following the power law, and by linear elasticity. Also, the edges of the plate are assumed to be simply supported.

An important additional simplification of the problem resulted from the replacement of the actual solid plate (Fig. 1) by an equivalent sandwich plate. A small element of this sandwich plate is shown in Fig. 2 in its distorted shape after loading. The principle involved in the replacement is that the cross-sectional area and the moment of inertia of the section be preserved; this is accomplished if one takes

$$h = (\tfrac{1}{2})h_{\text{real}}, \qquad D = (1/\sqrt{3})h_{\text{real}} \tag{1}$$

Here, h is the wall thickness of the cover plates of the sandwich, D is the distance between the centroidal planes of the upper and lower cover plates, as indicated in Fig. 2, and h_{real} is the real wall thickness of the real plate.

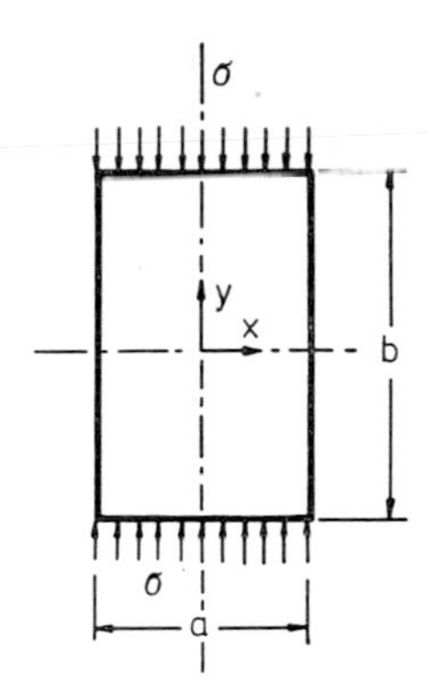

FIG. 1. Flat rectangular plate.

This model of the plate corresponds to the model of a beam, or column, known as the idealised I-beam, in which the entire cross-sectional area of the beam is concentrated in two mathematical points held apart a distance D by a perfectly shear resistant web of zero cross-sectional area. The justification for using this analytical model in calculations was supplied by a number of investigations, which are properly quoted in ref. 3.

It is useful to add here, however, that a somewhat more accurate value for D was proposed by Rabotnov [4] in 1964:

$$D = \left(\frac{n}{2n+1}\right)^{n/(n+1)} h_{\text{real}} \tag{2}$$

This relationship ensures that the rate of stretching and the rate of change of the curvature of the actual plate and its sandwich model are the same for pure tension and for the pure bending, respectively. Equation (2) naturally reduces to the second part of eqn (1) for the linear case when $n = 1$. However, the difference between the values of the numerical factor obtainable from eqns (1) and (2) is small. Rabotnov's formula yields a maximum of 0·578 for $n = 1$ and a minimum of $\frac{1}{2}$ for $n = \infty$, while the value calculated from the second part of eqn (1) is 0·578 for all values of n.

In the analysis of ref. 3, the equilibrium of the shell element of Fig. 2 was first established. One moment equation was automatically satisfied because S was taken to be the same on planes perpendicular to x and to y. The remaining two moment and three force equilibrium conditions were used to eliminate V, Q and S, and to obtain two new equations from which

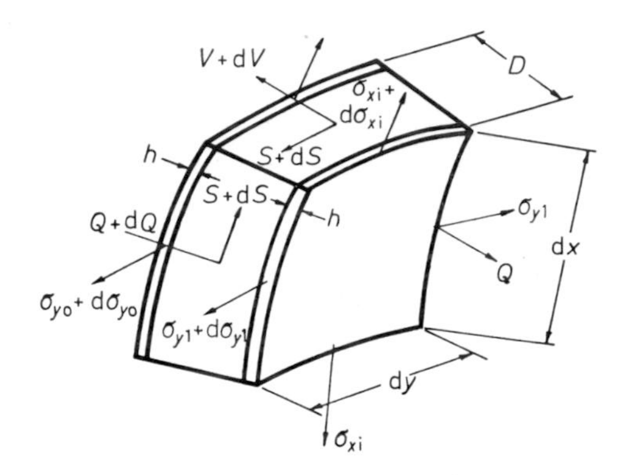

FIG. 2. Stresses and stress resultants on small idealised sandwich element.

all the shear stress resultants were missing. These two equations naturally do not suffice for the calculation of the four unknown normal stresses σ_{xo}, σ_{yo}, σ_{xi} and σ_{yi} and of the unknown lateral displacement rate $\dot{w}$. However, we can also write an overall condition of force equilibrium and two conditions of compatibility, as will be shown presently.

The analysis was carried out for the plate in its slightly distorted configuration which was assumed to be known at time t. The strain rates caused by arbitrary stresses were calculated from Odqvist's law

$$\dot{\varepsilon}_{ij} = CJ_2{}^m s_{ij} \tag{3}$$

where $\dot{\varepsilon}_{ij}$ is the i, j-th component of the strain rate tensor and s_{ij} is the i, j-th component of the stress deviator tensor. The latter can be calculated from the stress components σ_{ij} with the aid of the equation

$$s_{ij} = \sigma_{ij} - \delta_{ij}\sigma_{\text{ave}} \tag{4}$$

where σ_{ave} is the average normal stress and δ_{ij} is the Kronecker delta defined as

$$\delta_{ij} = \begin{cases} 1 & \text{when} \quad i = j \\ 0 & \text{when} \quad i \neq j \end{cases} \tag{5}$$

The second invariant J_2 is given by

$$J_2 = (\tfrac{1}{2}) s_{ij} s_{ij} \tag{6}$$

where the repeated subscripts indicate summation in accordance with Einstein's notation.

In ref. 3, the effect of the shear stresses on J_2 was disregarded. The argument was that the shear stresses vanish identically at the centre of the plate because of symmetry. Moreover, as long as the displacements are small, the shear stresses are small everywhere compared to the applied compressive stress σ. Also, the plate is so thin that it is justifiable to assume that a plane state of stress exists. On the basis of these considerations the expression for J_2 becomes

$$J_2 = \tfrac{1}{3}(\sigma_x{}^2 + \sigma_y{}^2 - \sigma_x \sigma_y) \tag{7}$$

and the constitutive equations for the x and y components of the strain rate can be written as

$$\dot{\varepsilon}_x = \tfrac{3}{2} k J_2 (2\sigma_x - \sigma_y), \qquad \dot{\varepsilon}_y = \tfrac{3}{2} k J_2 (2\sigma_y - \sigma_x) \tag{8}$$

The constant k is the one taken from the uniaxial steady creep law established for the material

$$\dot{\varepsilon} = k\sigma^n \tag{9}$$

and the creep exponent n was assumed to be 3.

When the creep strain rate $\dot{\varepsilon}_x$ is known, the difference between its value on the outer and inner faces is proportional to the rate of change of the curvature $\dot{\kappa}_x$. But this quantity can also be calculated in a simple manner from the expression assumed for the lateral deflections w, and the two calculations must yield the same result. This condition is one of the two compatibility conditions; the second is the analogous condition dealing with changes in the curvature $\dot{\kappa}_y$. The last equation needed is simply the statement that the integral of all the σ_y stresses over any section through the sandwich plate perpendicular to the y direction must be equal to the applied compressive load

$$P = 2ah\sigma \tag{10}$$

The total number of simultaneous equations to be solved is greater than would appear from this discussion. When the number of terms retained from the infinite series for the stresses and deflections is p, each equation will contain up to p trigonometric functions, each multiplied, as a rule, by some non-linear polynomial of the coefficients of the infinite series. Naturally, the equation will vanish identically only if each polynomial vanishes by itself. This fact renders the solution of the problem very laborious in the case when more than one or two terms of the infinite series representing stresses and deflections are retained.

In ref. 3, the lateral deflection w at time t was taken as

$$w = W_{11} \cos \alpha \cos \beta \tag{11}$$

and the expressions for the stresses were written as

$$\begin{aligned} \sigma_{yo} &= A_{00} + A_{11} \cos \alpha \cos \beta \\ \sigma_{xo} &= B_{11} \cos \alpha \cos \beta \\ \sigma_{yi} &= A_{00}{}^{*} + A_{11}{}^{*} \cos \alpha \cos \beta \\ \sigma_{xi} &= B_{11}{}^{*} \cos \alpha \cos \beta \end{aligned} \tag{12}$$

where

$$\alpha = \pi x/a, \qquad \beta = \pi y/b \tag{13}$$

The initial deviations from the ideal plane before load application were assumed to be

$$w_{00} = W_{00} \cos \alpha \cos \beta \tag{14}$$

Immediately upon load application these increase in consequence of the linear elasticity of the material to the value

$$w_0 = W_0 \cos \alpha \cos \beta \tag{15}$$

with

$$W_0 = W_{00} \frac{\sigma_E}{\sigma_E - \sigma} \tag{16}$$

Here, σ is the absolute value of the applied compressive stress and σ_E, the Euler stress, is the classical buckling stress of the simply supported plate in accordance with linear elasticity theory:

$$\sigma_E \simeq 3{\cdot}6E(h_{\text{real}}/a)^2 \tag{17}$$

The symbol E denotes Young's modulus of elasticity.

It was found in ref. 3 that the lateral displacement velocity $\dot{W}_{11}$ could be given in closed form as a linear function of the displacement W_{11} as long as the displacements were small. Similarly, a closed form solution was obtained for $\dot{W}_{11}$ as a cubic function of W_{11} at large values of the time t when the displacements were large. Next, it was shown that the sum of the two represented the displacement velocity at all times with sufficient accuracy for engineering applications.

The linear differential equation obtained for W as a function of t has a simple integral:

$$\frac{t}{t_E}\left(\frac{\sigma_E}{\sigma_E - \sigma}\right) = 0{\cdot}366 \ln \frac{\omega^2(1{\cdot}69 + \omega_0{}^2)}{\omega_0{}^2(1{\cdot}69 + \omega^2)} \tag{18}$$

where

$$\omega = \sqrt{3}(W_{11}/h_{\text{real}}), \qquad \omega_0 = \sqrt{3}(W_0/h_{\text{real}}) \tag{19}$$

Furthermore, the Euler time

$$t_E = \varepsilon_E/\dot{\varepsilon}_{\text{nom}} \tag{20}$$

and the Euler strain ε_E and the nominal strain rate $\dot{\varepsilon}_{\text{nom}}$ are defined by the equations

$$\varepsilon_E = \sigma_E/E, \qquad \dot{\varepsilon}_{\text{nom}} = k\sigma^3 \tag{21}$$

Equation (18) gives explicitly the time t that must lapse after load application in order to increase the non-dimensional initial deflection amplitude ω_0 to the value ω. Of course, ω_0 must be computed from the known value of $\omega_{00} = \sqrt{3}(W_{00}/h_{\text{real}})$ with the aid of eqns (14) to (16)

The critical time t_{crit} was defined as that value of t at which ω approaches infinity. It can be given as

$$\frac{t_{\text{crit}}}{t_E}\left(\frac{\sigma_E}{\sigma_E - \sigma}\right) = 0{\cdot}366 \ln\left[1 + 0{\cdot}564(h_{\text{real}}/W_0)^2\right] \tag{22}$$

The numerical coefficients in eqns (18) and (22) were calculated for that value of the side ratio b/a which yielded the highest deflection rate in the range of small deflections. If the initial imperfections of a long plate are of a random nature, that wavelength which develops most rapidly will be observed most often in experiments carried out with many specimens.

It is not known whether this wavelength will be preserved when the test is continued over such a long period of time that the deflections become large; such an assumption was made in the calculations. When the actual side ratio of a short plate is less than the value defined above, one can expect a critical time greater than that given by eqn (22).

THE EFFECT OF THE CREEP EXPONENT

In two doctoral theses written in the Department of Aeronautics and Astronautics of Stanford University, the creep buckling of simply supported rectangular plates was studied in the cases when n had the values 5 and 7. The calculations were carried out in the manner outlined above for $n = 3$. The panel characterised by $n = 5$ was investigated by Berke [5]. His creep law for plane stress was written in the form

$$\begin{aligned} \dot{\varepsilon}_x &= (9/2)kJ_2{}^2(2\sigma_x - \sigma_y) = (k/2)\sigma_{\text{equiv}}{}^4(2\sigma_x - \sigma_y) \\ \dot{\varepsilon}_y &= (9/2)kJ_2{}^2(2\sigma_y - \sigma_x) = (k/2)\sigma_{\text{equiv}}{}^4(2\sigma_y - \sigma_x) \end{aligned} \tag{23}$$

where J_2 is still given by eqn (7) and the equivalent stress σ_{equiv} is defined as

$$\sigma_{\text{equiv}}{}^2 = 3J_2 = \sigma_x{}^2 + \sigma_y{}^2 - \sigma_x\sigma_y \tag{24}$$

The lateral deflections and the stresses were written in the same manner as in eqns (11) to (14). Solution of the equations of equilibrium and compatibility in closed form was again possible for very small and very large displacements but the result obtained for large displacements appeared to be less accurate than in the case $n = 3$. Addition of the two displacement velocities and solution of the resulting differential equation yielded

$$\frac{t}{t_E}\left(\frac{\sigma_E}{\sigma_E - \sigma}\right) = 0{\cdot}172 \ln \frac{\omega^4(1{\cdot}67 + \omega_0{}^4)}{\omega_0{}^4(1{\cdot}67 + \omega^4)} \tag{25}$$

and

$$\frac{t_{\text{crit}}}{t_E}\left(\frac{\sigma_E}{\sigma_E - \sigma}\right) = 0{\cdot}172 \ln [1 + 0{\cdot}185(h_{\text{real}}/W_0)^4] \tag{26}$$

where

$$t_E = \varepsilon_E/\dot{\varepsilon}_{\text{nom}} \qquad \dot{\varepsilon}_{\text{nom}} = k\sigma^5 \tag{27}$$

The case of the seventh power creep law was treated by Honikman [6].

The creep law was written as

$$\dot{\varepsilon}_x = (27/2)kJ_2{}^3(2\sigma_x - \sigma_y) = (k/2)\sigma_{\text{equiv}}{}^6(2\sigma_x - \sigma_y)$$
$$\dot{\varepsilon}_y = (27/2)kJ_2{}^3(2\sigma_y - \sigma_x) = (k/2)\sigma_{\text{equiv}}{}^6(2\sigma_y - \sigma_x) \qquad (28)$$

The expressions for J_2 and σ_{equiv} can be found in eqns (7) and (24).

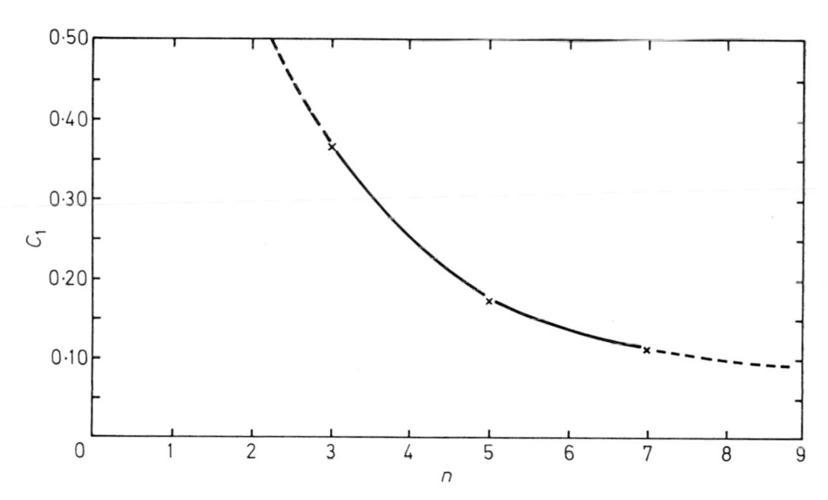

FIG. 3. Plot of C_1 versus n.

The main results of the analysis are the equations

$$\frac{t}{t_E}\left(\frac{\sigma_E}{\sigma_E - \sigma}\right) = 0{\cdot}111 \ln \frac{\omega^6(1{\cdot}38 + \omega_0{}^6)}{\omega_0{}^6(1{\cdot}38 + \omega^6)} \qquad (29)$$

and

$$\frac{t_{\text{crit}}}{t_E}\left(\frac{\sigma_E}{\sigma_E - \sigma}\right) = 0{\cdot}111 \ln\,[1 + 0{\cdot}0511(h_{\text{real}}/W_0)^6] \qquad (30)$$

where

$$t_E = \varepsilon_E/\dot{\varepsilon}_{\text{nom}}, \qquad \dot{\varepsilon}_{\text{nom}} = k\sigma^7 \qquad (31)$$

The critical times calculated for the three values of n can be presented in the unified form

$$\frac{t_{\text{crit}}}{t_E}\left(\frac{\sigma_E}{\sigma_E - \sigma}\right) = C_1 \ln\,[1 + C_2(h_{\text{real}}/W_0)^{n-1}] \qquad (32)$$

Plots of C_1 and C_2 versus n can be found in Figs. 3 and 4. The faired-in curves can be used to calculate the critical time of plates made of materials that follow the power creep law with n values different from 3, 5 and 7. In the drawing of the C_1 curve, use was made of the fact that the critical time does not exist, or, in other words, that infinitely large deformations are reached only after the passage of an infinitely long period of time, if the value of n is unity. This was shown for columns by Kempner and Pohle [7], and for plates by Lin [8].

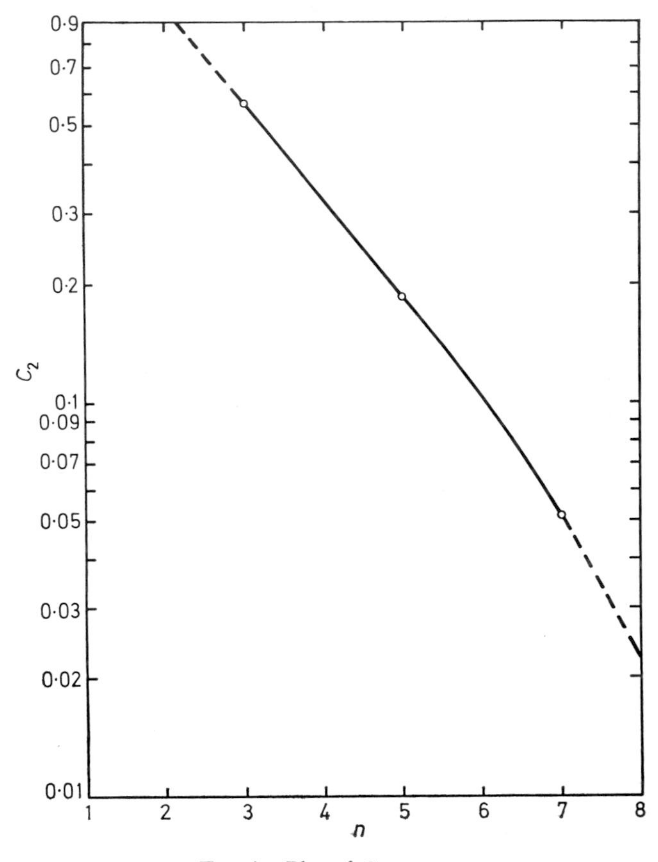

FIG. 4. Plot of C_2 versus n.

THE EFFECT OF THE SHEARING STRESSES

In ref. 3, σ_x and σ_y were taken as the principal stresses in the faces, or cover plates, of the sandwich. In reality, there are shear stresses also acting in the x–z and y–z planes and thus the direction of the principal stresses does not exactly coincide, in general, with the x and y directions. The effect of this approximation on the critical time was investigated by Levi [9] in a doctoral dissertation written at Stanford University.

In the more accurate analysis, the shear stress resultant S of ref. 3 was replaced by the shear stresses τ_{xyo} and τ_{xyi} (*see* Fig. 5). The assumption

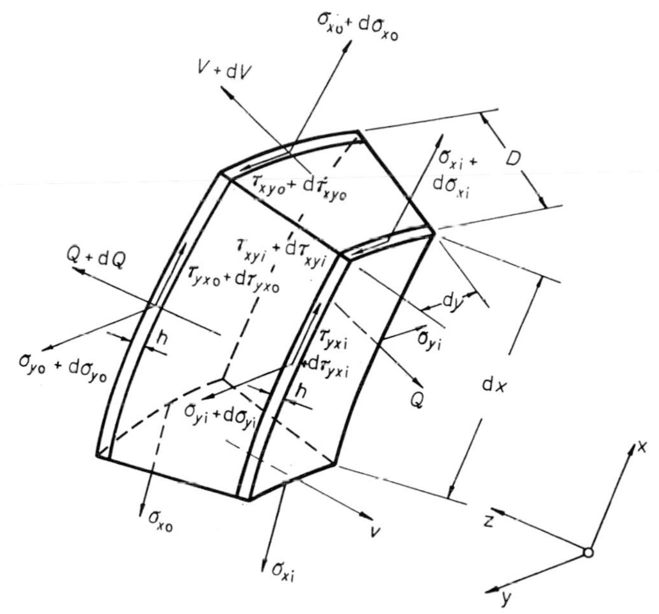

FIG. 5. Stresses and stress resultants on small idealised sandwich element for analysis of shear effect.

for w given in eqn (11) was retained and the assumptions contained in eqns (12) were augmented by adding the equations

$$\begin{aligned} \tau_{xyo} + \tau_{xyi} &= C_{11} \sin\alpha \sin\beta \\ \tau_{xyo} - \tau_{xyi} &= D_{11} \sin\alpha \sin\beta \end{aligned} \qquad (33)$$

The one new unknown introduced in this manner could be determined because a new equation became available when the equilibrium conditions were written separately for the two face plates.

The constitutive equations used earlier, eqns (8), retained their validity but the expression for the second invariant changed from that given in eqn (7) to

$$J_2 = (1/3)(\sigma_x^2 + \sigma_y^2 - \sigma_x\sigma_y + 3\tau_{xy}^2) \qquad (34)$$

The procedure used in ref. 3 was again followed and a closed form solution was obtained for the time t required to reach the non-dimensional displacement ω and for the critical time. The new solution is

$$\frac{t}{t_E}\left(\frac{\sigma_E}{\sigma_E - \sigma}\right) = 0{\cdot}368 \ln \frac{\omega^2(1{\cdot}26 + \omega_0{}^2)}{\omega_0{}^2(1{\cdot}26 + \omega^2)} \tag{35}$$

and

$$\frac{t_{\text{crit}}}{t_E}\left(\frac{\sigma_E}{\sigma_E - \sigma}\right) = 0{\cdot}368 \ln [1 + 0{\cdot}420(h_{\text{real}}/W_0)^2] \tag{36}$$

Comparison with eqns (18) and (22) shows that the numerical coefficient denoted C_1 in eqn (32) changed less than 1 per cent. Although the coefficient C_2 first introduced in eqn (32) was reduced by about 25 per cent, this coefficient appears only as one of two additive constants in the argument of the logarithmic function, and thus it influences only slightly the values of the time.

It can be concluded therefore that the simpler analysis of ref. 3 is accurate enough for engineering purposes.

NUMERICAL INTEGRATION OF THE DIFFERENTIAL EQUATION

It has already been mentioned that for both very small and very large deflections the simultaneous equations defining the problem can be reduced to a single differential equation of the first order in which the time derivative of the deflections is expressed as a function of the deflections. Summation of the functions obtained for small and for large deflections yielded an approximate expression valid for the whole range of the deflections. This simplification made it possible to integrate the differential equation in closed form.

To check the accuracy of this approximation, a computer program was worked out by Berke [5] which made possible a rigorous numerical solution of the equations without the necessity of resorting to the approximation just outlined. The equations mentioned under 'Résumé of the Earlier Solution' were rewritten in such a manner as to express all the unknowns in terms of the unknown coefficients A_{11}, B_{11}, the known quantity ω_0 and the quantity ω which is assumed to be known at the beginning of each step of the numerical procedure. In such a manner two algebraic equations

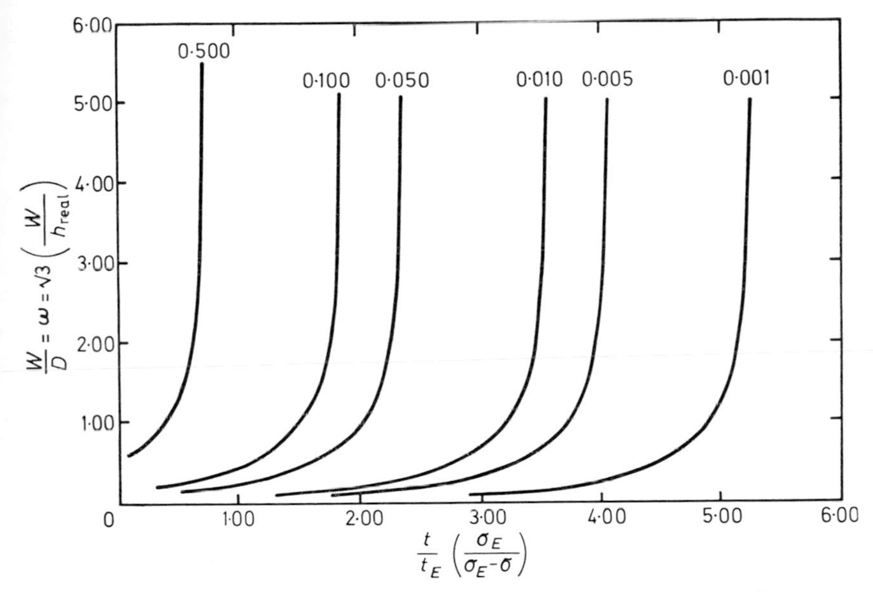

FIG. 6. Numerical solution for the midpoint deflection when $n = 3$.

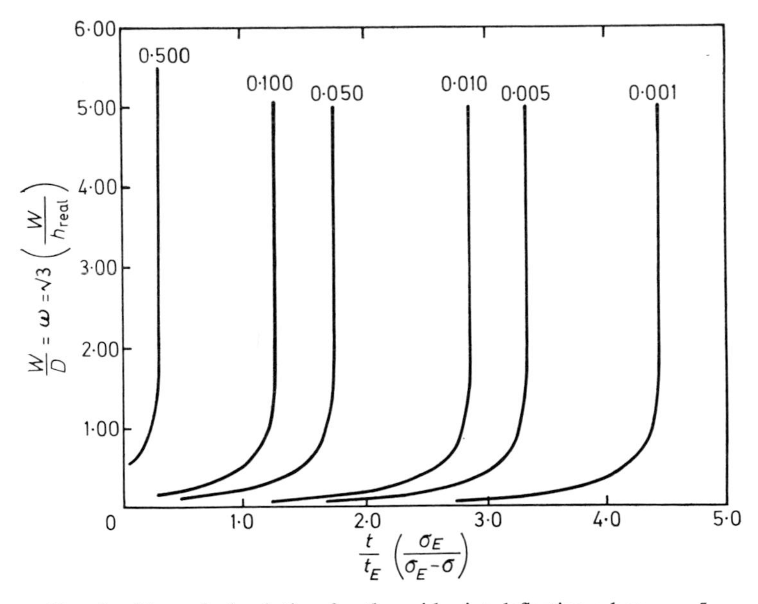

FIG. 7. Numerical solution for the midpoint deflection when $n = 5$.

(of which one is non-linear) and a first order ordinary differential equation were obtained for the three unknowns A_{11}, B_{11} and $\dot{\omega}$.

For a given value of the parameter ω_0 and a fixed value of ω, the two algebraic equations were solved for A_{11} and B_{11} by a Newton–Raphson iteration procedure. The results obtained for a sufficient range of the values of ω were then substituted into the expression for $\dot{\omega}$ which was integrated with the aid of Simpson's rule. This procedure, with the necessary modifications caused by the change in the value of n, was used by Berke [5] to obtain solutions for creep laws characterised by $n = 3$ and $n = 5$. Analogous calculations were carried out by Honikman [6] for the case $n = 7$.

The main results of this work are presented in Figs. 6, 7 and 8. They all contain curves of non-dimensional displacements ω plotted against non-dimensional time. The parameter is the non-dimensional initial deviation amplitude ω_0. Each curve shows the characteristic creep-buckling behaviour: at the beginning of the loading by the constant load the initial deviation amplitude increases very slowly, and at the end of the creep-buckling process the curve rises almost vertically. The transition between these two ranges becomes sharper with increasing values of n.

For comparison, values of the closed form solution are shown in Figs. 9 to 11. It can be seen that the difference between the closed form and the numerical solutions is negligible when $n = 3$. In this case, the closed form solution predicts very large deflections for 1 per cent and 7 per cent higher values of the time when $\omega_0 = 0{\cdot}001$ and $0{\cdot}500$, respectively, than the numerical integration. The difference increases with increasing values of n. When $n = 5$ or 7, the critical time obtainable from eqn (32) must be multiplied by about 0·9 and 0·5 in order to obtain the result of the numerical integration for $\omega_0 = 0{\cdot}001$ and $\omega_0 = 0{\cdot}50$, respectively.

An explanation for this behaviour is not difficult to find. For all values of n, the small deflection solution is accurate enough as long as the deflections are sufficiently small. But the deflection δ at which the seventh power term δ^7 alone is a sufficiently good representation of a polynomial $(1 + \delta)^7$ is much larger than the value of δ at which δ^3 represents in a satisfactory manner the expression $(1 + \delta)^3$. As a matter of fact, when $\delta = 7$, one has $\delta^3 = 343$ which is 67 per cent of $(\delta + 1)^3 = 512$. This can be a sufficiently good approximation because it was shown in ref. 3 that the lifetime is not sensitive to the behaviour at large deflections. When $n = 7$, one has to go to $\delta = 13$ to obtain a comparable accuracy (59 per cent). On the other hand, all the curves shown indicate that the buckling process becomes rapid at values of ω of 2, 3 or 4; the tangent to the curve is almost vertical,

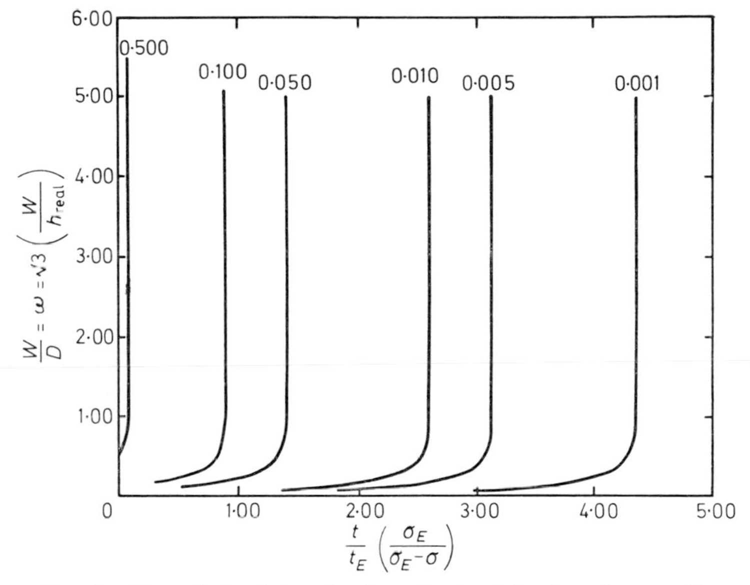

FIG. 8. Numerical solution for the midpoint deflection when $n = 7$.

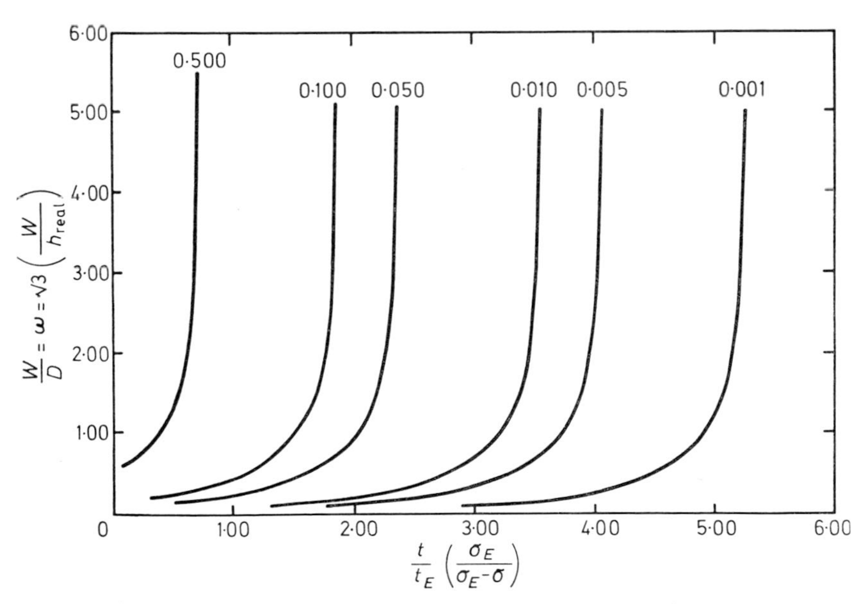

FIG. 9. Closed form solution for the midpoint deflection when $n = 3$.

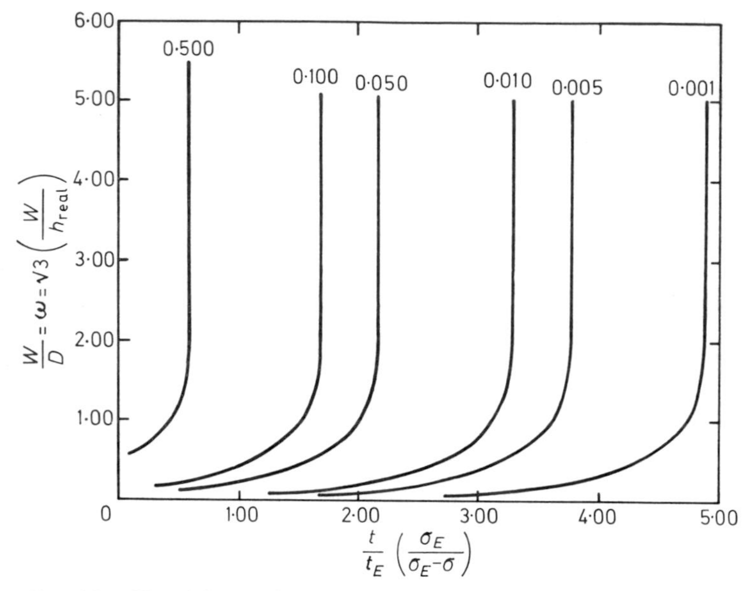

FIG. 10. Closed form solution for the midpoint deflection when $n = 5$.

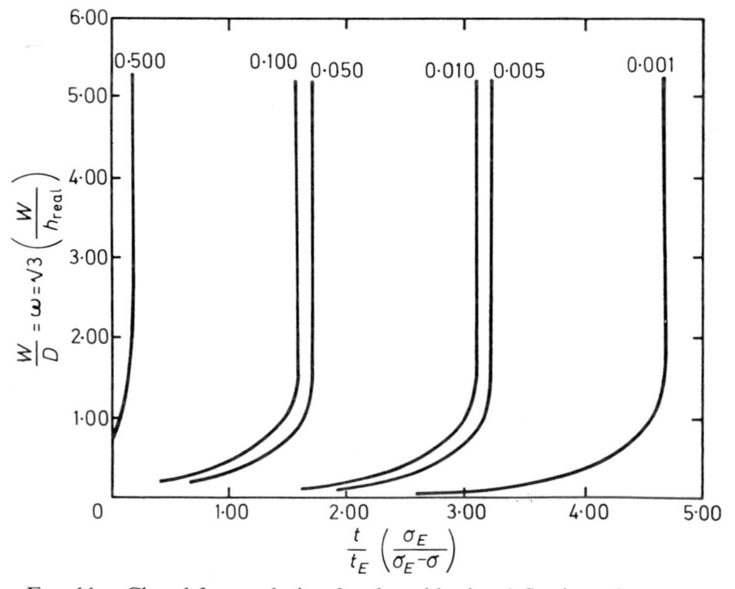

FIG. 11. Closed form solution for the midpoint deflection when $n = 7$.

and the change in time beyond these values of ω is very small. Hence, the creep-buckling process does not extend into the region of values of ω where δ^7 is a sufficiently good approximation to $(1 + \delta)^7$ (as was assumed in the calculations which resulted in the closed form solution).

In spite of these considerations, one sees that the closed form solution is satisfactory for engineering purposes even when $n = 7$ provided that ω_0 is very small. When the plate is inaccurately manufactured, it can buckle at one half or one third the critical time predicted by the closed form solution if $n = 5$ or 7.

THE EFFECT OF HIGHER HARMONICS

It was mentioned earlier that truncation of the infinite series representing the true solution of the problem introduces an unknown error in the calculations. The only reason why one can hope that the truncated series of eqns (11) and (12) will lead to an acceptable solution is past experience. First, the buckled shape of plates tested in the laboratory looks like the simple function given in eqn (11), and second, the effect of the higher harmonics was investigated theoretically by the senior author in his Wilbur Wright Memorial Lecture [10], at least for the case of the ordinary column, and was found to be unimportant.

The study of the effects of the higher harmonics was carried out by Berke [5]. He wrote, for the displacement and the stresses, the expressions

$$w = W_{11} \cos \alpha \cos \beta + W_{ij} \cos i\alpha \cos j\beta \tag{37}$$

and

$$\begin{aligned} \sigma_{yo} &= A_{00} + A_{11} \cos \alpha \cos \beta + A_{ij} \cos i\alpha \cos j\beta \\ \sigma_{xo} &= B_{11} \cos \alpha \cos \beta + B_{ij} \cos i\alpha \cos j\beta \\ \sigma_{yi} &= A_{00}^* + A_{11}^* \cos \alpha \cos \beta + A_{ij}^* \cos i\alpha \cos j\beta \\ \sigma_{xi} &= B_{11}^* \cos \alpha \cos \beta \times B_{ij}^* \cos i\alpha \cos j\beta \end{aligned} \tag{38}$$

Three combinations of values of i and j were investigated by Berke, namely

$$\begin{aligned} &\text{(a)}\quad i = 1 \quad j = 3 \\ &\text{(b)}\quad i = 3 \quad j = 3 \\ &\text{(c)}\quad i = 3 \quad j = 1 \end{aligned} \tag{39}$$

All the calculations were carried out for $n = 5$.

Because of the complexity of the non-linear equations governing the problem all the solutions were obtained with the aid of the digital computer. First, four simultaneous algebraic equations were solved for A_{11}, B_{11}, A_{ij} and B_{ij} for prescribed values of W_{11} and W_{ij}. Next, the results were substituted in expressions available for $\dot{W}_{11}$ and $\dot{W}_{ij}$ and the resulting two simultaneous first order differential equations were integrated by means of the Runge–Kutta forward integration method.

The solutions show that the effect of the higher harmonics on the development of the lateral deflections and on the critical time is small. Of the three harmonics taken into account, W_{13} was found to be slightly more important than the others.

CONCLUSIONS

The critical time of flat rectangular plates simply supported along their edges and loaded in uniaxial compression parallel to one pair of edges can be calculated from eqn (32). The values of the coefficients C_1 and C_2 can be read from Figs. 3 and 4. Strictly, the values are valid only for $n = 3$, 5, 7, but interpolation between these values and even extrapolation up to 9 and down to 2·5 should yield results acceptable for engineering purposes.

It was shown that neglect of the shearing stresses and the omission of higher harmonics did not introduce in the calculations inaccuracies of sufficient magnitude to make corrections to eqn (32) necessary. The same can be said of the simplification of the integration procedure achieved through a representation of the lateral deflection velocity of the plate as the sum of two terms, one valid for small, and the other for large deflections. However, this last statement is correct only when the creep exponent n is 3. As n increases, the inaccuracy increases and it can assume significant magnitudes when the non-dimensional initial deviation amplitude ω_0 is large. Thus, up to $n = 7$, no correction is required when $\omega_0 = 10^{-3}$, but the value of the critical time obtainable from eqn (32) should be multiplied by 0·5 when $n = 5$ or 7 and $\omega_0 = 0{\cdot}5$.

The critical stress of a perfectly elastic flat rectangular plate is not the maximal value of the stress that the plate can support; it is only the value at which large, visible deflections develop. In the same manner, a flat rectangular plate whose material creeps does not collapse and become useless when the critical time is reached. If the load is not released, it is redistributed in such a manner that the compression increases near the edges and decreases near the middle of the plate. After this redistribution the plate can support the load for a long period of time. The significance of the critical time of rectangular flat plates whose material creeps is that large, visible deflections develop when the critical value of the time is approached.

ACKNOWLEDGEMENT

A considerable part of the work reported here was done under Contract AF33 (615)–5115 with the US Air Force Flight Dynamics Laboratory, Structures Division.

REFERENCES

1. Hoff, N. J. (1958). 'A survey of the theories of creep buckling.' *Proc. of the Third US Nat. Congr. of Applied Mechanics*, p. 29. New York: American Society of Mechanical Engineers.
2. Rabotnov, Yu. N. and Shesterikov, S. A. (1957). 'Creep stability of columns and plates.' *J. of the Mech. and Phys. of Solids*, **6,** 27.
3. Hoff, N. J. (1968). 'Creep buckling of rectangular plates under uniaxial compression', p. 257. *Engineering Plasticity* (J. Heyman and F. A. Leckie (Eds.)). London: Cambridge University Press.
4. Rabotnov, Yu. N. (1964). 'Axisymmetrical creep problems of circular cylindrical shells.' *PMM*, **28** (6), 1040.
5. Berke, L. (1968). Two problems in structural stability, *Ph.D. Dissertation*, Stanford University.
6. Hoff, N. J., Honikman, T. C., Benoit, J. M. and Berke, L. 'Variation with the creep exponent, *n* of the critical time of flat plates in edgewise compression.' *US Air Force Flight Dynamics Laboratory Report* (to be issued).
7. Kempner, J. and Pohle, F. V. (1953). 'On the nonexistence of a finite critical time for linear viscoelastic columns.' *J. Aeronaut. Sci.*, **20** (8), 572–573.
8. Lin, T. H. (1956). 'Creep deflection of viscoelastic plate under uniform edge compression.' *J. Aeronaut. Sci.*, **23** (9), 883–887.
9. Levi, I. M. (1968). Creep Buckling of Plates. *Ph.D. Dissertation*, Stanford University.
10. Hoff, N. J. (1954). 'Buckling and stability.' The Forty-first Wilbur Wright Memorial Lecture. *J. Roy. Aero. Soc.*, **58** (517), 3.

Chapter 20

THE USE OF COMPUTERS IN THE CREEP ANALYSIS OF POWER PLANT STRUCTURES

C. H. A. TOWNLEY

SUMMARY

An essential step in the improvement of high temperature design methods is to establish stress and strain levels in structural components. Considerable progress has been made during the past 5 years in developing the necessary analysis techniques, using digital computers. While the results of such analysis are necessarily approximate, they are nevertheless helping to resolve many of the previous uncertainties about the behaviour of structures at high temperature. Four examples are considered: rotating discs under steady conditions, rotating discs under cyclic conditions, thick tubes under pressure and temperature transients, and butt welds in tubes.

NOTATION

σ Stress.
t Time.
T Temperature.
ε Creep strain.

INTRODUCTION

For a structure to give satisfactory service, it must be free from cracking and free from excessive distortion throughout its working life. It therefore follows that, at the design stage, predictions must be made about how the structure will respond to loads imposed on it in service, and adequate margins against failure must be demonstrated.

The integrity of a structure is largely determined by the performance of critical regions where geometrical discontinuities lead to stress and strain concentrations. Simple materials tests do not provide all the information which is needed to evaluate the design. The way in which stresses relax, and the way in which strains accumulate depend as much on the local geometry and the loading regime as on the material properties themselves.

In principle, it is possible to calculate the way in which local geometrical effects influence the stresses and strains throughout the structure, but much further development is needed before complete design calculations can be performed. Present computers are too small to deal with many of the complex shapes encountered in practice, and much more needs to be known about the deformation of materials under variable stress.

Nevertheless, considerable advances have been made in the field of creep calculations during the past 5 years. Although the computer programs are restricted to relatively simple shapes, and contain approximations to materials behaviour which are not completely realistic, they are proving of considerable value to the designer. They are helping to provide a scientific approach to creep design, and are helping to remove much of the empiricism which has previously been necessary. Four examples are given below, from investigations currently in progress at the author's laboratory which illustrate the way in which such calculations are being used to build up an understanding of the creep performance of structures. It will be noted that these investigations also serve to highlight the areas where present knowledge of materials performance is inadequate and to show the directions which future experimental work should take. This is particularly evident where cyclic loads are concerned. Most materials tests are carried out under strain controlled conditions, usually with zero mean strain. Calculations based on typical plant operating cycles show that conditions are in reality quite different. During the initial transient, stresses are usually quite large; during steady state operation, they fall to a lower level. In addition, the material in the critical region of the structure is often subjected to stress controlled, rather than strain controlled, conditions, and the mean stress is seldom zero.

ROTATING DISCS

Digital computers open up the way to rapid investigation of a wide range of fundamental problems in the design of turbine discs. With a given shape of disc, it is relatively simple to evaluate the stresses and strains brought about by a wide range of different operating regimes. Conversely, the effect of alterations in disc shape can be examined with fixed operating conditions.

The examples quoted below all relate to the first category, and apply to a parallel sided disc, with an outside diameter of 24 in (610 mm), and a bore of 4 in (102 mm) diameter. The PITT computer program was used throughout, and this is fully described in ref. 1.

The material was assumed to have a creep strain–time–stress relationship of the form

$$\varepsilon = e^{(0 \cdot 034T - 31 \cdot 1)}\sigma^{2 \cdot 3}t^{0 \cdot 34}$$

where

σ = Stress in ton/in^2

t = Time in hours

T = Temperature in °C

The rotational speed was taken to be 3000 rev/min.

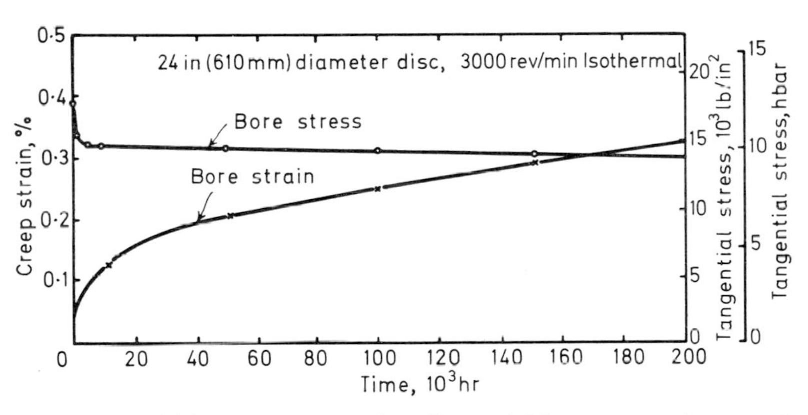

FIG. 1. Tangential bore stress versus time. Tangential bore creep strain versus time.

Figure 1 shows the bore stress and bore strain versus time relationships for a disc operating under isothermal conditions, assuming a strain hardening rule for the conversion of constant stress data to variable stress conditions. As would be expected, the bore hoop stress rapidly decreases from its initial elastic value of 17 800 lb/in^2 (12·3 hbar). After 5000 hr relaxation is virtually complete, and subsequently the stress remains almost constant at about 14 000 lb/in^2 (9·65 hbar). Bore strain rates are influenced partly by the stress redistribution, and partly by the primary creep characteristics of the material, and do not fall to relatively low values until 40 000 hr.

The calculations were carried out with all the material in the disc at a temperature of 500°C, and it is interesting to note that, provided isothermal conditions are assumed, the form of the stress–time curve is independent of temperature over a wide range.

It is also of interest to note that the almost identical relaxed stresses and cumulative creep strains are predicted at the bore when plane strain (thick disc) conditions are assumed, and when the time hardening rule is substituted for the strain hardening rule. The comparison between the different results is shown in Table 1.

TABLE 1

Creep relaxation of bore stress for 24 *in* (610 *mm*) *disc at* 3000 *rev/min isothermal conditions*

	Plane stress, strain hardening	Plane strain, strain hardening	Plane strain, time hardening
Initial elastic hoop stress at bore lb/in ^{2} (hbar)	17800 (12·3)	18500 (12·75)	18500 (12·75)
Bore hoop stress at 150000 hr lb/in ^{2} (hbar)	13900 (9·6)	15300 (10·55)	15000 (10·35)

The negligible difference in stress and strain predictions based on the two rules appears to be valid for a wide range of shapes of structure, provided the structure is subjected to constant load. On the other hand, the similarity between the plane stress and plane strain results is believed to be a feature of the disc shape.

Figure 2, line (b), shows the effect of introducing a temperature gradient into the disc. The temperature of the rim was taken to be 528°C and to decrease linearly with radius to a bore temperature of 500°C. The calculations were based on plane stress and strain hardening and in all other respects were identical to the previous example.

As would be expected, the effect of the temperature gradient is to increase the initial elastic stress at the bore, to a value of 20 800 lb/in^2 (14·3 hbar). What is at first sight surprising, is that the bore hoop stress does not relax to the same steady state value as before. The explanation can be found in line (c) of Fig. 2, which presents the results of a calculation, identical in all respects, except that the coefficient of expansion was set equal to zero. Owing to the higher temperature of the rim, the material here tends to creep faster than that at the bore, and it sheds some of its stress to the harder material. Similar considerations would apply even in the absence of a temperature gradient, where material properties varied throughout the disc, for example due to inhomogeneity.

It is often assumed by designers, particularly in the pressure vessel field, that creep is beneficial in relaxing out thermal stresses, and most

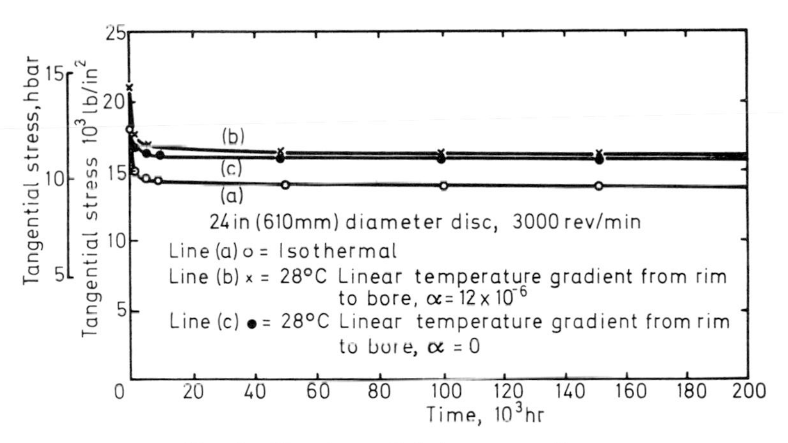

FIG. 2. Bore tangential stress versus time.

other types of secondary stress. While this statement is, in itself, true, these calculations make it quite clear that temperature differences can, at the same time, introduce other effects which are far from beneficial.

ROTATING DISCS UNDER CYCLIC LOADING

The PITT program has also been used to investigate the performance of a disc under a loading regime approximating to that which a turbine rotor experiences in service. Details of a typical loading cycle are shown in Fig. 3.

Figure 4 shows the tangential stress and the tangential strain at the bore of the disc for three successive operating cycles. Sharp peaks in stress and strain occur due to the transient conditions during start-up. However, once the transient has passed, stresses and strains fall to the levels predicted from the earlier steady load calculations. From the point of view of computation, it is therefore legitimate to superimpose the elastic

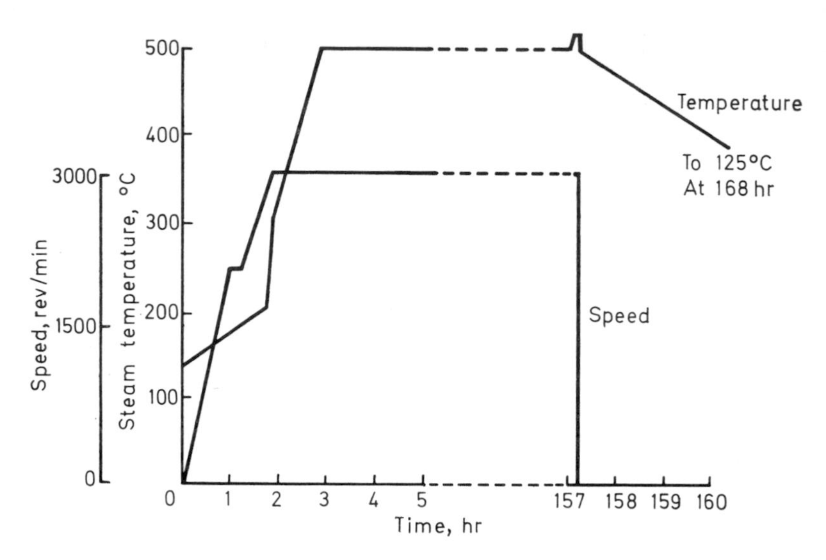

FIG. 3. Loading cycle for disc.

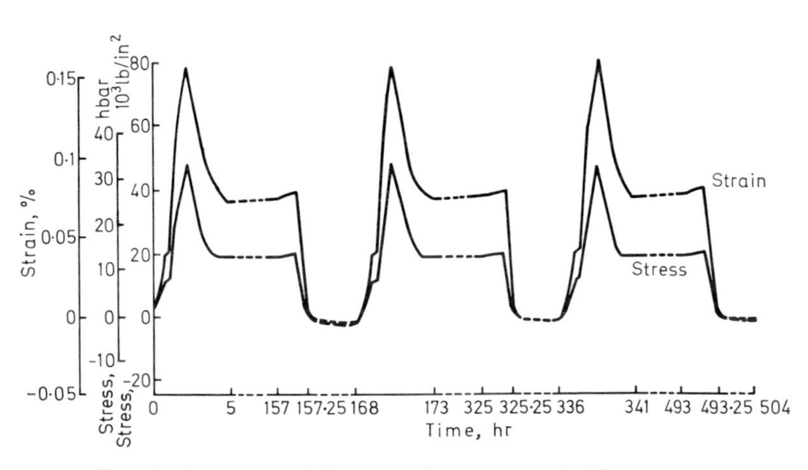

FIG. 4. Bore tangential stress and total strain, 2·75 hr start-up.

stresses at start-up and shut-down on the steady load creep curve to obtain predictions of stress levels later in the life of the structure.

The effect of an increased start-up rate has also been examined. Conditions identical to those given in Fig. 3 were assumed, except that the maximum speed was achieved after 0·15 hr instead of 1·5 hr, and the external temperature reached its steady value of 500°C after 0·25 hr instead of 2·5 hr. Because of the increased thermal transients, yielding occurs at the bore of the disc during the first start-up. However, shakedown takes place during the first cycle of loading, and thereafter all subsequent transients produce an elastic response. As far as subsequent history is concerned, stress levels can again be calculated by superposing the elastic stresses induced by the transients on the steady load creep curve. Indeed, the main effect of the plasticity during the first cycle is to accelerate the process of stress relaxation in the disc.

Very little information is available on the performance of materials at high temperature under cyclic conditions. What evidence there is tends to suggest that deformation rates are increased and that rupture lives are reduced.

Moderate increases in deformation rate will have little effect on the stress patterns in the disc, and on the long term changes in stress level. In general, they will tend to accelerate the approach to fully relaxed conditions. Large increases could lead to continued plastic cycling of the bore.

The bulk of the data on failure of low alloy steels, used in turbines, has been obtained from simple specimen tests, in which the material was cycled between fixed strain limits, with varying 'hold times' at the maximum strain levels. It is clear from Fig. 4 that strain levels are by no means constant through an operational cycle. Transient conditions during start-up lead to a peak in strain, which dies away as the disc settles down to steady operating conditions. It would be pessimistic to apply fatigue data based on the maximum strain range, and optimistic to ignore the peak completely, and this points to the need for materials tests under more realistic load cycles.

TRANSIENT BEHAVIOUR OF TUBES

The examination of tubes, subjected to pressure and temperature cycles, also provides information of general interest in the creep field. Typical results, using the PITT program, have been previously published [2]. For comparison with the calculations on rotating discs described above, the following example has been chosen.

The tube had an inner diameter of 87 in (2·03 m) and wall thickness of 6 in (152 mm). Plane strain conditions were assumed to apply, and the creep strain rate/stress relationship used was as shown in Fig. 5.

For relatively slow transients, no plasticity is involved, either during loading or unloading, but if the transients are of sufficient duration compared with the creep rate of the material some creep strain will occur under the influence of high transient stresses. Nevertheless, as in the case of the

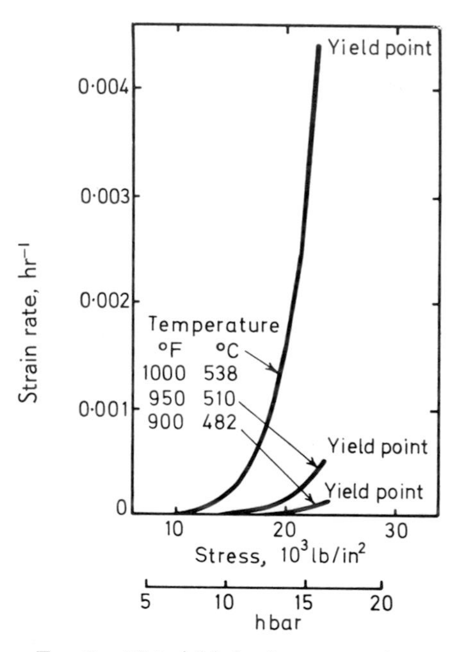

FIG. 5. Material behaviour, creep data.

turbine disc, a good approximation to long term behaviour can be obtained by superposing the elastically calculated transient stresses on the stress–time curve obtained from steady load calculations. However, this is not true of very fast transients, with high rates of temperature change, of which Fig. 6 is an example. The corresponding stress and strain history at the bore of the tube is shown in Fig. 7. It is no longer possible to make simple calculations to predict stress and strain time behaviour of the tube. A full cyclic calculation is necessary.

For comparison with the previous results, the loading regime of Fig. 6 was applied to an identical tube, using a material identical in every way,

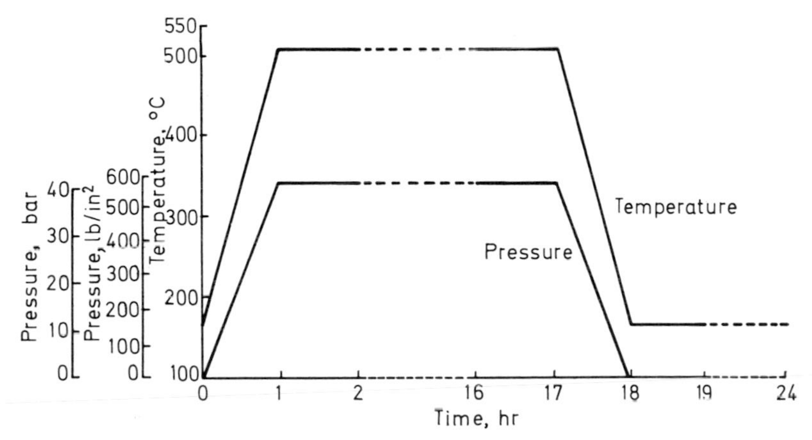

FIG. 6. Loading cycle for tube.

except that it had zero creep at all stress levels. The cyclic strain range was 40 per cent less than that in the previous example. This again shows that the assumption that creep is beneficial in relaxing secondary stresses must be treated with caution, especially where a structure is subjected to cyclic thermal loading.

As in the case of the turbine discs, the stress and strain cycles calculated for the tubes bear little resemblance to the conditions which apply in standard materials tests. However, once data are available from the computations, it is relatively simple to set up tests with more realistic loading conditions.

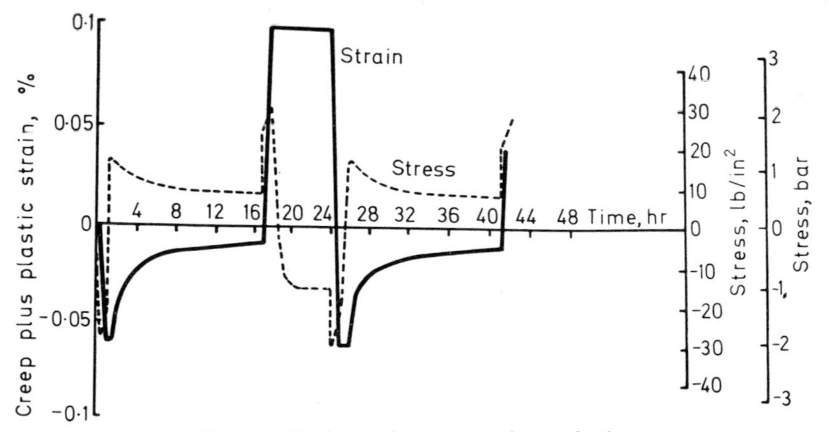

FIG. 7. Strains and stresses at bore of tube.

BUTT WELDS IN TUBES

The final example quoted in this chapter relates to another practical problem of current interest. With low and medium alloy steel piping, it is usual practice to make butt welds with filler materials which have higher creep rates and greater ductilities than the parent tube. Too large a difference in properties can lead to high stresses in the vicinity of the heat affected zone and of the weld interface, and this can be a contributory cause of failure. The aim of the calculations is to establish which weld

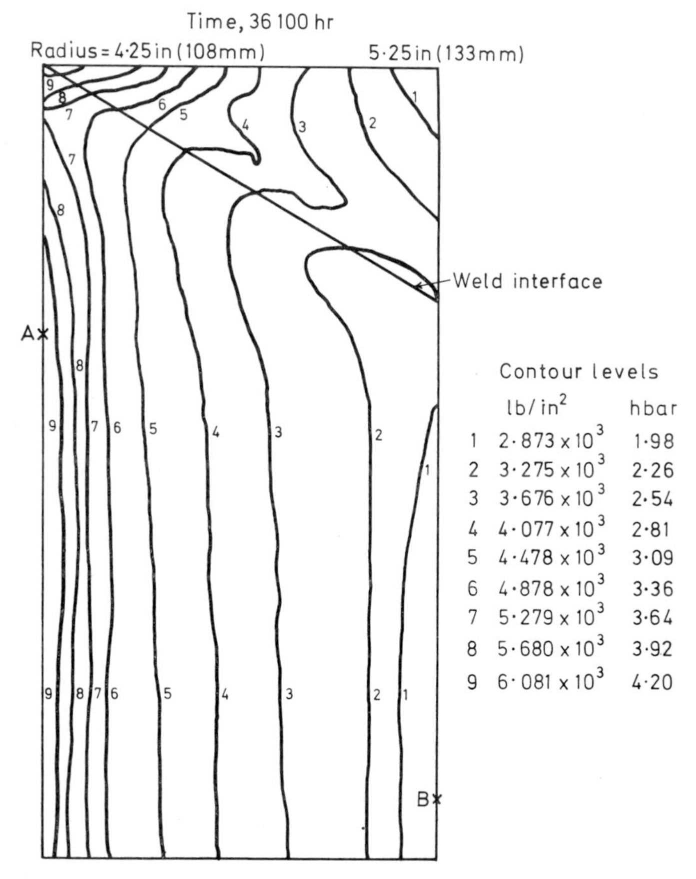

FIG. 8. Axial stresses in welded tube.

profiles, and which combinations of materials, give rise to the smallest stresses adjacent to the weld.

A finite element computer program, TESS, is being used for this investigation, and results are given below for a typical calculation.

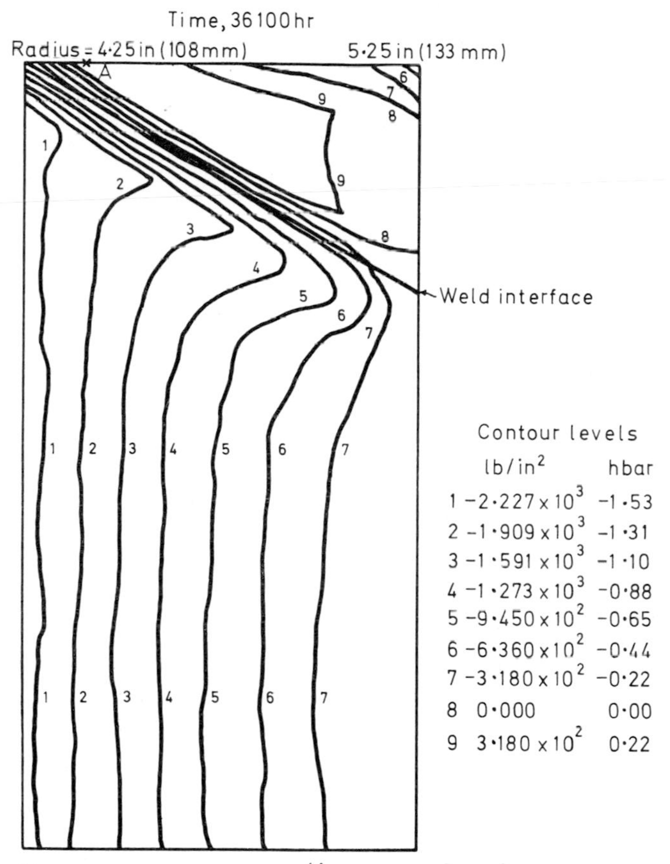

FIG. 9. Radial stresses in welded tube.

In this example, the pipe had an inner diameter of $8\frac{1}{2}$ in (216 mm) and a wall thickness of 1 in (25 mm). Isothermal conditions were assumed, at 565°C, in conjunction with an internal pressure of 2300 lb/in^2 (1·6 hbar).

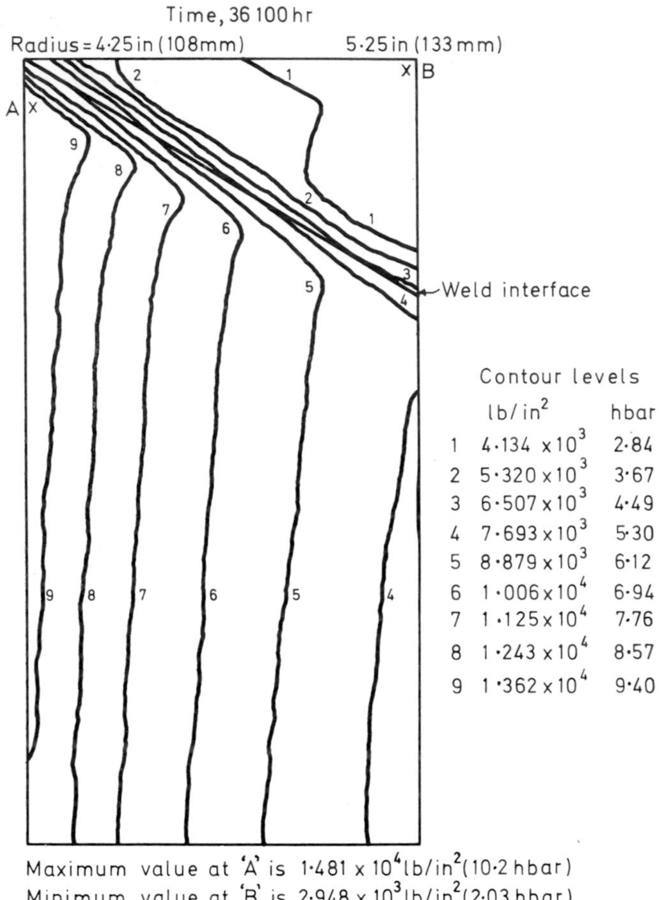

FIG. 10. Hoop stresses in welded tube.

The weld was of a 60° V form, and the creep properties were taken to be:

$$\varepsilon = 4{\cdot}11 \times 10^{-10}\,\sigma^{0{\cdot}63}t \qquad \text{for the pipe material}$$

and

$$\varepsilon = 4{\cdot}03 \times 10^{-12}\,\sigma^{1{\cdot}3}t \qquad \text{for the weld material}$$

where σ is the stress in pounds per square inch and t is the time in hours.

Since the materials have approximately the same Young's modulus, the initial stresses everywhere conform to the Lamé distribution. However, as creep proceeds, the difference in creep rates of the two materials gives rise to an additional system of stresses, necessary to maintain compatibility of the structure. The changes in the stresses local to the weld can be observed as this process proceeds. Figures 8, 9 and 10 are plots of stress contours after 36 100 hr, when the structure is approaching steady conditions.

In assessing the rupture behaviour of such welds, the properties of the parent tube and of the weld metal are clearly of little importance. The highest stresses occur in the region of the interface and of the heat affected zone, where metallurgical properties generally tend to be poor. The calculations show that data must be obtained from creep tests on composite welded specimens, loaded in such a way as to achieve the correct ratio of shear stress to direct stress.

CONCLUSIONS

The examples described in this chapter illustrate the scope of current theoretical stress and strain analysis in the creep range. Although it is still only possible to deal with simple shapes, and fully reliable constitutive equations are not yet available to describe material deformation, such analyses are proving to be of considerable value. On the one hand they provide an important insight into the interaction between geometry, loading system and materials behaviour. On the other, they serve to highlight the areas where further research is needed on materials failure criteria.

ACKNOWLEDGEMENT

This chapter, based on work carried out at Berkeley Nuclear Laboratories, is published by permission of the Central Electricity Generating Board.

REFERENCES

1. Frederick, C. O., Chubb, E. J. and Bromley, W. P. (1966). 'Cyclic loading of a tube with creep, plasticity and thermal effects.' *Appl. Mech. Convn* (1966). *Proc. Instn Mech. Engrs*, **180** (3I), 448–461.
2. Townley, C. H. A. and Chubb, E. J. (1967). 'Computation of strain history of a thick tube during thermal and pressure cycling.' *Proc. Internat. Conf. on Thermal and High Strain Fatigue*, 15–27. London: Institute of Metals.

BIBLIOGRAPHY

1. Crossland, B. (1965). 'Torsion testing machines.' *Proc. Instn Mech. Engrs*, **180** (Pt 3A), 243–254.
2. Penny, R. K. and Leckie, F. A. (1968). 'The mechanics of tensile testing.' *Int. J. Mech. Sci.*, **10** (4), 265–273.
3. Johnson, A. E. (1950). 'A high sensitivity torsion creep unit.' *J. Sci. Instrum.*, **27,** 74–75.
4. Finnie, I. (1958). 'Creep buckling of tubes in torsion.' *J. Aero. Sci.*, **25,** 66–67.
5. Ohji, K. and Marin, J. (1963). 'Creep of metals under non-steady conditions of stress.' *Thermal Loading and Creep in Structures and Components. Proc. Instn Mech. Engrs*, **178** (Pt 3L), 126–134.
6. Warren, J. W. L. (1966). 'A survey of the mechanics of uniaxial creep deformation of metals.' *Aero Res. Counc. C. P.* No. 919. London: HMSO.
7. Everett, F. L. (1931). 'Strength of materials subjected to shear at high temperatures. *Trans. Amer. Soc. Mech. Engrs*, **53,** APM-53-10.
8. Skelton, W. J. and Crossland, B. (1967). 'Results of high sensitivity tensile creep tests on a 0·19 per cent carbon steel used for thick-walled cylinder creep tests.' *Conf. High Pressure Engng. Proc. Instn Mech. Engrs*, **182** (Pt 3C), 151–158.

SECTION IV

REVIEW OF A. E. JOHNSON'S RESEARCH

Chapter 21

THE CONTRIBUTION OF DR A. E. JOHNSON TO THE DEVELOPMENT OF ENGINEERING DESIGN FOR CREEP

A. I. SMITH AND J. HENDERSON

The work on which Dr Johnson was engaged for over a quarter century has been internationally recognised as a unique and significant contribution to the determination of the basic relationships between complex stress, strain, temperature and time, essential to the development of a theory for the rational creep design of components and structures. Dr Johnson's main objective was the establishment of laws of deformation and fracture under multiaxial stress systems, including time dependent (creep) behaviour, time independent plastic strain relations, creep under changing stress conditions, creep relaxation, and the effect of vibrational stress and creep prestrain. His extensive publications between 1931 and 1965 provide an interesting résumé of the historical progress of experimental and theoretical knowledge of creep and its importance in high temperature engineering design.

EARLY WORK ON UNIAXIAL CREEP

When Dr Johnson joined the Engineering Division, National Physical Laboratory, in 1929, he became Assistant to H. J. Tapsell, who had commenced creep testing at NPL about 1920 (and whose influence largely determined the lines that British work in this field were to follow for the next quarter century). Advances in temperatures in steam power plant and chemical plant resulted in creep becoming technically important in the 1920s and Dr Johnson's earliest papers, published with Tapsell as co-author, reported the results of some of the first systematic determinations of creep properties of engineering materials for use at temperatures at which creep might be significant.

In 1931, these authors published the results of creep tests [1], undertaken on behalf of the British Electrical and Allied Industries Research Association (BEAIRA) on a 0·3 per cent C cast steel for steam chests and 0·4 per cent C forged disc steel 'chosen as representative of those used in construction of steam turbines'. These steels had given satisfaction up to 400°C, but in view of the requirement to increase working temperatures to achieve higher thermal efficiency, their creep properties above 400°C were required. The longest tests were only of a few thousand hours' duration but they extended over a range of stress and temperature in the modern manner. Similar programmes of tests were undertaken on copper alloys for the British Non-Ferrous Metals Research Association [2], and a series on various steels and non-ferrous alloys for the chemical industry [3].

In 1930, NPL commenced what was to prove to be a 20 years' study of the creep properties of carbon steels, and Dr Johnson was one of the investigators associated with this notable research, again undertaken on behalf of BEAIRA. In a paper presented to the 1936 World Power Conference [5], the causes of scatter of creep properties and of the abnormally high creep rates of certain carbon steels were discussed; the improvement in creep properties afforded by steels containing $\frac{1}{2}$ per cent Mo was illustrated, but the tendency of the latter steels to rupture with intercrystalline cracking at high testing temperatures was emphasised.

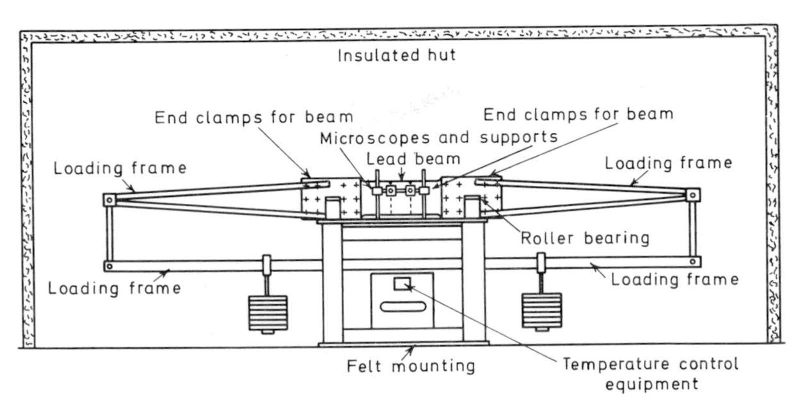

FIG. 1. Apparatus for creep testing lead beam in bending. (From ref. 4.)

PIONEER WORK ON COMPLEX STRESS CREEP

In 1935, Tapsell and Dr Johnson published the results [4] of one of the earliest experimental studies (1932) of the creep deformation of a component having a stress gradient across its section—in this instance, a beam under uniform bending moment, Fig. 1. For experimental simplicity, a lead beam was tested at room temperature. It was established that plane sections remained plane during creep, that redistribution of stress occurred relatively quickly in these tests, and that the creep deformation of the beam could be calculated from simple tensile creep data. In 1940, these authors published a paper [6] which was a milestone in Dr Johnson's career—the beginning of his studies of creep under multiaxial stress systems. The work

FIG. 2. View of complex stress creep laboratory at NEL showing tension/torsion test units.

had commenced several years earlier, and the experimental technique of using combined tension and torsion creep tests on thin wall tubular test pieces was based on the test approach used by Taylor and Quinney† in time independent tests (1931); in an appendix to this paper, Dr Johnson presented the theoretical analysis of the experimental results. This classic paper marked the beginning of a long time basic research, the object of which was the development of a comprehensive theory of creep deformation and fracture under three-dimensional stress systems. During the following two decades, Dr Johnson's was the only laboratory, Fig. 2, in the world in which a continuing programme of systematic research in this field was pursued.

Although by 1941 the special demands of the war were changing, to some degree, the emphasis in Dr Johnson's work, in that year he published (with Tapsell as co-author) the first comprehensive presentation [7], for design purposes, of 'long time' creep properties of power plant steels, based on the results of a series of 3 year tensile creep tests on several carbon steels. This paper was followed by the results of the first detailed experimental and analytical study [8] of recovery in tensile creep; this research was undertaken to explore the partial recovery of primary creep strains during periods of unloading.

TURBINE DISC RESEARCH

The importance of aircraft development in wartime is reflected in a joint paper (1942) with Cox [9], for the Aeronautical Research Council, on the theory of torsion of a symmetrical frustrum of a regular tetrahedron, and a paper [10] with Schofield (1943) for the Ministry of Aircraft Production on the effect of age hardening on creep of duralumin at 100° to 200°C. In 1941, the Ministry of Aircraft Production appointed a Disc Panel to study the material properties and stressing of jet engine discs, and Dr Johnson was Secretary of this Panel from 1941 to 1949. He undertook much of the experimental and analytical work for it, and submitted a series of reports which were later published, in summarised form [43], in 1956.

The Disc Panel considered, in detail, the problems of disc design, including methods of calculating elastic stresses, and the effects of plastic

† Taylor, G. I. and Quinney, H. (1931). 'The plastic deformation of metals.' *Phil. Trans. R. Soc.*, **230**, 323–362.

strain, due to yielding and to creep, on stress distributions in the disc. A large volume of experimental data on the properties of early jet disc materials was determined, and included in the programme of work were some of the earliest tests on 'fir tree' blade root fixings under creep conditions, Fig. 3; the latter tests were undertaken by Dr Johnson at NPL. These reports provided the analytical bases of design of the turbine discs of early British jet engines.

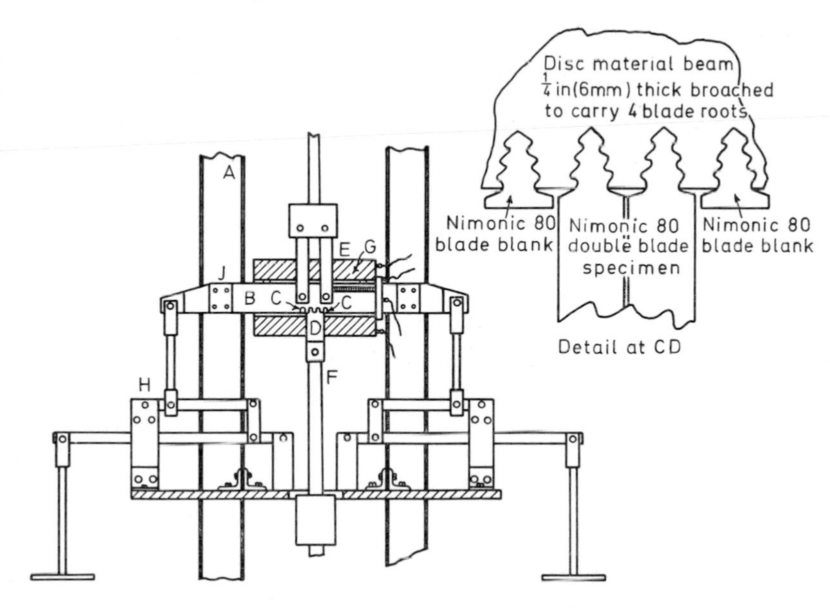

FIG. 3. Apparatus for combined tensile and hoop stress creep tests on blade roots. (From ref. 5.)

A. Framework of existing 5 ton (50 kN) creep unit; B. Beam of disc material 19 in (480 mm) × $2\frac{3}{4}$ in (70 mm) × $\frac{1}{4}$ in (6 mm) broached centrally on lower edge to carry 4 blade roots; C. Blade root blanks (Nimonic 80); D. Double blade root specimen in Nimonic 80; E. Double link adaptor; F. Existing creep M/C adaptor; G. Furnace with platinum resistance control; H. Lever loading mechanism magnifying 21:1; J. Joint to loading mechanism.

WORK ON STEAM POWER PLANT MATERIALS

At the end of the war, interest swung back to materials for high temperature use in power stations, and Dr Johnson undertook several further investigations for the BEAIRA. Designers required a reliable method of estimating long term properties from short term test data, and during the previous two decades, many short time creep limits had been proposed by workers in the creep field, but the validity of these 'creep limits' as indications of long time properties was in doubt. In 1945, Dr Johnson and Tapsell published a detailed assessment [11] of these short time measures of creep strength in relation to the long time creep properties of a series of carbon steels, followed by a similar paper in 1948 on a series of molybdenum. steels [14]; the limitations of all the short time criteria were demonstrated.

In an exhaustive study, also for the BEAIRA, of the tensile stress relaxation properties of carbon steels [22] and a chromium–molybdenum steel [23], it was found that, for the particular test condition and type of material, creep recovery did not have a highly significant influence in the stress relaxation range tested. From the results of this investigation, Dr Johnson concluded that the stress relaxation curves were not directly predictable from creep curves on the basis of either a time hardening or a strain hardening theory.

In 1951, as a member of the team of engineers and metallurgists who had extensively studied the high temperature properties of carbon steels, Dr Johnson was one of the authors of a voluminous report [28], submitted to the BEAIRA, on the effects of manufacturing variables and heat treatment on creep of carbon steels. An abridged version of this report was subsequently published by the Iron and Steel Institute†.

Although Dr Johnson's attention was increasingly turning to his basic research on creep under multiaxial stress systems, he continued to publish occasional papers on special aspects of conventional properties. For example in 1948 he published the results [15] of an experimental study, sponsored originally by the Ministry of Aircraft Production, of the stresses and strains, under repeated cycles of uniform bending moment, of a magnesium alloy at room temperature. In 1953, he reported [36] (with Frost as co-author) the results of an investigation of the effects of cyclic stress and temperature on the creep properties of several power plant steels; it was demonstrated that fair predictions of the effects of minor superimposed cycles of stress or temperature could be obtained on the fractional life hypothesis, but the effects of large fluctuations (*e.g.* 50 per cent overstress or 50 deg F overheating) could not be adequately predicted from the existing knowledge of creep properties.

† Jenkins, C. H. M. and Tapsell, H. J. (1952). 'Factors influencing the creep resistance of wrought carbon steels.' *J. Iron & Steel Inst.*, **171**, 359–371.

PIPE FLANGES RESEARCH

Dr Johnson undertook considerable work for the Pipe Flanges Research Committee of the Institution of Mechanical Engineers, Fig. 4, the results of which were published in 1955; in this, the third and Final Report [41] of the Committee, comprehensive analyses of the stress distribution and

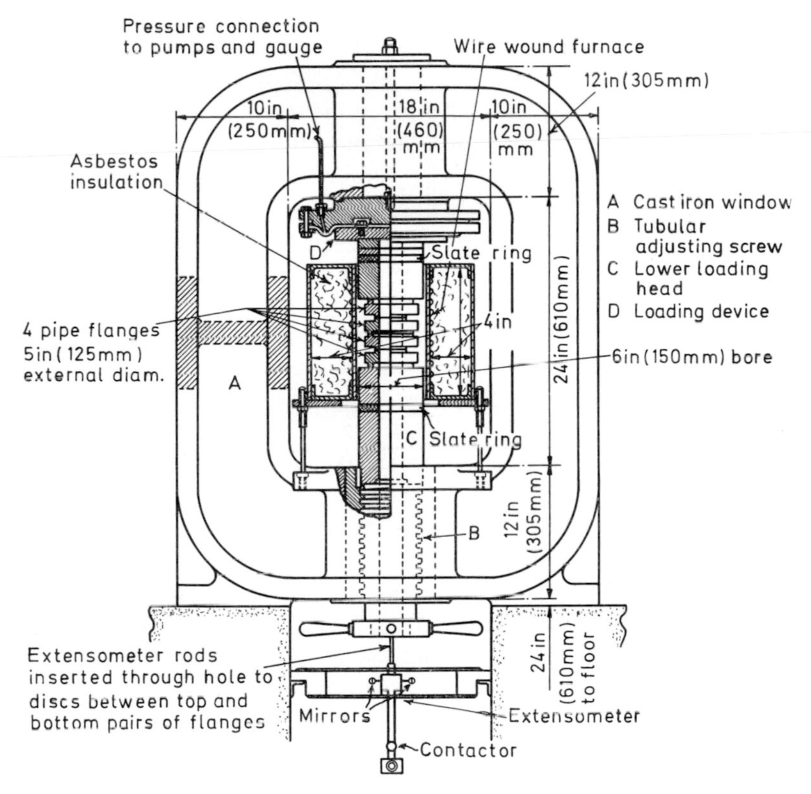

FIG. 4. Apparatus for creep relaxation tests on model pipe flanges. (From ref. 41.)

creep deformation of flanged joints were presented. The three reports of the Pipe Flanges Research Committee provided the main technical basis for the design of BS bolted flanged joints in high temperature pipe systems. In 1946, he had published a theoretical treatment [12]—more rigid than earlier attempts—of the deflection of a thick plain ring under a uniformly distributed couple; expressions were derived for radial and axial deflections, and for the axial, shear, radial and circumferential stresses. This analysis was directly relevant to the work of the Institution of Mechanical Engineers Pipe Flanges Research Committee.

COMPLEX STRESS CREEP DEFORMATION RESEARCH

In the last 15 years of his life, Dr Johnson concentrated, almost completely, on his work on multiaxial creep—'complex stress' creep, as he called it. The programme of combined tension and torsion creep strain tests on an isotropic carbon steel, which had been interrupted by wartime activities, was completed and the results published in 1949 [16], together with results of a similar programme on a cast aluminium alloy [18]; results for a magnesium (2 per cent Al) alloy [24] and Nimonic 75 [30] followed in 1950 and 1951, respectively. Criteria for the onset of plastic yielding and primary creep were established and principal creep rate stress equations derived, Fig. 5. In general, at low and moderate stresses, the creep rate stress equations were of the St Venant, von Mises type, involving a simple power function of stress. However, at high stresses, anisotropy tended to develop, possibly due to plastic strain on loading. All creep obeyed, quite closely, the von Mises criterion of plastic flow.

In Dr Johnson's complex stress work, the experimental technique was to use combined tension and torsion tests on thin walled tubular specimens at elevated temperature, together with the two limiting cases of this type of test: pure torsion and pure tension, Fig. 2. This method of investigating the effect of general stress systems depends on the validity of the assumption that the effect of the addition or subtraction of a suitable hydrostatic stress on any complex (three dimensional) stress system produces an equivalent combined tension and torsion stress system which would have similar creep characteristics. In addition, it was assumed that isotropy prevailed—that the principal axes of stress and creep rate remained coincident, at least for moderate creep strains, and that volumetric changes due to creep were negligible.

In his complex creep research, Dr Johnson's policy was to choose test materials either because they were representative of a class of high temperature materials, or because they exhibited a particular pattern of creep behaviour. For example, in tests at room temperature, magnesium gave creep curves of similar shape to those of many heat resistant alloys at high temperature, thus offering a simple means of studying the creep of model structures, Fig. 6.

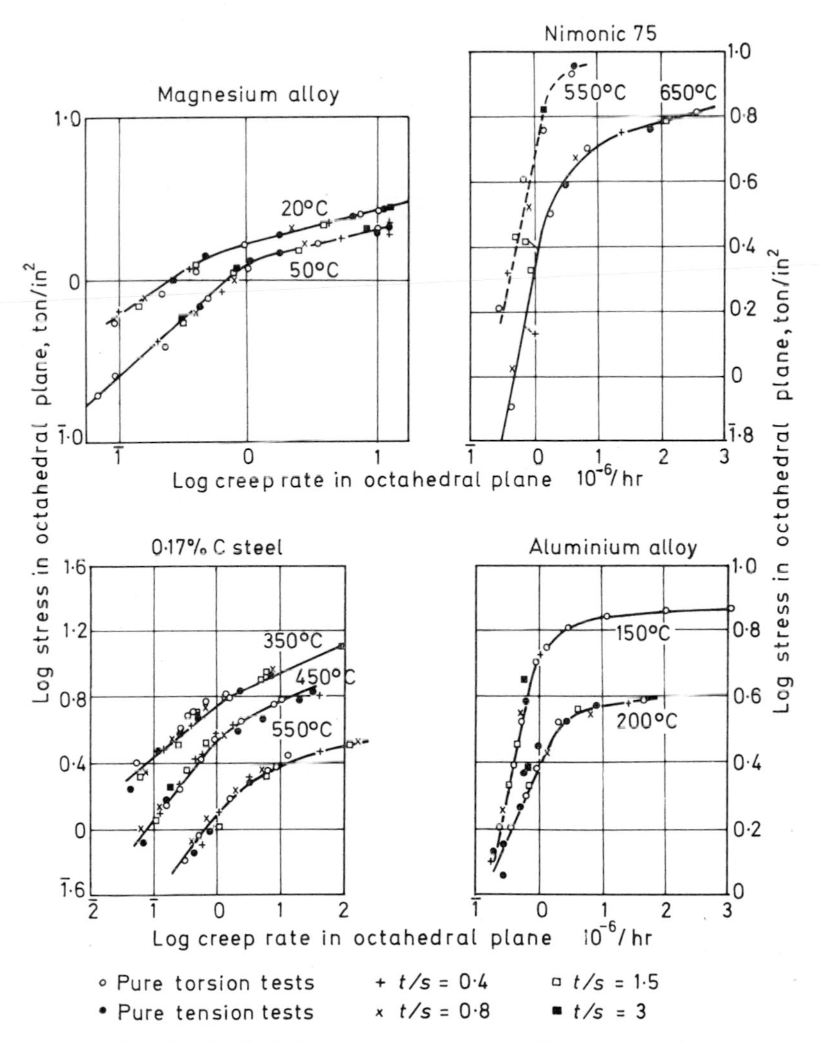

FIG. 5. Octahedral stress versus creep rate for four materials.

Dr Johnson's experimentally confirmed complex stress primary creep equation represented the multiaxial counterpart of the well established uniaxial creep rate equation

$$C = A\sigma^n t^m$$

A particular advantage of this equation was its relatively simple transformation to the multiaxial form. Thus the constants derived from uniaxial

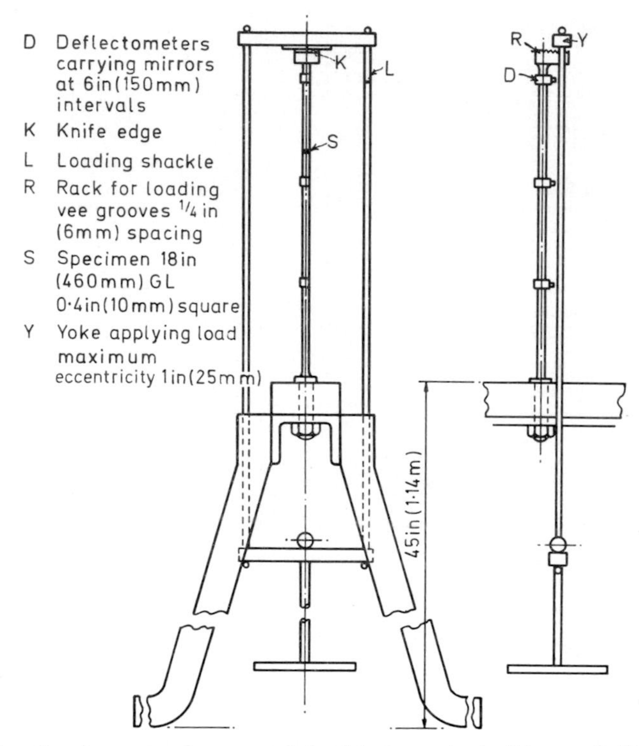

FIG. 6. Apparatus for eccentric load tests on struts. (From ref. 47.)

creep tests on a given material and at a particular temperature could readily provide the corresponding complex stress equation for these conditions.

In 1952, Dr Johnson changed his form of presentation of the creep rate equation from that involving either the octahedral stress or shear stresses to the tensor notation (see p. 168):

$$C_{ij} = F(J_2)S_{ij}\varphi(t)$$

where S_{ij} denotes the components of the stress deviation tensor, J_2 is the second invariant of the stress deviation tensor, and $\varphi(t)$ is a function of time.

Among the needs to employ the tensor notation, apart from its shorthand convenience, were that such presentation made for ready comparison with the increasing number of theoretical approaches that were being proposed internationally. The major reason, however, was that Dr Johnson wished to emphasise that the primary creep equation satisfying his results represented a particular case of the general relation and expression proposed by Prager† for plastic deformation and which could be reasonably expressed only in tensor form.

COMPLEX STRESS CREEP FRACTURE RESEARCH

Up to 1955 only low deformation creep strains under multiaxial stress systems had been considered by Dr Johnson, these being of comparable magnitude to the elastic loading strains. In many high temperature components, however, such as steam pipes, considerably higher creep deformations are permissible, and Dr Johnson therefore turned his attention to the study of tertiary creep and fracture. The first metal tested to fracture under combined stresses was a $\frac{1}{2}$ per cent Mo steel [42], and this series of tests was undertaken specifically to check the hypotheses of Siegfried‡ that below the equicohesive temperature the criterion of rupture is the stress deviator, but above that temperature the hydrostatic stress component is the criterion of failure. The tests proved Siegfried's hypothesis invalid since the critical criterion was proved to be the maximum principal stress. Tests on a commercially pure copper [42] supported these findings, Fig. 7a. However, corresponding researches on an aluminium alloy RR 59 [50] revealed an octahedral stress criterion for creep fracture time, Fig. 7b. Systematic study of these and other materials at elevated temperature enabled Dr Johnson to conclude that the appropriate stress criterion of failure was related to the type of cracking preceding rupture: materials exhibiting general cracking failed in a given time at a critical value of the maximum principal stress, whereas those which failed with the development of a single crack were governed by the octahedral stress.

† PRAGER, W. (1945). 'Strain hardening under combined stresses.' *J. Appl. Phys.*, **16**(12), 837–840.

‡ SIEGFRIED, W. (1943). 'Failure from creep as influenced by the state of stress.' *J. Appl. Phys.*, **10**(4), 202–212.

Dr Johnson continued to investigate the primary creep range for the more complicated condition of complex stress creep under changing stress systems, since in a component in service the loading and therefore the ratio of principal stresses may change with time. Several metals were investigated in this work to test the validity of existing theories for these conditions. The experimental technique consisted of adding increments of torsional stress at 168 hr intervals, in combined tension and torsion tests,

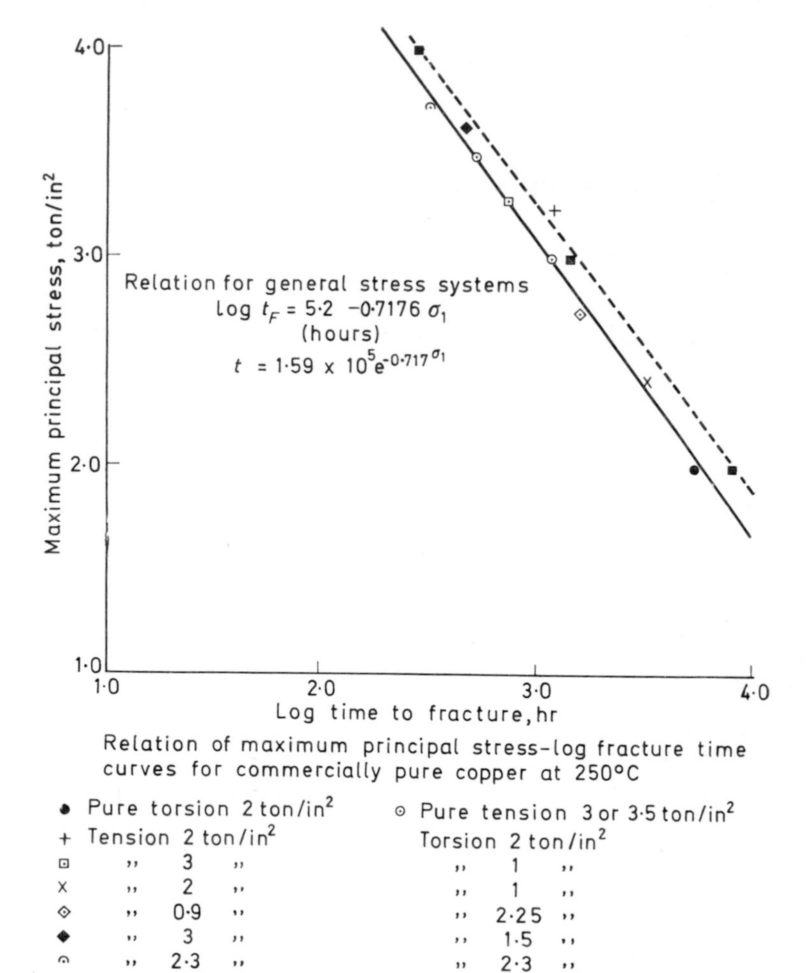

FIG. 7(a). Maximum principal stress versus creep fracture time for copper.

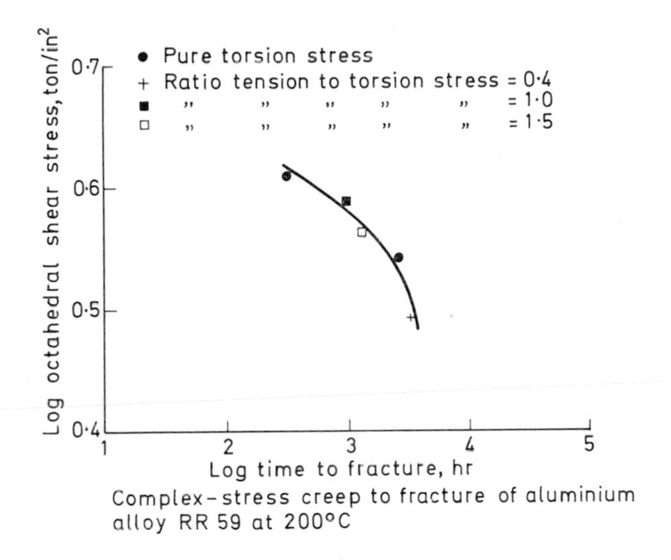

FIG. 7(b). Octahedral shear stress versus creep fracture time for aluminium.

and observing the corresponding axial and shear creep strains. The results were adequately represented by the inclusion of a strain history term in the stress function of the equation for simple loading conditions:

$$C_{ij} = [AF(J_2) - f(I_2^{\frac{1}{2}})]S_{ij}t^m$$

where I_2 is the second strain invariant.

In the field of complex stress creep fracture a further complication is the effect of vibratory or fatigue stressing [53] on life to fracture, and Dr Johnson's experimental study revealed that, depending on the material and test conditions, superimposition of vibratory stress might either reduce or increase the time to rupture.

In certain practical applications, material is subjected to conditions approximating more closely to constant strain, rather than constant stress (load) conditions, and the stress is reduced as creep strain replaces part of the original elastic strain. Again, Dr Johnson was the first to undertake a systematic investigation of stress relaxation under multiaxial stress, and he found that an age hardening theory adequately fitted his results [48].

CREEP OF COMPONENTS

In 1961, Dr Johnson considered that his laws of primary/secondary creep under conditions of multiaxial stress were sufficiently well established to justify their application to calculation of the creep deformation of simple components, or model structural elements typical of the shapes and materials used in high temperature engineering. There then followed a

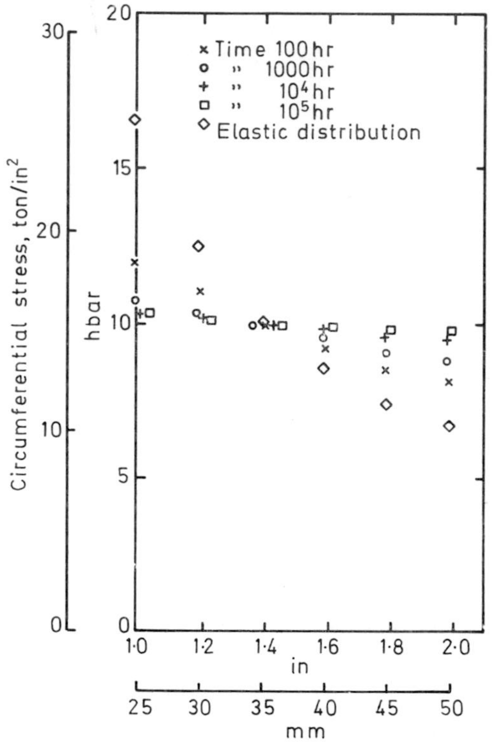

FIG. 8. Redistribution of stress in thick cylinder. Inner radius 1 in (25 mm); outer radiuf 2 in (50 mm); internal pressure 10 tons/in^2 (1·54 hbar).

series of papers on the creep deformation of various components, including tubes under internal or external pressure [51]; thick walled spheres under internal or external pressure [54], solid bar or thick walled tube under combined bending, torque and axial load [57]; thick walled spherical [61] and cylindrical pressure vessels [63] subjected to pressure and radial thermal gradient; and circular bars under torque [65]. A unique feature of these analyses was that they predicted the stress changes with time, from the initial condition to the completely redistributed state, Fig. 8.

In 1962, Dr Johnson and his collaborators in later years (J. Henderson and B. Khan) summarised the results of more than 20 years' study of creep, stress relaxation and rupture under multiaxial stresses, in a monograph [55] published by HMSO; this, and a review [52] of work on this subject in British and overseas laboratories are likely to remain major references in this field for many years. His experimental programme was continued, with special emphasis on the effect of changing stress systems [71], and on fracture under multiaxial stress [69, 70], and he was still actively engaged on these at the time of his death and had drafted, only a few months previously, his research programme for the following 5 years.

SIGNIFICANCE OF DR JOHNSON'S WORK

Dr Johnson's research is a classic example of the achievements of an individual scientist who, with only one or two assistants, concentrated on pioneer work in a highly specialised field which presented considerable theoretical and experimental difficulties. His long series of published papers have gained recognition in many countries, and his techniques and approach have been emulated by several workers overseas. Although the assimilation of his results into design procedures will take some years, his work constitutes one of the few major contributions to the advancement of knowledge of creep in relation to design of components and structures for elevated temperature service.

ACKNOWLEDGEMENT

This chapter is published by permission of the Director, National Engineering Laboratory, Department of Trade and Industry.

BIBLIOGRAPHY OF DR A. E. JOHNSON'S PAPERS

1. TAPSELL, H. J. and JOHNSON, A. E. (1931). 'The strength at high temperatures of a cast steel as used for turbine construction.' *Dept. Sci. Industr. Res. Engng Res.* Special Report No. 17. London: HMSO. Reviewed in *Engineering* (1932), **133**(3448), 191–192.

2. TAPSELL, H. J. and JOHNSON, A. E. (1932). 'The properties of copper in relation to low stresses. Part II, Creep tests at 300 and 350°C of arsenical copper and silver arsenical copper.' *J. Inst. Metals*, **48**(1), 89–96.
3. TAPSELL, H. J., JOHNSON, A. E. and CLENSHAW, W. J. (1932). 'Properties of materials at high temperatures. 6 The strength at high temperatures of six steels and three non-ferrous materials. *Dept. Sci. Industr. Res. Engng Res.* Special Rept. No. 18. London: HMSO.
4. TAPSELL, H. J. and JOHNSON, A. E. (1935). 'An investigation of the nature of creep under stresses produced by pure flexure.' *J. Inst. Metals*, **57**(2), 121–140.
5. JENKINS, C. H. M., TAPSELL, H. J., MELLOR, G. A. and JOHNSON, A. E. (1936). 'Some aspects of the behaviour of carbon and molybdenum steels at high temperatures.' *Trans. Chem. Engng. Cong. World Pwr. Conf.*, **1,** 122–162.
6. TAPSELL, H. J. and JOHNSON, A. E. (1940). 'Creep under combined tension and torsion. Part 1 The behaviour of a 0·17 per cent C steel at 455°C.' *Engineering, Lond.*, **150,** 24–25, 61–63, 104–105, 134, 164–166.
7. TAPSELL, H. J. and JOHNSON, A. E. (1941). 'Properties of steels as a basis for design for high temperatures. Part 1 Carbon steels.' *Proc. Instn Mech. Engrs*, **144**(3), 97–106.
8. JOHNSON, A. E. (1941). 'The creep recovery of an 0·17 per cent C steel.' *Proc. Instn Mech. Engrs*, **145,** 210–220.
9. COX, H. L. and JOHNSON, A. E. (1942). 'Torsion of a symmetrical frustrum of a regular tetrahedron.' *ARC Report Strut.* 597. Teddington, Middlesex: Aeronautical Research Council.
10. SCHOFIELD, T. H. and JOHNSON, A. E. (1943). 'The effect of heat treatment on the creep properties of Duralumin.' *ARC Report A*.436. Teddington, Middlesex: Aeronautical Research Council.
11. JOHNSON, A. E. and TAPSELL, H. J. (1945). 'A comparison of some carbon steels on the basis of various creep limits.' *Proc. Instn Mech. Engrs*, **153,** 169–181. (Emergency War Issue No. 6.)
12. JOHNSON, A. E. (1946). 'The deflection of a thick plain ring under a uniformly distributed couple.' *Proc. Instn Mech. Engrs*, **155,** 101–111. (Emergency War Issue No. 16.)
13. British Intelligence Objectives Subcommittee (1946). 'Report of a visit to Germany and Austria to investigate alloys for use at high temperatures.' *BIOS Report* No. 396, London: HMSO.
14. JOHNSON, A. E. and TAPSELL, H. J. (1948). 'A comparison of some carbon molybdenum steels on the basis of various creep limits.' *Proc. Instn Mech. Engrs*, **159,** 163–171.
15. JOHNSON, A. E. and HERBERT, D. C. (1948). 'The stress strain characteristics of a magnesium alloy beam.' *Aircraft Engng*, **20**(237), 330–334.
16. JOHNSON, A. E. (1949). 'The behaviour of nominally isotropic, 0·17 per cent C cast steel under complex stress systems at elevated temperatures.' *Proc. Instn Mech. Engrs*, **161,** 182–186.
17. JOHNSON, A. E. (1949). 'Behaviour of cast steel at elevated temperatures. Creep behaviour at 350, 450 and 550°C.' *Engineer, Lond.*, **188,** 126–128, 138–141, 165–168, 189–191.

18. JOHNSON, A. E. (1949). 'The creep of a nominally isotropic aluminium alloy under combined stress systems at elevated temperatures.' *Metallurgia*, **40**(237), 125–139.
19. JOHNSON, A. E. (1949). 'The behaviour of metals under multi-axial stress systems; the relationship between the applied tensile or direct stress and the torsion or shear stress.' *Aircr. Engng*, **21**(247), 284–286.
20. JOHNSON, A. E. (1949). 'The relaxation test in terms of creep and creep recovery.' *Metallurgia*, **39**(234), 291–297.
21. JOHNSON, A. E. (1949). 'The plastic creep and relaxation properties of metals.' *Aircr. Engng*, **21**(239), 2–8, 13.
22. JOHNSON, A. E. (1949). 'Creep and relaxation of metals at high temperatures.' *Engineering, Lond.*, **168**(4362), 237–239.
23. JOHNSON, A. E. (1949). 'The relaxation of chrome molybdenum bolt steel at elevated temperatures.' *British Electrical & Allied Industries R.A. Report* J/T 144. Reviewed in *Engineering, Lond.* (1949), **168**, 237–239.
24. JOHNSON, A. E. (1950). 'The creep of a nominally isotropic magnesium alloy at normal and elevated temperatures under complex stress systems.' *Metallurgia*, **42**(252), 249–262.
25. JOHNSON, A. E. (1950). 'High-sensitivity torsion creep unit.' *J. Sci. Instrum.*, **27**(3), 74–75.
26. JOHNSON, A. E., TAPSELL, H. J. and CONWAY, H. D. (1951). 'Creep tests of 150 hours duration at 100, 150 and 200°C on some cast magnesium alloys.' *R and M* No. 2675, pp. 1–5. Teddington, Middlesex: Aeronautical Research Council.
27. JOHNSON, A. E., TAPSELL, H. J. and CONWAY, H. D. (1951). 'Comparison of the creep properties of three cast magnesium alloys based on tests of 1000 hours duration.' *R and M* No. 2675, pp. 6–16. Teddington, Middlesex: Aeronautical Tesearch Council.
28. JENKINS, C. H. M., TAPSELL, H. J., BECKER, M. L., SCHOFIELD, T. H. and JOHNSON, A. E. (1951). 'The relationship of composition, heat treatment, grain size, and microstructure to the creep strength of carbon steels.' *BEAIRA Report* No. J/T 139. Leatherhead: British Electrical and Allied Industries Research Assoc.
29. JOHNSON, A. E. and FROST, N. E. (1951). 'Fracture under combined stress creep conditions of a 0·5 per cent Mo steel.' *Engineer, Lond.*, **191**(4967), 434–437.
30. JOHNSON, A. E. (1951). 'Creep under complex stress systems at elevated temperatures.' *Proc. Instn Mech. Engrs*, **164**(4), 432–447.
31. JOHNSON, A. E. and FROST, N. E. (1952). 'The temperature dependence of transient and secondary creep of an aluminium alloy to BS 2L42 at temperatures between 20 and 250°C and at constant stress.' *J. Inst. Metals*, **81**(2), 93–107.
32. JOHNSON, A. E. and FROST, N. E. (1952). 'Stress and plastic strain relations of a magnesium alloy.' *Engineer, Lond.*, **194**, 713–719.
33. JOHNSON, A. E. and FROST, N. E. (1952). 'Rheology of metals at high temperatures.' *J. Mech. Phys. Solids*, **1**(1), 37–52.
34. JOHNSON, A. E. and FROST, N. E. (1953) 'Creep properties of steels for power plants.' *Engineering, Lond.*, **175**, 25–28, 58–60, 249–250.

35. JOHNSON, A. E. and FROST, N. E. (1953). 'Equipment for compression creep tests at high temperatures.' *Engineering, Lond.*, **176,** 28–29.
36. JOHNSON, A. E. and FROST, N. E. (1953). 'Creep properties of steels for power plants.' *Engineering, Lond.*, **175,** 820.
37. JOHNSON, A. E. and FROST, N. E. (1953). 'Theory of complex stress creep.' *Engineer, Lond.*, **196**(5103), 632–633.
38. JOHNSON, A. E. and FROST, N. E. (1954). 'Combined tension and torsion machine for relaxation tests.' *Engineer, Lond.*, **198**(5160), 834–835.
39. JOHNSON, A. E. and FROST, N. E. (1954). 'Some aspects of the behaviour of metals under complex stresses at high temperatures.' British Soc. of Rheology, Summer Meeting, East Kilbride.
40. JOHNSON, A. E., FROST, N. E. and HENDERSON, J. (1955). 'Plastic strain and stress relations at high temperatures.' *Engineer, Lond.*, **199**(5173), 366–369 (5174), 403–405 (5175), 457–458.
41. JOHNSON, A. E. (1954). Third report of the Pipe Flanges Research Committee. *Proc. Instn Mech. Engrs*, **168**(15), 423–445.
42. JOHNSON, A. E. and FROST, N. E. (1956). 'Note on the fracture under complex stress creep conditions of a 0·5 per cent Mo steel at 550°C and a commercially pure copper at 250°C.' *Creep and Fracture of Metals at High Temperatures*, pp. 363–382. London: HMSO.
43. JOHNSON, A. E. (1956). 'Turbine disks for jet propulsion units.' *Aircr. Engng*, **28**(328), 187–195 (329), 235–243 (330), 265–272 (331), 325–332 (332), 348–356.
44. JOHNSON, A. E., HENDERSON, J. and MATHUR, V. D. (1956). 'Combined stress creep fracture of commercial copper at 250°C.' *Engineer, Lond.*, **202**(5248), 261–265; **202**(5249), 299–301.
45. JOHNSON, A. E., MATHUR, V. D. and HENDERSON, J. (1956). 'The creep deflection of eccentrically loaded magnesium alloy struts.' *Aircr. Engng*, **28**(334), 419–425.
46. JOHNSON, A. E., HENDERSON, J. and MATHUR, V. D. (1958). 'Pure torsion creep tests on magnesium alloy (2 per cent Al) at 20°C and on 0·2 per cent C steel at 450°C, at low rates of strain (10^{-6} to 10^{-9} per hour).' *Metallurgia*, **58**(347), 109–117.
47. JOHNSON, A. E., HENDERSON, J. and MATHUR, V. D. (1958). 'Creep under changing complex stress systems.' *Engineer, Lond.*, **206,** 209–216, 251–257, 287–291.
48. JOHNSON, A. E., HENDERSON, J. and MATHUR, V. D. (1959). 'Complex stress creep relaxation of metallic alloys at elevated temperatures.' *Aircr. Engng*, **31**(361), 75–79 (362), 113–118.
49. JOHNSON, A. E., HENDERSON, J. and MATHUR, V. D. (1959). 'Note on the prediction of relaxation stress–time curves from static tensile stress data.' *Metallurgia*, **59**(335), 215–220.
50. JOHNSON, A. E., HENDERSON, J. and MATHUR, V. D. (1960). 'The complex stress creep fracture of an aluminium alloy.' *Aircr. Engng*, **32**(376), 161–170.
51. JOHNSON, A. E., HENDERSON, J. and KHAN, B. (1961). 'The behaviour of cylindrical pressure vessels and tubes under internal and external pressure at elevated temperatures.' *Proc. Instn Mech. Engrs*, **175**(25), 1043–1069.
52. JOHNSON, A. E. (1960). 'Complex stress creep of metals.' *Metall. Rev.*, **5**(20), 447–506.

53. JOHNSON, A. E., HENDERSON, J. and KHAN, B. (1961). 'Complex stress creep fracture of copper at 250°C under vibratory stress.' *Engineer, Lond.*, **212**(5509), 304–308.
54. JOHNSON, A. E., HENDERSON, J. and KHAN, B. (1961). 'Metallic thick-walled spherical pressure vessels at elevated temperatures.' *Engineer, Lond.*, **212**(5527), 1078–1080.
55. JOHNSON, A. E., HENDERSON, J. and KHAN, B. (1962). *Complex Stress Creep, Relaxation, and Fracture of Metallic Alloys.* Edinburgh: HMSO.
56. JOHNSON, A. E. (1962). 'Research on complex stress creep of metals and structures at some centres in the USA.' *NEL Report* No. 20. East Kilbride, Glasgow: National Engineering Laboratory.
57. JOHNSON, A. E., HENDERSON, J. and KHAN, B. (1962). 'Creep of a solid metallic bar or thick-walled tube of circular section at elevated temperatures, when subject to various combinations of uniform bending moment, torque and axial load.' *Int. J. Mech. Sci.*, **4,** 195–203.
58. JOHNSON, A. E., HENDERSON, J. and KHAN, B. (1962). 'Elastic thick plain ring under a couple due to loads along the circumferences of two particular circles. *NEL Report* No. 22. East Kilbride, Glasgow: National Engineering Laboratory.
59. JOHNSON, A. E., HENDERSON, J. and KHAN, B. (1962). 'Creep stress distribution in rectangular beams of various materials under a uniform bending moment.' *NEL Report* No. 36. East Kilbride, Glasgow: National Engineering Laboratory.
60. JOHNSON, A. E., HENDERSON, J. and KHAN, B. (1963). 'Pure torsion creep tests on aluminium alloy to specification BS 2L42 at 200°C, and at low rates of strain (10^{-7} to 10^{-9} per hour).' *Metallurgia*, **67**(402), 173–177.
61. JOHNSON, A. E. and KHAN, B. (1963). 'Creep of thick-walled spherical vessels subject to pressure and radial thermal gradient at elevated temperatures.' *Int. J. Mech. Sci.*, **5**(6), 507–532.
62. JOHNSON, A. E., HENDERSON, J. and KHAN, B. (1963). 'Multiaxial creep strain/complex-stress/time relations for metallic alloys with some applications to structures.' *Jt. Int. Conf. on Creep. Proc. Instn Mech. Engrs*, **178**(Pt 3A), 2-27–2-42.
63. JOHNSON, A. E. and KHAN, B. (1964). 'Creep of metallic thick-walled cylindrical vessels subjected to pressure and radial thermal gradient at elevated temperatures.' *Proc. Instn Mech. Engrs*, **178**(Pt 3L), 29–42.
64. JOHNSON, A. E., HENDERSON, J. and KHAN, B. (1964). 'Stress and total strain rate distribution in spherical pressure vessels.' *Engineer, Lond.*, **217**(5648), 729–739.
65. JOHNSON, A. E., HENDERSON, J. and KHAN, B. (1964). 'Creep stress distribution in circular bars of various metallic materials under pure torque.' *Appl. Mater. Res.*, **3**(1), 45–54.
66. JOHNSON, A. E. and KHAN, B. (1964). 'Creep of metallic thick-walled cylindrical vessels subject to internal and external pressure and torque at elevated temperatures.' *J. Mech. Engng Sci.*, **6**(2), 191–201.
67. JOHNSON, A. E. and KHAN, B. (1965). 'Thermal gradient in cylindrical pressure vessels.' *Engineer, Lond.*, **219**(5705), 924–930.

68. JOHNSON, A. E. and KHAN, B. (1965–66). 'A biaxial stressing creep machine and extensometer.' *Proc. Instn Mech. Engrs*, **180**(3A), 318–323.
69. JOHNSON, A. E. and KHAN, B. (1965). 'Complex-stress creep fracture of Nimonic 90 at 750°C.' *Metallurgia*, **72**(430), 55–66.
70. JOHNSON, A. E. and KHAN, B. (1965). 'Complex-stress creep fracture of magnesium–2 per cent aluminium alloy at 50°C.' *NEL Report* No. 198. East Kilbride, Glasgow: National Engineering Laboratory.
71. JOHNSON, A. E. and KHAN, B. (1965). 'Creep under changing complex stress systems in copper at 250°C.' *Int. J. Mech. Sci.*, **7**, 12, 791–810.
72. JOHNSON, A. E. (1967). 'Some progress in creep mechanics.' B. BROBERG (Ed.), *Recent Progress in Applied Mechanics* (The Folke Odqvist volume), pp. 289–327. London: John Wiley & Sons.

INDEX